The Future of Food

Also Available from Bloomsbury Academic

Encyclopedia of Spices and the Spice Trade: History, Cuisine, and Culture
Victoria R. Williams

Food Shortage Crisis: Origins and Global Impact
Dawn M. Drake

Food and World Culture: Issues, Impacts, and Ingredients
Linda S. Watts and Kelty Clark-Mahoney

The Future of Food

Issues and Innovations in the 21st Century and Beyond

Honor May Eldridge

BLOOMSBURY ACADEMIC
NEW YORK • LONDON • OXFORD • NEW DELHI • SYDNEY

BLOOMSBURY ACADEMIC
Bloomsbury Publishing Inc, 1359 Broadway, New York, NY 10018, USA
Bloomsbury Publishing Plc, 50 Bedford Square, London, WC1B 3DP, UK
Bloomsbury Publishing Ireland, 29 Earlsfort Terrace, Dublin 2, D02 AY28, Ireland

BLOOMSBURY, BLOOMSBURY ACADEMIC and the Diana logo are trademarks of Bloomsbury Publishing Plc

First published in the United States of America 2026

A catalog record for this book is available from the Library of Congress.

ISBN: HB: 979-8-7651-2985-2
ePDF: 979-8-7651-2987-6
eBook: 979-8-7651-2986-9

Typeset by Newgen KnowledgeWorks Pvt. Ltd., Chennai, India
Printed and bound in the United States of America

For product safety related questions contact productsafety@bloomsbury.com.

Contents

Acknowledgments

Thank you to:

My amazing research assistants—your passion and support for this book make me confident that the future of food is in good hands.

My husband, son, and daughter—your love and respect made this book possible

Preface

When I started this book, I did my research—as any good author would. I read much of the existing literature on the future of food, and I came across a curious trend. The vast majority of books, even the most academic, opened their texts with an anecdote about their children. Author after author, academic after academic, disclosed intimate stories about the food that their children eat, their dietary preferences, their allergies, and their first foods. Perhaps, it is because the subject of food is inherently linked to our sense of nurture, of connection, of childhood. Food is central to our existence as humans. It is at the nucleus of our survival as a species—basic and essential. It is rooted in some primal. Thinking about the future of food is intertwined with thinking about the future of society. Perhaps even the future of humanity itself.

I am no different. My writing on the future of food is deeply linked to my child and my experience as a mother. I handed in the manuscript for my first book, and two days later, my daughter was born. Contemplating my second book was done against a backdrop of mashed bananas, half-eaten cucumber sticks, and rejected rusks. It was finished as she ran around the garden outside as a busy, chaotic toddler, demanding ice cream and cookies, pregnant with my second.

While my first book looked backward and contemplated the history of food, this one looks forward and considers how food will change and evolve in the coming years—just as she continues to change and evolve. What will my daughter be eating when she is grown? What will she feed her children? Will it be protein bites, grown in a sterile lab? Lettuce grown in underground bunkers, filled with UV lights and smart hydroponic systems? As a middle-class white woman, living in the Global North, will she continue to have access to whatever food she wants, with disregard for natural systems or seasons? Or will climate change have proved a leveler to our food system with all global citizens adjusting their diet to face a new reality? Will she learn to grow her own food in her backyard and value the complex ecological systems that allow a seedling to grow? Or will building connection with the soil seem alien to her urban reality?

These questions fascinated me, and I set out, through this book, on a quest to find answers. What I found were more questions. The future of food will be decided by the wider forces within the world. It cannot be separated from other economic and political drivers. And so the questions I asked evolved. What will the different global responses to the climate crisis mean for the future of food? If we embrace a technological fix for emissions, what path will that take our food futurc down? If we exceed three degrees, can food production adapt? How does the rise in geopolitical instability ripple out in the food system? What repercussions will future food security experience as a result of isolationism and a new era of tariffs? The reality is that the future of food will be defined by the wider context. It is impossible to say what will happen with certainty. Consequently, I have tried to allow the alternative futures of food to coexist through these pages to present the range of possibilities that might unfurl. This book does not claim to predict the future but rather to map the terrain of what might be—so that we, as eaters, parents, citizens, and stewards of the planet, can begin to make sense of where we are headed and what kind of food future we want to help create.

Abbreviations

AFDB	African Development Bank
APHIS	USDA Animal and Plant Health Inspection Service
BEAD	Broadband Equity, Access, and Deployment Program
BSE	Bovine Spongiform Encephalopathy
CAFO	Concentrated Animal Feeding Operations
CAP	Common Agricultural Policy
CAT	Catastrophic Risk Protection
CDC	Centers for Disease Control and Prevention
DNA	Deoxyribonucleic Acid
ECM	Extracellular Matrix
EFSA	European Food Safety Authority
EGEIS	European Glyphosate Environmental Information Sources
EIG	Economic Innovation Group
EIT	European Institute of Innovation and Technology
EPA	Environmental Protection Agency
FAA	Federal Aviation Authority
FAO	Food and Agriculture Organization of the United Nations
FBS	Fetal Bovine Serum
FCIP	Federal Crop Insurance Program
FDA	Food and Drug Administration
FSIS	Food Safety and Inspection Service (USDA)
GE	Gene Editing / Edited
GHG	Greenhouse Gas
GM	Genetically Modified / Genetic Modification
GMO	Genetically Modified Organism
GRAS	Generally Recognized As Safe
GURT / GURTs	Genetic Use Restriction Technology / Technologies
IoT	Internet of Things
IPCC	Intergovernmental Panel on Climate Change
IRRI	International Rice Research Institute
LANGEBIO	National Laboratory of Genomics for Biodiversity (Mexico)
LCA	Life Cycle Assessment

LMO	Living Modified Organism
LUC	Land Use Change
MARA	Ministry of Agricultural and Rural Affairs (China)
MAS	Marker Assisted Selection
MPCI	Multiple Peril Crop Insurance
MSC	Mesenchymal Stem Cells
NCD	Noncommunicable Diseases
NGO	Non-Governmental Organization
NGT	New Genomic Techniques
NOAA	National Oceanic and Atmospheric Administration
NRDC	Natural Resources Defense Council
OCDH	Observatorio Cubano de Derechos Humanos (Cuban Human Rights Observatory)
OGG	Open Green Genomes Initiative
PETA	People for the Ethical Treatment of Animals
PPE	Personal Protective Equipment
PRC	People's Republic of China
RAS	Recirculating Aquaculture Systems
RNA	Ribonucleic Acid
ROS	Reactive Oxygen Species
RTC	Reduced to Clear
R&D	Research and Development
SDGs	Sustainable Development Goals
SFA	Singapore Food Agency
SSA	Sub-Saharan Africa
TVP	Textured Vegetable Protein
UAV	Unmanned Aerial Vehicle
UCSB	University of California, Santa Barbara
UFA	Upcycled Food Association
UN	United Nations
UPF	Ultra-Processed Food
USAID	United States Agency for International Development
USDA	United States Department of Agriculture
WFP	World Food Programme
WHO	World Health Organization
WRI	World Resources Institute
WTO	World Trade Organization
WWF	World Wide Fund for Nature

Introduction

Climate change has rightfully placed the food system in the global spotlight. If current trends continue, the planet's average temperature is projected to rise by 2 degrees Celsius above preindustrial levels by 2030.[1] Rising temperatures, shifting rainfall patterns, and increasingly severe weather events are already disrupting how and where we grow food. If global greenhouse gas (GHG) emissions continue unchecked, crop yields could decrease by up to 23 percent by the end of the century.[2] Our current food production model is only making matters worse. Modern industrial agriculture plays a significant role in driving climate change. Agriculture contributes about 30 percent of GHGs, with livestock and fisheries the largest contributors.[3] Given this, it is clear that we must urgently reconsider our food production system. If we are to ensure that it becomes sustainable and resilient in the face of climate change, fundamental changes need to be made.

Climate is just one of many converging pressures reshaping the way we eat. Beneath the surface lies a deeper, more tangled web of challenges that cannot be solved by focusing on climate-sensitive agriculture alone. Ecological degradation from farming and the associated biodiversity loss need to be urgently addressed. Industrial farming systems, designed to maximize yield and minimize cost, have accelerated soil loss, polluted waterways, and driven biodiversity's steep decline. Forests are cleared to produce cheap meat and grow grain for feed. Fish stocks are depleted to serve an ever-growing global appetite. With each passing year, we erode the very natural systems that make food production, and our very existence on this planet, possible.

While these environmental crises of climate change and biodiversity loss have been raging, another has been brewing more insidiously—the public health crisis of obesity and associated noncommunicable diseases. Since the 1970s, hypertension rates have nearly doubled, while diabetes prevalence has increased more than fourfold worldwide.[4] This is linked to our global agriculture system that aims at calories (and profit) over health and nutrition. Ultra-processed foods (UPF) that are full of salt, fat, and sugar have surged in popularity and affordability. Rising inequality means that, while some people have the luxury of shopping for organic kale and oat milk, others cannot afford basic staples. Many are economically limited to buying the cheapest food options, which are predominantly nutritionally poor but calorie-rich. Globally, billions still lack access to healthy, nutritious food.[5] Even in wealthy countries, many communities face growing food insecurity and are forced to rely on ultra-processed snacks and fast food because it's cheap and accessible. Sugary drinks dominate in the poorest communities—heavily promoted and deeply entrenched. By consequence, health systems are buckling under the strain of diet-related diseases—from obesity and diabetes to heart disease and malnutrition.

The world is producing more food than ever before.[6] Yet, despite the vast quantity, farmers have not become rich. Large agrifood corporations might have made billions in the past century, but our small family farmers have lost out. Many have sold up. Those remaining struggle to stay economically viable. This is true the world over, from the subsistence farmer in Kenya to the independent rancher in Montana, and the rice farmer in South Korea. In many countries, the traditional model of the local smallscale farmer may disappear in the next century. Instead, agriculture will be dominated by large corporate farms that can ship food across oceans to feed consumers a continent away. This globalized model of food production comes with its own risks. Despite global economies being more intertwined than ever before, our world is facing growing levels of political instability and rising violence. This is disrupting trade and having a significant impact on our food system that rests on our global interconnectivity. Tariffs are increasing costs and restricting exchange, while conflicts, such as in Ukraine, are destroying entire agricultural supply chains, as well as causing unrolled devastation and destruction to people's lives.

With all these competing tensions influencing and driving our food system, when we talk about the "future of food," it's essential that we don't view it solely through the lens of climate change. Yes, climate is a critical factor—but we must confront the growing interconnected crises of producing and consuming food in an increasingly unstable and unpredictable environment. Focusing only on climate risks misses the bigger picture. If our goal is to build a better future, we need to widen our lens and look beyond carbon emissions and genuine resilience.

People have already attempted to wrestle with how to make the future of food more sustainable. The primary example of this is the 2019 EAT-Lancet Commission report. Developed by an international panel of leading scientists, it is widely recognized as a landmark, evidence-based framework on how to transform food systems in the face of climate change, biodiversity loss, and rising diet-related disease.[7] It offers a powerful vision for the future of food, demonstrating how dietary shifts in high-income nations could significantly benefit both human and planetary health. According to the report, if wealthier countries transitioned to healthier, more plant-forward diets, agricultural greenhouse gas emissions could be reduced by up to 61 percent.[8] This shift would primarily come from decreased demand for resource-intensive foods like red meat and dairy, which are major contributors to methane and nitrous oxide emissions. Additionally, the dietary transition would free up vast amounts of agricultural land—land that could then be restored to natural ecosystems or reforested. This change has the potential to sequester up to 98 gigatonnes (Gt) of CO_2, providing a critical carbon sink and helping to mitigate climate change.[9] These numbers underscore the profound position that food systems hold in climate solutions. The EAT-Lancet findings make it clear: aligning diets with planetary boundaries is not only possible but also necessary for a sustainable future—and high-income nations have an outsized role to play.

Yet such a transformation isn't just about emissions; it also demands significant changes in cultural norms, economic incentives, and agricultural policy—none of which is easy to achieve. The EAT-Lancet Commission's 2019 report is often described as a milestone in setting out what a healthy and sustainable future for food could look like for the whole world. But there's a big gap between that vision and the messy realities of how our food systems actually work—and where they seem to be headed. For one thing, the report's recommendations are based on global averages, while food cultures, farming systems, and nutritional needs vary hugely from place to place. What's ideal on paper may not fit easily into local diets or farming practices. Our current food systems, from industrial farming to ultra-processed foods, are shaped by decades of habits, business models, and marketing. Without big, coordinated action, the future of food is far more likely to follow these existing patterns than to snap into line with the EAT-Lancet vision. Delivering against the EAT-Lancet objectives would mean major changes to markets, subsidies, trade, and jobs—changes that face strong political and economic resistance, especially in the current political climate.

The real challenge is figuring out how to shape and ameliorate our global food system in a way that works for different cultures, economies, and communities around the world, recognizing the recalcitrant nature of people's dietary choices. Solving the crisis in the food system requires us first to understand it fully—only by grappling with its complexity can we begin to identify solutions that will truly deliver long-term, meaningful change. Humanity's tendency to reach for quick fixes without fully understanding the complexity of the problem is clearly illustrated by past efforts to "solve" the future of food. The Green Revolution emerged in response to fears of overpopulation and widespread hunger. In the face of this threat, the solution appeared to lie in technology and agrochemicals. High-yield seeds, synthetic fertilizers, and pesticides were promoted as the answer to feeding a growing global population. And, on the surface, they delivered: crop yields rose dramatically, and rates of hunger fell in many regions. But this success came at a steep, and largely unacknowledged, cost. The singular focus on boosting production overlooked the broader implications of such an intensive approach. The widespread use of synthetic chemicals led to soil degradation, water pollution, and a dramatic decline in insect populations and biodiversity. Monoculture farming practices, encouraged by the Green Revolution's model, further undermined ecological resilience and long-term sustainability. In attempting to solve one problem, we created another—arguably even more dangerous—set of challenges.

Once again, fear over how to feed a growing global population is mounting—and once again, the response is a surge of technological optimism. As we consider the future of food, solutions must be assessed in the context of the systems they are meant to change. As we look to the future of food—particularly in the search for quick substitutes and fast, scalable innovations—the example of the Green Revolution reminds us of the need for critical scrutiny, transparency, and humility in our approach to change within our food system. When we treat symptoms without addressing root causes, we risk entrenching the very problems we aim to fix or creating new ones that will haunt us for years to come. New tools—synthetic meat, precision agriculture, and climate-resilient crops—are often heralded as saviors. But too much faith in innovation can distract from more grounded, systemic change. The future of food demands more than innovation; it requires reflection, humility, and a deeper understanding of the systems we hope to transform.

It is impossible to offer a definitive prediction of the "future of food." Instead, this book explores the various factors that will influence and shape its trajectory. As a famous futurist, Jim Dator, aptly puts it, "The future cannot be predicted—but preferred futures can be imagined."[10] Rather than trying to forecast a singular outcome, this book delves into the possibilities, examining the forces at play that will impact the food systems of tomorrow. It challenges us to consider not only what's possible but also what's desirable. The future of food is complex, influenced by a multitude of societal, technological, environmental, and economic factors.

We are entering a period when decisions made today will determine the direction our food systems will take—how food is produced, distributed, consumed, and understood. However, as we look ahead, it's important to ask: whose future are we building, and for what purpose? The choices we make now are not just about what works but about what we value and the kind of world we want to create.

This book invites readers to imagine a range of possible futures for our food system. Could society transition to one that is more equitable, with food justice and livelihoods at its core? Might it become sustainable and resilient, capable of withstanding global challenges? Or should the focus remain narrowly on hunger, emphasizing yields and production above all else? Each of these visions carries its own challenges and trade-offs, and this book acknowledges the difficulty of turning them into reality. Will society embrace technologies that promise convenience but risk harming public health or the environment? Will we prioritize true food security for all, or allow corporate interests to continue shaping the system in their favor? These questions are central to the discussion, and they help us imagine futures that are not just feasible but worth striving for.

Technology will surely play a role. The rapid advancement of technology holds exciting promise for transforming the food system of the future. From precision agriculture and artificial intelligence (AI) that optimize resource use to biotechnology that enables nutrient-rich crops and alternative proteins, innovation is reshaping what and how we eat. Digital supply chains and blockchain systems can reduce waste and enhance transparency, while smart packaging and data-driven logistics promise greater efficiency from farm to fork. At the consumer level, wearable health monitors and personalized nutrition platforms could align diets more closely with individual well-being and environmental sustainability. Together, these technologies open the possibility of a food system that not only feeds a growing population but also does so in ways that protect ecosystems and promote healthier lives.

However, technology cannot substitute for the deeper work of reckoning with why the system is failing in the first place. Too often, innovations are created in search of a problem, locking us into narrow solutions and overlooking the deeper cultural, economic, ecological, and political foundations of food systems. A fix that ignores these roots may ease symptoms while creating new risks. Truly meaningful progress requires more than innovation. It requires reflection, honest assessment, and the willingness to hold space for complexity before jumping to solutions. If we fail to define the challenge, we risk repeating past mistakes—patching problems with products, chasing innovation while missing transformation. This book offers a framework for moving beyond quick fixes, inviting readers to interrogate assumptions, clarify priorities, and imagine pathways toward food systems that are not only innovative but truly just, resilient, and transformative.

This book embraces complexity: solving the food crisis requires asking hard questions, listening to diverse voices, and grounding solutions in context. It is an invitation to engage with the nuances within the future of food. It challenges readers to think critically about the role of food in our lives and to consider what kind of future we want to build together.

Part 1

Food System in Crisis

Introduction

This book explores the future of food, examining a range of exciting innovations and emerging technologies with the potential to disrupt our current food system. We'll delve into these developments in Parts 2 and 3. But before we look ahead, it's important to first understand the state of the food system today—and why its trajectory matters. In Part 1, it is worthwhile taking the time to define the problems. In the chapters in this part, we will contemplate the challenges that we face. We will dive deep into each of those categories of crisis in turn. Only when the problem is defined can we begin to identify potential solutions. Unless we are clear about those vulnerabilities, we cannot expect to come up with remedies or solutions. Without a clear definition of the problem, we risk merely creating a Band-Aid solution and the problems will remain, just hidden from view under the surface. They will likely bubble up again at a future date—and there is a risk that they will be even more devastating by that point having been allowed to fester. Once defined, we can consider what future of food we want and what choices will lead us closer to that future.

Our environment has been pushed to the limit. As a result of industrialization, greenhouse gases (GHGs) have increased global temperatures. If climate change continues on its current track, it is anticipated that by 2030, the planet's average temperature will have increased 2 degrees Celsius from preindustrial levels.[1] Every aspect of society will experience an impact from this. For agriculture, climate change has had a devastating impact already—and the consequences are likely to worsen. Yields could plummet in the coming years as the effects of the climate crisis begin to bite. Extreme weather events are increasingly common, while reduced rainfall and sustained periods of drought have killed off a huge number of crops.[2] As well as being impacted by climate change, agriculture is also a driver of the climate crisis. Since the invention of agrochemicals, we have been increasingly farming outside of natural systems. GHG emissions from agriculture account for about 30 percent of global warming; of these emissions livestock accounts for over 30 percent.[3] Current practices are highly carbon-intensive with food being produced in industrial farming systems and shipped around the world. Livestock, especially cows, release huge amounts of methane into the atmosphere, which is twenty-eight times more potent than carbon dioxide.[4] The methane then traps heat in the Earth's atmosphere, raising the global temperature. But GHG emissions are only one of the environmental consequences caused by modern farming practices. Biodiversity is declining rapidly. There has been a catastrophic 73 percent decline in the average size of monitored wildlife populations in the last fifty years, according to the World Wildlife Fund Living Planet Report 2024.[5] This is largely driven by agriculture and the associated land use change. It is estimated that agriculture is the primary driver of species decline for 86 percent of the species that are currently threatened.[6]

Not only is our food system damaging the planet, but it is also failing the public. While hunger around the world has reduced in the past fifty years, another problem has taken its place.[7] Obesity has risen globally. By extension, noncommunicable diseases, such as diabetes and heart disease, have increased by 39 percent since 2000.[8] Public diets are in crisis. Processed food (aka junk food) consumption has risen globally since 2000.[9] The rise in the availability of ultra-processed foods means that many people are consistently eating foods that are high in fat, sugar, and salt. What is deeply concerning is that these foods are often the cheapest to produce and, therefore, the cheapest to buy. As a result, today, the highest consumption rates of sugary drinks are in low-income communities.[10] Yet, these communities also face the highest obesity and noncommunicable diseases (NCD) rates by extension, while individuals with higher incomes have better diets and better health outcomes.[11] This creates a cycle that condemns people to the comorbidities of ill-health and poverty. Many of those who have the unhealthiest diets are children. Child obesity in the United States is 19.3 percent.[12] Unless this dynamic is addressed urgently, the next generation will be doomed to repeat the same cycle.

These two challenges of environment and public health might have been more manageable had our world not been facing elevated levels of geopolitical conflict and disruption. Our food supply today is globalized. This means that disruptions in one corner of the world lead to impacts on the other side. Over the last two decades, we have seen a rise in authoritarian populist politicians with President

Putin dominating Russia, President Trump returning for a second term in the United States, and Prime Minister Modi firming up his power base in India.[13] This has happened against a backdrop of a strengthening China. Experts predict that China will become the largest economy (overtaking the United States) in the 2030s.[14] Due to the unsettled nature of global politics, there has been a rise in conflict and violence around the world from skirmishes within regions (such as the Sudanese civil war) to global conflict (such as the ongoing war in Yemen).[15] This has disrupted food supply. Many farmers have been caught in the middle of these conflicts—whether that is the farmer who had to leave his wheat fields in Ukraine in 2022, the Mexican avocado producer struggles to be profitable when high tariffs are applied, or the fisherman in England who can no longer rely on Spanish canneries after the UK's exit from the EU.[16]

Given all of these factors, it is not surprising that farming is in crisis. We see a record number of farmers leaving the industry. In the United States, there has been a decline in farms from its peak of approximately 6.8 million in the 1930s to 1.9 million in 2022.[17] The decline has predominantly been the closure of small, family farms. In Europe, this trend is also present. Between 1960 and 2020, France declined from nearly 1.9 million farms to under 400,000.[18] There is a lack of new entrants coming in to take their place. Instead, young people prefer to move to the city. It is estimated that 60 percent of urban populations will be under eighteen by 2030.[19] In the past, family farms would have passed to the next generation. Today, many farms are being put on the market by farmers nearing retirement who do not have a successor. The next generation is reticent to join the industry because few consider it a lucrative or viable career. They are not wrong. The farms are quickly snapped up; they either get consolidated into larger transnational companies that benefit from economies of scale and political reach or developed into housing with its huge profit margins. Consequently, rural communities are dying. This decline only exacerbates the problem since the lack of opportunity and dynamism makes rural areas even less appealing for young people. The economics of farming today make it hard to make a living—particularly as a smallscale producer. In the United States, the average farm income was projected to be around $81,000 in 2023. This is significantly lower than pre-1970s, adjusting for inflation.[20] Adjusting for inflation, real farm income has been on a downward trend since the 1950s. This pattern is mirrored around the world. The Food and Agriculture Organization of the United Nations (FAO) reported in 2020 that farmers' share of food prices has been declining globally, meaning that farmers receive a smaller percentage of the final price consumers pay for food products.[21] Some years can be particularly challenging.

Shocks to the market can come thick and fast. For example, due to extreme flooding in 2023 and a change in government support payments, arable farm incomes in the UK plummeted by 73 percent in one year.[22]

The weaknesses in our food system sit in multiple quadrants: climate, geopolitical, health, and economic. Sadly, the reality of our current food system is fairly bleak. Climate change is causing crop failure.[23] Obesity is rising globally.[24] Supply chains are at risk of breaking down. And in response, farmers are abandoning the industry in record numbers.[25] Simply put, our food system is in crisis and nearing the brink of collapse. Given the level of exposure on multiple fronts, there is a real likelihood of a shock to the system coming in the immediate or near future. Worse still, it raises the chance that shocks could come simultaneously with dire consequences for food supply. Urgent action is needed in order to address the vulnerabilities. With a clearly defined problem and an ambition for what we want to create, we can then select solutions that help us inch closer to that model.

1

Environmental Challenges

Every day, humans burn fossil fuels for energy purposes, cut down trees, and consume meat from methane-producing livestock. These factors, among others, lead to rising sea levels and temperatures, which in turn increase the frequency and severity of extreme weather events like flash floods, droughts, and unpredictable hurricanes. If climate change continues on its current track, it is anticipated that by 2030, the planet's average temperature will have increased 2 degrees Celsius from preindustrial levels, with an additional 250,000 people predicted to die each year due to climate change-related causes.[1] Increased migrations are also expected as people flee areas that are destroyed by extreme weather that have made land become inhospitable. These displacements are already occurring at an unprecedented rate. In 2022, a record 32.6 million domestic migrants were displaced by disasters, with 98 percent of those disasters resulting from extreme weather events.[2] Given the devastating impacts, addressing the climate crisis is of utmost importance. As UN Secretary-General António Guterres puts it, "We need an exit ramp off the highway to climate hell."[3] In 2015, the United Nations adopted the Sustainable Development Goals to promote the cause of protecting the planet and ensuring that by 2030 all people enjoy peace and prosperity.[4] These seventeen goals include explicitly climate-related aims, like affordable and clean energy and climate action. However, these goals indicate aspirations, and we need urgent action to mitigate the impacts of climate change. With the climate continuing to change, the future of the planet and of the human population hangs in the balance.

Modern agricultural production is a key contributor to climate change and environmental degradation. Greenhouse gas emissions from agriculture account for about 30 percent of global warming.[5] Within this, crop production accounts for 27 percent of global agricultural emissions, land use accounts for 24 percent, and supply chain activities, like transport and processing, account for 18 percent.[6] However, the maintenance of livestock and fisheries is responsible for the majority, comprising 31 percent of total emissions from agriculture. This includes activities like pasture and farmland management, emissions produced by the animals themselves, and fuel usage. Livestock, especially cows, release huge amounts of methane into the atmosphere. Microbes in their digestive tract break down fiber into methane, which is then belched out.[7] Methane, which is twenty-eight times more potent than carbon dioxide, then traps heat in the Earth's atmosphere, raising the global temperature.[8]

Part of what makes agricultural production environmentally damaging is that it requires lots of space. A total of 38 percent of global land is used for agriculture. Increasing demand for productive land spurs deforestation; in fact, the expansion of agricultural enterprises is responsible for close to 90 percent of all global deforestation. Peatlands are drained and converted into arable fields while forests are cleared for crops and farms, destroying habitats that provide essential services to the ecosystem. The rearing of livestock in particular requires significant land space and deforestation; livestock production takes up 70 percent of total agricultural land usage and covers 20 percent of Earth's surface.[9] This has been driven by increasing demands for meat and is a trend that will likely continue since there is a projected 14 percent growth rate in meat consumption from 2020 to 2030.[10] These farms are not solely used for human consumption. About 36 percent of all crops go to feeding livestock.[11] Deforestation for increased meat production accelerates the rates of global warming from GHGs as trees that usually store carbon are cut down, releasing the gas into the atmosphere. The drastic environmental impacts of deforestation are particularly pronounced in numerous countries in the Global South. Brazil, Indonesia, Tanzania, Mozambique, Bolivia, and Paraguay all rank among the states with the highest deforestation rates.[12]

While agriculture is accelerating the rate of climate change, climate change is in turn wreaking havoc on farming and food production. While certain regions in the temperate zones now have climates to support new crops (e.g., Madagascar can now grow rice) or longer growing seasons, that is not true the world over. Overall, crop yields are expected to decrease by up to 23 percent by the end of the century if global GHG emissions continue on their current track.[13] Hotter climates mean thirstier crops but also mean less water. Worse still, a warming climate strips the soil of its moisture, as water evaporates at a faster rate. Increased and more extreme droughts may further destabilize crop production as rich, fertile topsoil that farmers rely on dries out. Soil may seem to be in endless supply, but its fertility is slowly being depleted, turning it to dust as it loses organic matter and carbon. By 2050, around 90 percent of the Earth's topsoil could be gone.[14] Consequently, food production will suffer. Topsoil is threatened by rainfall and floods as well, which will increase in intensity and frequency as the climate warms. Intense precipitation and flooding wash it away (especially when it has lost its health), leaving plants' roots exposed. These exposed roots make plants more fragile and susceptible to rot and disease. Given that 95 percent of our food is grown in topsoil, preserving it in the face of these climate change-induced threats is critical.[15] A warming world is also expected to see weeds and pests thrive and spread. Some of these pests are likely to expand their ranges, infringing on areas that have little experience containing them. For other insects, the changing climate will create more favorable conditions for their spread. Globally, farmers already spend around $60 billion USD on pesticides and $33 billion USD on herbicides, with both markets expected to grow by an average of around 5 percent annually.[16]

The impact of disrupted and declining harvests will be most felt by those in the Global South—despite the fact that emissions are predominantly driven by the Global North. These communities are on the frontlines of climate change and disproportionately experience climate change's damaging and destructive consequences. Rising sea levels and extreme weather events threaten homes, livelihoods, and food supply. Yet, they often have the least resources to fight against it or invest in adaptation. Farmers and fisherfolk from Africa to Central America and Asia are facing ever-declining yields and an inability to feed their families, prompting migration.[17] Ultimately, the unequal distribution of climate impacts reveals a deep injustice at the heart of the global food system—one that demands urgent and equitable solutions.

Planetary Boundaries

When considering the impact that humans have wrought on our environment, the primary focus tends to be on the impact of carbon dioxide and other greenhouse gases driving up global temperatures. However, that is not the only area where human activity, and for the purposes of this discussion agriculture, is altering the planet. In 2009, a group of internationally renowned scientists at the Stockholm Resilience Centre identified a set of Planetary Boundaries within which humanity can continue to develop and thrive for generations to come alongside these natural systems.[18] While one of these is carbon, it is only one among others. These nine boundaries are as follows: biological diversity, freshwater use, land use change, climate change, ocean acidification, biochemical flows (nitrogen and phosphorus cycles), stratospheric ozone depletion, novel entities (modified organisms), and atmospheric aerosol loading (air pollution). When we think about the future of our environment, we need to think beyond carbon. This framework starts to recognize the complexity of our natural systems, and scientists are clear that crossing these boundaries will generate abrupt or irreversible biophysical changes and reduce the planet's ability to support life. We have done serious damage to all nine, but it is biodiversity loss and the nitrogen cycle (in addition to climate change) that pose the greatest threat.

Without equivocation, we are facing an ecological crisis. Naturalists have dubbed it the Sixth Great Extinction since we are seeing huge declines in biodiversity. There has been a catastrophic 73 percent decline in the average size of monitored wildlife populations in the past fifty years, according to the World Wildlife Fund Living Planet Report 2024.[19] This rapid rate of loss is estimated to be between 1,000 and 10,000 times higher than the natural extinction rate.[20] Despite attempts to curb this destruction, more and more species are threatened with extinction. The UN's IPBES report outlines that more than 1 million plants and animals—a shocking 25 percent of all remaining global species—are threatened with extinction over the next decades as a result of human activity.[21] Agriculture is one of the biggest drivers of biodiversity loss. Too frequently native ecosystems are destroyed to make way for farmland. It is estimated that agriculture is the primary driver of species decline for 86 percent of species that are currently threatened.[22] One category that has seen significant declines is pollinating insects, such as bees and butterflies. These insects are critical to ecosystem health, but their populations are under threat, particularly due to the application of insecticides. This has a knock-on impact on food security. Three-quarters of crops require pollination. With fewer pollinators, crop yields suffer. Many well-established farms use artificial pollination or transport pollinators to farms at key moments in the growing cycle. However, other farms (particularly in the Global South) do not have backup options. Recent research suggests that global pollinator losses cause 500,000 human early deaths a year by reducing the supply of healthy food.[23] Biodiversity loss is also happening below the soil line. Perhaps it doesn't seem as important as a tiger or elephant, but soil microbes are going extinct. These little organisms ensure healthy, well-functioning soil. Through applications of fungicides and pesticides, the toxicity of the soil has damaged the ecosystem that exists under our feet.[24] There is still limited research into how these microbes contribute to soil health, but their destruction will likely have an outsized effect on our food.

Case Study A: Bite Back—Fighting Invasive Species One Meal at a Time

Fish accounts for 17 percent of global animal protein consumption and provides at least 20 percent of the daily protein intake for more than 3 billion people.[25] China by far led global fish and seafood consumption in 2022, followed by India, Indonesia, and the United States.[26] Popular choices, such as certain types of tuna or salmon, are overfished in some world regions, putting wild populations at risk. This has led to calls to eat less seafood and choose a wider variety. Eating more types of species can reduce pressure on the most popular ones and may also benefit ecosystems as a whole. One way to diversify consumer choices could be by incorporating invasive species into diets. Invasive species are non-native organisms that have been (either intentionally or accidentally) introduced to a new area and often cause economic or ecological harm in their new environments. One example is the lionfish. Native to the tropical waters of the South Pacific and Indian Oceans, lionfish began appearing in Florida, United States, in the 1980s and reached the Caribbean in the early 2000s.[27] It preys on smaller native fish and ultimately disrupts the balance of the reef ecosystem. Overall, lionfish are thought to have caused an 80 percent drop in the survival of small reef fish across the Caribbean.[28] Given that some coastal communities around the world rely on fish for up to 70 percent of their daily protein intake, why not use lionfish to help meet local needs?[29] In Curaçao, a Caribbean island, the invasive lionfish has caused destruction since it first arrived there. To combat the problem, local restaurants are preparing lionfish dishes such as "fruity, zesty ceviche with the lionfish fillets" or "lionfish tacos and fried lionfish and chips," aiming to help reduce its population and curb its spread.[30] A similar strategy has been implemented in the United States to manage invasive nutrias, a species of rodent, in wetlands along the Gulf Coast, in the Pacific Northwest, and in the Southeastern United States.[31] As part of National Invasive Species Awareness Week in February 2025, the US Fish and Wildlife Service released a list of invasive species that Americans can hunt, catch, and cook to help control their population.[32]

The biodiversity of our crops is also suffering. Our drive to industrialize farming since the 1950s has seen a drastic decline in heritage and landrace varieties. These are traditional, locally adapted varieties of a crop (or sometimes a breed of livestock) that have developed over time through natural selection and farmer selection in a specific geographic region. Global diets have become more and more consistent, relying on the same staple crops. Corn, rice, and wheat dominate our grain consumption, even when there are thousands of other types that humans could eat. Beef, chicken, lamb, and pork are the primary types of protein that humans choose to eat. There are many more species that we could consume, which we will discuss in Part 2, "The Future of Protein." Even within those commonly consumed species, varieties are bred for uniformity, and diversity is lost in the process.

With the loss of diversity comes the rise of threats since diversity brings resilience into a system. Genetic resilience in farming strengthens crops and livestock against pests, diseases, and climate extremes, reducing the risk of catastrophic losses. It supports long-term food security by maintaining diverse traits that can adapt to changing conditions. By relying on a wider genetic base, farming systems become more sustainable, less input-dependent, and better able to withstand shocks. With lower levels of biodiversity in our crops, there is an inherent weakness that pests and diseases can exploit. A virus will spread like wildfire through an industrialized monoculture since there are no natural systems in place to stop it. Such was the case with the Gros Michel banana (one of the earliest examples of monocropping) that was almost entirely wiped out by the Panama Disease in the 1920s. Similarly, invasive species can take hold when there are no natural predators. This has happened in California, where sea urchins have taken over huge stretches of the coast. Due to the decline in the sunflower sea star, which has been listed as critically endangered by the IUCN since 2020, the sea urchins have no natural predators and have killed off many of the kelp forests.[33]

Nitrogen in Our Food System

Nitrogen is essential for farming. In the air we breathe, it exists as N_2, which is stable and harmless because the two atoms are tightly bound together. But once that bond is broken, nitrogen can join with oxygen or hydrogen and become reactive. Nitrate (NO_3-) and ammonium (NH_4+), along with phosphorus, potassium, and other trace elements, are vital for plants and encourage vigorous growth. While all plants are able to take up nitrogen from the soil in the form of ammonium or nitrate, only legumes are able to "fix" their own nitrogen due to a relationship with rhizobia bacteria in their root nodules. The bacteria live symbiotically with the plant and can readily convert atmospheric nitrogen into ammonia that the plant can use directly. Drawing on this process of "biological fixation," farmers can use legumes (either as cover crops or in rotation) to add nitrogen to their soils without using manufactured fertilizers. However, not all plants have this ability to fix nitrogen, and so synthetic fertilizers are often applied to provide the crops with additional nitrogen so that they grow. The use of nitrogen fertilizer in agriculture has been one of the key drivers of yield increases over the past century. The most widely used forms of nitrogen fertilizer are ammonium nitrate (34 percent N) and urea (46 percent N). Due to the direct relationship between nitrogen fertilizer application and crop yields, farmers across the globe have been encouraged to use nitrogen fertilizer to achieve food security. In fact, many developing countries subsidize fertilizer. Worldwide, nitrogen fertilizer consumption reached 109.17 million tons by the end of 2022 with the fertilizer market estimated to be worth $202.2 billion USD today and growing.[34]

The manufacture of nitrogen fertilizer creates a major problem because it is a significant source of GHG emissions due to its energy-intensive production. Manufactured nitrogen is produced through the Haber-Bosch method that uses natural gas as its source of hydrogen and requires a huge amount of energy. The production of all manufactured fertilizers is estimated to contribute 1.2 percent of the world's GHG emissions, and the production of nitrogen fertilizer alone was estimated to release 0.8 percent of total global GHG emissions.[35] However, the production of nitrogen-based fertilizer is not the end of its emissions footprint. Emissions also arise from the storage and transportation. Factoring in the entire process, the total emissions per lb of fertilizer are 1.5–3.0 kg CO_2-eq (including transport and storage).[36] But our reliance on synthetic fertilizer goes beyond a carbon problem. The Stockholm Resilience Centre has calculated that we have already put three and a half times more reactive nitrogen into the environment than planetary ecosystems can manage.[37]

Our atmosphere is damaged by nitrogen pollution. Ammonia can create fine particulate matter (particles less than 2.5 micrometers). Fine particulate matter can cause asthma, respiratory disease, and cancers and also penetrate deeply into the lung to impair lung function and damage the cardiovascular system. Nitrogen oxide gases (nitric oxide, nitrous oxide, and nitrogen dioxide) can cause problems when breathed in directly. Nitrous oxide, an important GHG, also contributes to climate change by breaking down ozone. It has a greenhouse effect that is 300 times greater than carbon dioxide; thus a tiny amount contributes significantly to climate change. In total, 36 percent of global Nitrous Oxide results from human activity, and of this, global agriculture accounts for 67–80 percent.[38]

One of the ways that nitrogen fertilizers damage our environment is in our waterways. Nitrogen fertilizers are highly soluble and so easily run off the fields into rivers and streams. Nitrate leaching from soils can cause eutrophication in waterways. The increase in the supply of nutrients to an ecosystem—typically nitrogen and phosphorus—can lead to excessive growth of algae, known as "blooms." When the algae die, it causes a reduction in oxygen levels. A serious consequence of eutrophication is the decrease in oxygen levels in deep waters. Without oxygen, other species cannot survive, which can lead to the loss of biodiversity. The bacteria decompose the organic matter but also deplete the available oxygen, creating a state of hypoxia. Both invertebrates and fish can be killed extensively as the area of suitable habitat for food, growth, and reproduction becomes significantly reduced. Coral reefs, seagrass beds, fish, and shellfish can be particularly sensitive to eutrophication and oxygen reduction. The negative impacts can also move up the food chain to seabirds, mammals, and other marine animals. This is a growing problem worldwide with hypoxic conditions occurring in rivers, lakes, reservoirs, estuaries, and coastal waters. This leads to major losses in biodiversity. Coastal hypoxia can lead to "dead zones" where biological life is not supported. The number of dead zones has approximately doubled each decade since the 1960s, and they now affect more than 150,000 square miles globally.[39] The largest recorded dead zone of 9,000 square miles (equivalent to the size of Luxembourg) is off the coast of the United States, caused by fertilizer run-off washing out from the Mississippi.[40] Nitrates in water also damage human health and are associated with cancers of the digestive tract, bladder and breast, and thyroid disease.[41]

Water and Irrigation Impacts

Run-off from fertilizer is not the only issue with water that agriculture poses. Farming requires a huge amount of water—70 percent of the world's freshwater.[42] Livestock production makes up 27 percent of the global human water footprint.[43] Crop farming also requires a lot of water—on average, it takes about twenty-two–thirty gallons of water to grow one pound of tomatoes and a shocking 3,000 to 5,000 gallons of water per pound of rice.[44] Nuts have notoriously high water demands with almonds and pistachios having the highest consumption. These water footprints are getting worse since these crops increasingly face drought conditions and require more and more water. Roughly half of the globe lives under "highly water-stressed conditions" for at least one month a year, so farmers increasingly need to irrigate since rainfall is insufficient.[45] If they don't irrigate, crops will likely fail, but modern irrigation methods don't always function in harmony with natural systems. Higher levels of irrigation often deplete natural aquifers that then do not have sufficient time to refill as they might have in the past. This means that water is becoming more and more valuable, which has led to questions over rights to water and ownership of water sources where it was originally taken for granted. For example, Chilean law considers water rights akin to property rights, and there was even a 2022 attempt to embed the right to water in the constitution.[46]

The intensity of water stress exacerbates inequities around the world.[47] By 2099, the global supply of freshwater per head will fall by 29 percent from what it was in 2000. Africa will see a massive 67 percent decline, while, as glaciers melt, Europe's freshwater will rise by 28 percent.[48] The role of embedded water is also key to understanding the impact of globalized agriculture and water. Ultimately, the global trade of food is the trade of embedded water. Embedded water can be thought of as virtual water. For example, as an avocado grows in Mexico, it absorbs water. By the time it reaches maturity, that individual fruit has absorbed 320 liters of water.[49] This water has been extracted from the Mexican landscape and is embedded within the fruit. Once that fruit is shipped, the embedded water goes with it. This means that when you buy an avocado, a tomato, or a lettuce, you are also buying the water used to produce it. The majority of the time when it is imported from a Global South country, the water has been taken out of a water-stressed environment and will not be returned. Such is the case with all exported crops, but it is particularly problematic with crops that have high levels of embedded water that are produced in water-stressed regions. For example, countries such as Morocco, South Africa, India, and Peru are facing high or extremely high levels of water stress but are still exporting horticultural products to Global North nations.

While water stress in the Global South has thus far primarily been framed as an environmental issue, there is evidence to suggest that it is also a geopolitical security question too. Many warn that water could spark the next resource war.[50] According to the Carnegie Endowment for International Peace, "from the Great Ethiopian Renaissance Dam dispute to border skirmishes between India and China, and political unrest in the North Africa and the Middle East, it is clear that water-related competition only escalates in scale and intensity as climate change further restricts access to this ever-vital resource."[51] Lake Chad provides an example of what can happen when water becomes scarce for at-risk communities. Water vulnerability has been directly linked to the rise in extremism and the recruitment into terror organizations, such as Boko Haram.[52] Disaffected youth cannot find employment in the traditional industries of fishing and agriculture and have turned to these groups for purpose and employment. Secondly, displacement caused by climate migration has bred resentment and hostility in communities already pressured and constrained. Water stress has similarly impacted the economic security of farmers in Afghanistan and driven many to extremism. Afghanistan has experienced a 70 percent deficit in precipitation in 2018, resulting in 2 million food-insecure people.[53] One farmer said, "I would do anything to save my son from hunger. I would join Daesh or the Taliban," which demonstrates the direct security impacts of water stress.[54]

Case Study B: A Sweeter Future—Honey without Bees

Can science help our food innovate to survive the mounting pressures of ecological collapse? As beekeepers face colony collapse, can we imagine a future of honey without bees? "10 years from now, honey from bees will be an expensive artisanal gift only for special occasions," says MeliBo cofounder and CEO Darko Mandich.[55] Global demand for honey continues to rise as honeybee populations decline due to viruses, parasites, bacterial and fungal pathogens, and climate change.[56] In response, MeliBio, a food-tech innovator from California, has developed a plant-based honey that mimics the flavor of traditional honey. Using precision fermentation and plant science, the company aims to engineer the same enzymes found in a honeybee's stomach, offering an authentic alternative while eliminating the environmental concerns associated with industrial beekeeping. Although honey production is a natural process, modern commercial beekeeping raises several concerns. Bees are often confined in artificial hives, restricting their natural behaviors. Furthermore, commercial honey production depends primarily on a few types of bees bred for high productivity, which can result in a narrowed gene pool, outcrowding of native pollinators, ecological disruptions, and the spread of disease.[57] Many North American farmers rely on pollination from non-native European honeybees, which can outcompete native bees for resources and spread diseases. This growing imbalance underscores the need for sustainable, bee-free alternatives. By eliminating bees from the equation, MeliBio's innovation supports biodiversity while ensuring a stable and scalable "honey" supply without contributing to habitat destruction. Such innovations demonstrate that sustainability and tradition can coexist, preserving beloved flavors without harming the environment. At the same time, honoring the beekeeping tradition remains vital; preserving this knowledge ensures that communities that have long practiced beekeeping are not left behind in this shift toward innovation.

Soil Health and Fertility

While not specifically recognized as a Planetary Boundary, soil health is critical to keeping the world functioning and staying within the boundaries. Soil is the foundation upon which all terrestrial life depends. A report published by the United Nations in 2022 estimates that up to 40 percent of all soils worldwide are moderately or severely degraded—a figure that could rise to 90 percent by 2050 if deforestation, overgrazing, intensive cultivation, urbanization, and other harmful practices persist.[58] Soil loss due to excessive plowing occurs in nearly all countries. Plowing exposes the soil, leaving it bare to be easily eroded and the soil organic matter lost. Around the world, soil is being swept and washed away ten to forty times faster than it is being replenished, at an estimated cost of up to $500 billion USD per year.[59] Once it is lost, humanity's most valuable agricultural resource is gone forever. Historically, fields would have been farmed in rotation, allowing a fallow period to ensure that the soil had an opportunity to recover. As discussed earlier, legumes would also have been used in the rotation to bring natural nitrogen into the ground. However, the invention of synthetic fertilizers meant that fields could be continually cultivated with the fertility added through applications of agrochemicals. This worked in the short term, but in the long term, the natural fertility of the soil has steadily declined. As soil health has declined, the rate of erosion has increased. A total of 75 billion tons of fertile soil is removed every year from the global soils by erosion.[60] With poor quality soils, more synthetic fertilizer needs to be applied to get the same yield—trapping farmers into a dangerous cycle. This degradation has been exacerbated by climate change and the increased frequency of extreme weather events. Heavy rainfall can result in topsoil being washed away. Deforestation has also exacerbated the loss of healthy topsoil. When trees are cut down to expand farmlands, the land loses a natural windbreak, increasing exposure and soil loss. Root systems die when trees are cut, leaving little to hold the soil together and in place. The decrease in and eventual loss of fertile soil starts a dangerous cycle with old farmland being abandoned and new territory pushed into, compounding the destruction.

Conclusion

There can be no doubt that we are facing both climate and ecological crises. According to the UN, human-induced global warming has spurred changes to the Earth's climate that are unprecedented in recent human history. There has been a 68 percent average decline of birds, amphibians, mammals, fish, and reptiles since 1970.[61] There are many causes for these changes, but one of the key ones is farming. Farming is both contributing to the climate and biodiversity crisis but also suffering from its impacts. Its role in driving other environmental degradations such as declining soil health and worsening air and water quality are also incontrovertible. Ultimately, biodiversity loss, the climate crisis, and the wider destruction of our ecosystems will continue to accelerate, unless we change the way we produce food. This is why we need to consider what the future of our food system will be. If we continue as we currently are, our ecosystems, and perhaps the planet more widely, will collapse. In part, the problem is that the demand for food outstrips what our natural systems can sustain. In the Global North, we want more and more food—and we want it all year round. This demand is met by the Global South, which grows huge quantities for export, even when it degrades or destroys the immediate environment. At a global level, this dynamic pushes our planet to the breaking point. It is essential that society conceptualizes a new model of food production that can deliver healthy food in sufficient quantities to feed the planet while staying within the Planetary Boundaries and working in harmony with nature. Dutch scientists, for example, found that "if the UK and other high-income European nations simply ate the nationally-designated 'healthy diet,' their GHGs would fall by a startling 17%."[62] Wired journalist, Matthew Reynolds wrote in his book, *The Future of Food: How to Feed the Planet without Destroying It*, "We know it's possible to grow enough food to feed the planet, but we haven't yet worked out how to produce food without tipping out of balance the precarious ecological systems that we rely on."[63] Yet, is mitigation enough? Given the scale of the problem, should we be doing something more radical than merely trying to limit the damage? Do we need to create an entirely new approach to food production that functions within Planetary Boundaries and in harmony with the planet? This presents a huge challenge and is the subject of this book.

2

Public Health in Crisis

We are eating ourselves into crisis, and the systems that feed us are crumbling under the strain. Over the past two decades, public diets have been worsening around the world.[1] According to the Noncommunicable Diseases Alliance, poor diets are characterized as featuring a high intake of salt, sugar, and trans fats but a low intake of fruit, vegetables, fibers, grains, nuts, seeds, and, therefore, low micronutrients.[2] While not directly the same, poor diets are often associated with higher levels of ultra-processed food consumption. UPFs are industrially manufactured products made mostly from refined ingredients and additives, designed for convenience and taste but often low in nutritional value and linked to negative health outcomes. High consumption of UPF is causing obesity to rise, and it is projected that the number of adults living with obesity will rise from 0.81 billion in 2020 to 1.53 billion in 2035—a shocking 15 percent of the global population.[3] Yet, this is not a problem that is limited to the Global North. Obesity is rising across the world. While extreme hunger still exists, the primary issue facing the Global South is poor nutrition, not a lack of calories. Within the next decade, 79 percent of the adults worldwide who are overweight or obese will live in the Global South and 88 percent of all overweight or obese children.[4] Higher levels of obesity have a knock-on effect since obesity is associated with a higher frequency of noncommunicable diseases, such as diabetes and heart disease, putting huge pressure on health services around the world.[5] According to the American Action Forum, heart disease, type 2 diabetes, obesity, and thirteen different types of cancer associated with dietary ill-health have cost an estimated $7.6 trillion USD in direct health care costs in the United States between 2011 and 2020.[6] Between 2021 and 2023, these diseases were estimated to cost $2.9 trillion USD in healthcare costs.[7] There needs to be urgent action to tackle and (hopefully) reverse these trends—if not, it will have a profoundly detrimental impact on future generations.

The Damaging Effects of HFSS Foods

People across the world are eating "bad" food with ever greater frequency. Almost everyone knows that these foods with high fat, salt, and sugar (HFSS) are bad for your health, but few understand how they truly impact your body and its functioning. High levels of salt, or sodium, intake promote water retention; immediate symptoms include increased thirst, bloating, puffiness, and headaches. The kidneys are usually able to regulate these in the short term, but consistent overconsumption of salt can lead to more serious risks. Excess salt leads to a narrowing of blood vessels, causing the heart to pump more blood throughout the body. This puts the individual at risk for high blood pressure, heart disease, or stroke—cardiovascular disease is the deadliest noncommunicable disease in the world.[8] Other potential issues include kidney disease, osteoporosis, stomach cancer, and kidney stones.[9] Excess sugar intake is widely known for posing a higher risk of obesity and diabetes. This, in turn, increases the risk of heart disease. A 2014 JAMA study found that those who got 17 percent to 21 percent of their calories from added sugar had a 38 percent higher risk of dying from cardiovascular disease compared with those who consumed 8 percent of their calories as added sugar.[10] High amounts of sugar overload the liver and lead to fat accumulation, posing a risk for diabetes.

A diet high in saturated fat can lead to several health risks, mainly from high cholesterol.[11] Saturated fats have no double bonds between carbon atoms in their fatty acid chains, making them solid at room temperature, and are typically found in animal products like butter and meat. Excess saturated fat intake can lead to obesity, heart disease, and even cancer.[12] That said, there are fats that play a positive role in our diets. Unsaturated fats are considered "healthy" and can help lower blood cholesterol levels.[13] These unsaturated fats contain one or more double bonds, which create kinks in the chains, keeping them liquid at room temperature; they are common in plant oils, nuts, and fish. While saturated fats are linked to higher levels of low-density lipoprotein (LDL), which is a "bad" cholesterol. Unsaturated fats—especially polyunsaturated and monounsaturated—are considered healthier for your heart. Trans fats are a type of unsaturated fat that has been chemically altered through a process called hydrogenation, which makes them more solid and extends their shelf life. Commonly found in processed foods, baked goods, and fried items, trans fats are a particularly damaging type of fat. Trans fats are harmful because they raise LDL cholesterol that deposits in artery walls, raising the risk of heart disease. Trans fats also lower HDL ("good") cholesterol, which helps remove cholesterol from the bloodstream and protect arteries. Due to these serious health concerns, many countries have taken steps to restrict or ban artificial trans fats altogether.

There are lessons to be learnt from our early embrace of trans fats. At the turn of the twentieth century, German chemist Wilhelm Normann discovered a process to convert liquid vegetable oils into solid or semisolid fats—hydrogenation. These trans fats were initially marketed as a healthier alternative to saturated fats like butter and lard and gained huge popularity in the 1950s and 1960s. Partially hydrogenated oils offered food manufacturers a cheap, stable fat that extended shelf life and created appealing textures in processed foods—from margarine to baked goods. Beyond this, in the 1970s and 1980s, medical advice started encouraging people to reduce saturated fat intake—further boosting their popularity. At the time, trans fats were celebrated as progress in public health, offering a plant-based substitute to the animal fats associated with heart disease. However, science soon caught up with the reality: trans fats are far more harmful than originally thought. Consumed over time, they significantly increase the risk of coronary heart disease, stroke, and type 2 diabetes. The World Health Organization (WHO) estimates that trans fat consumption is responsible for around 500,000 premature deaths globally each year. The trans fat story is a cautionary tale about unintended consequences. It highlights the danger of being dazzled by a fresh innovation and adopting "quick fixes" in the name of health or innovation without fully understanding the long-term impacts. We must evaluate impact, not just intention.

Once lodged in our bodies, fat can have different health impacts. Within the body, there are two different types of fat: subcutaneous fat and visceral fat. Subcutaneous fat is the fat you can see: the fat that builds up on the body. Visceral fat lies deeper, out of sight, wrapping around the abdominal organs. For those who are obese, both types of fat are present in large amounts. However, visceral fat is regarded to be much more harmful than subcutaneous fat. In general, subcutaneous fat (stored right under your skin) releases anti-inflammatory molecules that boost metabolism and regulate energy balance. However, visceral fat (the kind that builds up around your organs) produces inflammatory chemicals that may cause heart disease and diabetes.[14] In fact, indulging in high-fat foods for just five days can impair the body's ability to properly oxidize glucose and respond to insulin, thus increasing risk for diabetes.[15]

If you compound all of these risks together into diets that are high in fat, salt, and sugar, the risks are escalated, making them more damaging to the body. Yet, despite this damaging impact, these foods have become more and more popular. In part, this is because of how your body responds to them. HFSS foods have addictive qualities, such as the sensation in the mouth and the brain chemical response. Consumption of fatty or sugary foods triggers dopamine release in the brain, which is part of the body's chemical reward system. In studies with rats, foods high in sugar and fat have induced signs of addiction—both behavioral and biological. This includes building tolerance, withdrawal symptoms, and chemical changes in the brain that increase cravings.[16] No wonder it is difficult for people to cut out junk food—even if they really want to make steps to improve their diets. Consuming these carefully curated foods can have an even more profound impact when it starts young, creating an addictive response. As Pen Vogler wrote in *Stuffed: A Political History of What We Eat and Why It Matters*, "The masterstroke is that locking immature palates into a pattern of eating ultra-processed foods creates Peter Pans who never grow out of them."[17]

Ultra-Processed Foods

Linked to the rise of HFSS foods is the rise of UPFs—such as sodas, breakfast cereals, and instant noodles. Numerous studies show that high UPF consumption is associated with higher rates of obesity, type 2 diabetes, cardiovascular disease, various cancers, and mental health problems.[18] While UPF commonly overlaps with HFSS foods, they are not automatically synonymous. HFSS and UPF are two different categories of "damaging" foods that have their own distinct impact on the body, and recent evidence suggests the harms of UPF cannot be explained by poor nutrient content alone. Chris van Tulleken wrote *Ultra-Processed People: Why Do We All Eat Stuff That Isn't Food and Why Can't We Stop?* and pointed to the rise of UPFs as having an effect on public health that goes beyond the HFSS diet that had come to dominate public food consumption.[19] Many UPFs are also high in fat, salt, and sugar, but there are also those that aren't. For example, protein drinks are UPF but tend to be low in salt, fat, and sugar. Some UPFs, like artificially sweetened "diet" drinks or some packaged breads and yogurts, may meet nutrient thresholds to avoid HFSS classification, but yet still exhibit risks associated with ultra-processing. Conversely, some HFSS foods, such as homemade desserts, are high in fat and sugar but not necessarily ultra-processed. In the UK, analysis suggests that about 16 percent of UPFs fall outside current HFSS-based policies, highlighting gaps in regulation.[20] Scientists are only just beginning to understand the way in which UPFs act in the body, their impact on public health, and how they differ from HFSS foods. A systematic review by researchers at University College London found that even after adjusting for fat, sugar, and salt, UPFs remained strongly associated with poor health outcomes.[21] This points to other mechanisms at play, possibly linked to how these foods are designed to be hyper-palatable, energy-dense, and rapidly consumed. Indeed, a randomized controlled trial by the National Institute of Health (NIH) showed that participants on a UPF diet consumed around 500 more calories per day and gained weight, despite matched nutrients and calorie availability between diets.[22] Factors such as reduced water content (making foods more energy-dense) and the fast-eating behavior they induce all appear to drive overconsumption. Further controlled studies are underway to untangle these mechanisms.

How are UPF foods defined? Van Tulleken defines them as anything wrapped in plastic with ingredients one wouldn't find in a domestic kitchen.[23] In 2009, epidemiologist Carlos Augusto Monteiro led a team of researchers at São Paulo University to develop a more academic classification of processed foods. In response to the poor public health in many parts of Brazil, as well as the increasing prevalence of obesity, he developed the NOVA food classification system to distinguish different types of food.[24] They segmented foods into four categories. Group 1 is unprocessed or minimally processed foods, for example, an apple. Group 2 is processed culinary ingredients, for example, butter or sugar. Group 3 is processed foods (with the addition of salt or sugar), for example, breads and cheeses, or canned fish. Group 4 is the group that gained the most attention in recent years—ultra-processed foods. Under NOVA's definition, UPFs are simply put, "industrial formulations typically with 5 or more and usually many ingredients."[25] They continue, "Besides salt, sugar, oils, and fats, ingredients of ultra-processed foods include food substances not commonly used in culinary preparations … and additives whose purpose is to imitate sensorial qualities of unprocessed or minimally processed foods and their culinary preparations or to disguise undesirable qualities of the final product."[26] This would therefore include commonly consumed foods such as soft beverages, cookies or muffins, spreads, and prepackaged meats like sticks or nuggets. They are cheap, mass-producible, and palatable, making them extremely popular and widespread. According to the NIH, "their ultra-processing makes them highly profitable, intensely appealing and intrinsically unhealthy."[27]

The definition and classification of UPFs expressed by NOVA could be open to multiple interpretations. The US Food and Drug Administration (FDA) is in the process of establishing a federally recognized definition, while the European Prospective Investigation into Cancer (EPIC) has opted to develop its own classification.[28] The EPIC definition of food processing includes three categories: highly processed foods, moderately processed foods, and unprocessed foods.[29] Since EPIC is primarily concerned with the effects of food processing on cancer, they developed a more detailed and precise classification system. For example, NOVA determines vegetables to be "processed" if they are treated to extend shelflife or enhance taste. However, EPIC dives deeper into food processing techniques, using more specific vocabulary like "salting, pickling, concentration, fermentation, drying, and canning in a commercial sauce or in fat."[30] Overall, EPIC developed three detailed, specific categories while NOVA developed four broader categories of food classification. The two systems serve different purposes: NOVA focuses on nutritional data, while EPIC focuses on technology data and the specifics of food processing techniques. It has been mentioned that a standardized, single classification system would be less confusing for the general public. However, no such movement has been made to streamline the NOVA and EPIC definitions into one comprehensive system.[31]

Case Study C: Nanotechnology in Food—a Growing Concern

Emerging innovations like nanotechnology are raising new concerns for public health. Used in food production for decades, nanotechnology can enhance appearance, texture, and flavor. Titanium dioxide (TiO_2), one of the most common nanoparticle additives, is used as a whitener in products like sugar, yogurt, and milk, as well as in candies like M&Ms and Skittles. However, recent research has raised red flags about its safety. The US Food and Drug Administration (FDA) still considers TiO_2 safe up to 1 percent of a food's weight.[32] However, the European Food Safety Authority (EFSA) concluded in 2021 that it could no longer be considered safe as a food additive.[33] This led to a ban in the European Union in 2022.[34] The FDA maintains that if a traditional food additive is deemed safe, its nanoparticle form is as well—but critics argue that nanoparticles behave differently due to their size and should be evaluated independently.[35] Unlike the EU, which required labeling of nanotechnology in products as early as 2011, the United States does not mandate disclosure on packaging.[36] With thousands of nanotech-containing products already on store shelves, this lack of transparency and regulatory oversight is increasingly problematic. EFSA's concerns included the potential genotoxic effects of nanoparticles—their ability to damage DNA and potentially cause mutations, abnormal chromosomes, and cancer.[37] Because of their size, nanoparticles can pass through biological barriers like the blood-brain barrier and accumulate in tissues and organs, compounding harmful effects over time. Ingestion through food or migration from packaging poses a clear route of exposure. Studies in mice have shown that nanoparticles can disrupt immune function, cause lung and gut inflammation, and interfere with the gut microbiome.[38] A particularly harmful effect is the generation of reactive oxygen species (ROS)—unstable molecules that can damage cell membranes, organelles, and disrupt normal cell functions, including in the nucleus. These disruptions may lead to oxidative stress, chronic inflammation, and tissue damage. Despite their widespread use, there is currently no reliable way to detect nanoparticles in food, and long-term human studies are lacking. As the EU has warned, precaution is necessary until more definitive safety data is available.[39]

Cheap Food, Bad Diets

The reality is that highly processed foods are cheap—and price is often the motivating factor behind why consumers choose these foods. As a result, central to the crisis of diets is the question of accessibility and social justice. Often, UPFs or HFSS foods are the most affordable and accessible foods for low-income communities to purchase because fresh produce is too expensive. In the Global North, healthy foods cost, on average, twice as much per 1,000 calories than processed food.[40] This question of affordability creates a two-tier food system where wealthy individuals can afford healthy food, and low-income communities are stuck with damaging diets. American households earning less than $10,000 a year buy twice as many sugar drinks as those earning $100,000.[41] But more than just price, UPFs are popular because they are convenient. They are easy to eat on the go if you're working multiple jobs or are simply exhausted after a long day. If you need to stay awake on a night shift, a can of soda, full of sugar, hits the spot, or you can just grab some fries to eat in the car on your way home, ready to collapse. Many processed meals (such as instant ramen) only need boiling water added. They do not require kitchens to prepare—an essential consideration if your accommodation does not have an oven—and don't require you to use much electricity, which might be an unaffordable luxury. Furthermore, they last a long time, so you don't have to worry that they will rot before you have a chance to eat them. You can buy them on payday and be confident that they will last through the month. They taste good and are full of energy, which is appealing after a long day. After all, no one wants to eat a kale salad after a twelve-hour day of delivering packages or working on an assembly line. This does not mean that the people who buy UPFs do not know that the foods are unhealthy or that they are environmentally damaging. It is just the reality that ready access to cheap food is often too convenient to question the method of production or the long-term consequences, especially when you're struggling financially. Pen Vogler rightly points out when discussing the tension, "Hard times push people into the toughest choices: whether to go hungry to let your children eat; whether to buy cheap food which will satisfy the first pangs of hunger, but not give the nutrients the body needs."[42]

Higher consumption of UPFs often intersects with food insecurity, and despite rising obesity hunger is still a major issue worldwide. Food insecurity, at the household level, is when people cannot access enough food to lead an active, healthy life.[43] This concept has evolved over the years but primarily exists in conjunction with hunger. Yet, while hunger might be experienced on a single day or week, food insecurity is a state of being. It exists in opposition to food security—"when all people, at all times, have physical and economic access to sufficient safe and nutritious food that meets their dietary needs and food preferences for an active and healthy life."[44] A total of 18 million households in the United States are food insecure, while 733 million people globally are food insecure.[45] Almost one in ten people in the world do not have enough to eat.[46] The issue is not that we cannot produce enough food to meet this gap. We already produce enough food to feed the world.[47] The problem is the inequities of our society. As Hannah Witan, Professor of Land and Food Systems at the University of British Columbia, said, "The problem of food insecurity is a problem of poverty—it's not a problem of insufficient food at a global level."[48]

For many (whether in the Global North or Global South) with limited resources to purchase food and other everyday items, price rapidly becomes the deciding factor, and UPFs are cheaper. UPFs make up a significant portion of the diets of low-income communities in the Global North. As a result, low-income communities have the highest obesity and NCD rates, while individuals with higher incomes have better diets and better health outcomes.[49] This creates a negative feedback loop where poor diet limits an individual's economic opportunity, which, in turn, hinders the ability to afford a healthy diet. Economists Sue Horton of the University of Waterloo and Richard Steckel of Ohio State University estimated that inadequate diets cost the world about 6 percent of GDP in lost output and double that in developing countries.[50] Chronic illness (such as type 2 diabetes) impacts the ability to work, in particular to take on physically intense jobs. It also hides malnutrition since diets with high levels of UPFs tend to lack the necessary nutrients for healthy body functions.

Dietary ill-health in children is a profound problem—particularly for children of low-income families that struggle to access healthy food. According to the WHO, over 39 million children under five years old are overweight or obese with another 300 million children under eighteen. This is not limited to the Global North, although prevalence is higher. "In Africa, the prevalence of overweight or obese children under 5 years has increased by nearly 24% since 2000. Almost half of the children under 5 years who were overweight or living with obesity in 2024 lived in Asia."[51] Dietary ill health hurts their future. Chronic illness can restrict their academic attendance, causing them to miss out on lessons and key learning, but more than that, poor diets can also restrict their ability to focus. This combination lowers their attainment and limits lifelong earning potential. A study found that children who were obese performed worse on standardized tests and had lower grade point averages, compared to peers with healthy weights, controlling for socioeconomic factors.[52] This can create a cyclical situation where their children face similar struggles to access healthy food choices. Yet, it doesn't have to be this way. The Leeds HENRY study in the UK showed that upward trends in early-age obesity can be reversed by putting large efforts into preschool child support.[53] Moreover, it seems more socially acceptable and effective to intervene in the diets of children than it does in the diets of adults. For example, India's Integrated Child Development Services (ICDS), launched in 1975, is the world's largest early childhood nutrition and development program. Aimed at children under six and their mothers, ICDS delivers supplementary nutrition, plus growth monitoring, immunization referrals, and preschool education. In states like Uttar Pradesh and Bihar, where child malnutrition was widespread, ICDS scaled up significantly in the 2000s. Uttar Pradesh expanded centers and improved food supply chains, while Bihar introduced take-home rations to reach remote families. These efforts showed measurable impact. Between 2005 and 2015, stunting in children under three fell from about 52 percent to 45 percent in these states.[54] However, challenges have persisted. Food quality and delivery can be inconsistent, and many critics note that the program often prioritized quantity over quality, with insufficient focus on nutrient adequacy and behavior change.[55] Despite these shortcomings, ICDS remains a landmark program. Its integrated, community-based approach has reached millions of children, improving nutrition and early development. It also underscores an important lesson: combating child malnutrition requires not just food but also education, health care, and community engagement.

Healthy and nutritious school meals are also critical for addressing poor diets in children. Oftentimes, school meals might be the only proper, hot meal that a child gets. It is important that these meals are healthy and full of nutrients that growing bodies need. There has been much work done to improve the quality and nutritional value of school food. The starting point is updating the nutritional guidelines. This has successfully been done by New Zealand, which amended its guidelines to "eating a diet with more whole, low or minimally processed foods."[56] Improved guidelines can lead to better meals for children. The leading example is Japan's Kyushoku School lunch program. Kyushoku is Japan's nationally coordinated school lunch program, with roots dating back to 1889, and provides healthy, balanced meals to elementary and middle school students nationwide. Beyond simply feeding children, the program also promotes education and awareness around health, culture, and community responsibility. The Japanese government encourages schools to source ingredients from local producers, helping to reduce waste and pollution from mass transportation. Kyushoku has delivered significant health benefits as well. Children who eat school-provided lunches tend to consume more fiber and less fat and salt compared to those who bring meals from home.[57] They also eat more protein-rich foods like fish and meat, along with greater amounts of vegetables and fruit, contributing to better overall nutrition. For other countries trying to implement school meals, logistics of delivery is a marathon—especially in somewhere like India. Each day, India serves 100 million hot lunches for kids that are tailored to the dietary requirements of each demographic group.[58]

Market Failure?

Poor diet is not a question of "bad" choices by the individual. As we have discussed, it is often dictated by economic status. Yet, an individual's freedom to make their own dietary choice is limited not only by economic situation—it is also a direct result of the power and influence of food corporations. These companies are worth around $10.5 trillion USD and have a vested interest in ensuring that their products, however damaging, are purchased.[59] A recent Chatham House paper showed that the more society produces, the cheaper food becomes, the more society consumes.[60] Companies have incentive to encourage and cajole consumers into buying more and more. For example, according to food expert Professor Tim Lang, "Coca-Cola spends about $4bn a year on ads worldwide. This is twice the total budget of the UN World Health Organization and twelve times the WHO's health promotion budget stream."[61] There is little hope of a public health message cutting through that noise. The food industry is a massive economic powerhouse, with far-reaching influence over consumer choice and public health. Much like the tobacco industry, these food corporations have vast resources to push back against any attempts to regulate their products. The global food and beverage industry is a colossal sector, valued at approximately $7.4 trillion in 2025, with projections to reach $9.4 trillion by 2029—more than three times larger than the global pharmaceutical industry.[62] This gives them tremendous power.

One key tactic is their heavy investment in research, which allows them to produce studies, reports, and articles that either promote their products or undermine criticisms about their health impact. Renowned nutritionist Marion Nestle has been a vocal advocate against this influence, consistently challenging the industry's ability to shape public opinion with biased research. In 2015, she exposed how the Coca-Cola Company funded research through the Global Energy Balance Network—a so-called independent scientific group that minimized the role of sugary drinks in obesity and instead promoted exercise as the primary solution.[63] This strategy effectively redirected the public discourse on obesity away from the company's products. Beyond influencing research, the food industry wields significant financial power in lobbying and legal challenges. They have the means to sway policy decisions and delay regulatory measures that could benefit public health. The industry spent over $50 million on lobbying in the United States alone to influence public policy and prevent regulation and restrictions on sugary drinks, mandate clearer labeling, or impose taxes on unhealthy food products.[64] This combination of research, lobbying, and legal clout has created an incredibly powerful opposition to any meaningful effort to improve public diets, making the battle for better nutrition and food policy a highly contentious and uphill fight.

This corporate power and marketing prowess have contributed to the perception of UPFs in the Global South as being an indicator of glamour and sophistication. For example, Lays Gourmet chips are marketed as a luxury good in India with a tagline "it's crafted, not cooked," with endorsements from Bollywood star Saif Ali Khan and a television advert featuring a banquet in a European palace. People want to be seen eating these foods since it conveys to their peers that they are wealthy enough to afford them. This is an example of conspicuous consumption. Conspicuous consumption, a term coined by sociologist Thorstein Veblen in 1899, refers to a social phenomenon in which people acquire goods and other tangible measures of wealth to display their status to others.[65] The acquisition of Lays chips goes above and beyond what is necessary for survival and comfort—it is specifically for show and to relay a message of prosperity and wealth to one's community. This mindset leads to competition among peers, where people will acquire more and more things to outdo each other. In areas of extreme poverty, the possession of goods has a larger marginal value than in wealthier areas. This has resulted in people from the Global South emulating Global North diets and eating processed snack foods publicly as a way of demonstrating wealth. Consequently, it is more common to see conspicuous, overt overconsumption of UPFs in particularly underserved areas such as the Global South.[66] As a result, many international UPF snacks have replaced traditional local foods. An example of this is the rise in the consumption of ultra-processed snack foods in Mexico over healthier options, such as guacamole or vasos de fruta. Corinna Hawkes, a leading UK academic on food policy, has termed this the "globalization/diet nexus." Her term aims to highlight that globalization exacerbates uneven dietary choices between Global North and Global South nations, writing, "lower-income groups may well experience convergence towards poor quality obesogenic diets."[67] Conspicuous consumption is also contributing to the growing popularity of international restaurant chains, such as McDonald's and Starbucks. These brands often use aspirational consumer psychology to promote their brands with advertising that features professional urban couples and celebrity endorsements. For the outlets themselves, they select locations in luxury shopping malls and utilize more stylish interiors, featuring those with Wi-Fi and air conditioning—some even feature table service.

As Global South countries become wealthier and their GDP rises, citizens typically purchase more processed and calorie-dense foods, which leads to an increase in obesity rates. For example, Mexico in the early 1980s had a relatively low obesity rate of around 10 percent of adults. However, as the country's economy grew and processed food consumption increased in the 1980s and 1990s, this number skyrocketed. By 2018, over 36 percent of Mexican adults were classified as obese.[68] This shift in dietary patterns is closely linked to higher disposable income and greater access to cheap, ultra-processed foods. However, this relationship is not linear. At a certain economic threshold, the trajectory reverses. Once nations increase their GDP and transition into high-income countries, obesity rates begin to decline. For example, Japan now boasts one of the lowest obesity rates among OECD countries at just 4.3 percent (compared to the United States' 40 percent).[69] This reversal is largely due to increasing public awareness of health and wellness among the population, more individual disposable income for healthier options, and government interventions such as food policies, nutrition programs, and robust healthcare systems that prioritize prevention and education. This pattern is known as the Obesity Kuznets Curve.[70] It illustrates the idea that, as nations progress economically, they experience a rise in obesity, followed by a decline as affluence increases and better health policies are enacted. The curve underscores the importance of tailoring nutrition policies to a country's specific stage of economic development. To effectively tackle rising obesity rates, interventions must reflect the unique socioeconomic and cultural contexts of each nation.

Public Policy Interventions

Governments have incentive to fix this crisis and reduce the percentage of their populations facing dietary ill-health. Beyond being the right thing to do, obesity and NCDs cost them huge amounts of money in public health and social programs. In the United States, UPFs and the link to poor health outcomes gained increased public attention with Robert F. Kennedy Junior's tenure as Secretary of Health and Human Services, who committed to "Making America Healthy Again" (MAHA). He has been very vocal about the damage that these foods can cause, saying "sugar is poison," that UPFs are driving the "obesity epidemic," and that Americans are being "unknowingly poisoned" by seed oils.[71] He even went as far as criticizing the food industry, saying, "The problem is, industry is making money on keeping us sick."[72] He has expressed a personal commitment to updating the national dietary guidelines and reducing the amount of added sugar allowed. He is lobbying to remove artificial food coloring from US food—predominantly used in UPFs. He announced that he had reached "an understanding" with major food retailers to eliminate these colorings—although that has not been confirmed by the brands themselves.[73] Some of these colorings have already been banned in the EU for their potential carcinogenic and hyperactivity qualities, and states across America (both red and blue) have been enacting regional restrictions.[74] Many public diet experts have welcomed these interventions—although with caution, mindful of his more controversial stances on other aspects of health. Marion Nestle has said, "On some issues, he is completely off the wall, but then on chronic disease … childhood obesity, he's fantastic. I've never heard anybody at this level of government talk about these issues in this way. It's thrilling."[75] Many experts, beyond Nestle, have highlighted that the decline in personnel at the FDA through government rollbacks might limit his ability to successfully deliver on these promises.

Attempts to influence diet have seen a rise in restrictions on food aid programs. In the United States, the Supplemental Nutrition Assistance Program (SNAP) provides low-income households with funds to purchase food, delivered through an Electronic Benefit Transfer (EBT) card that functions like a debit card. The Special Supplemental Nutrition Program for Women, Infants, and Children (WIC) offers targeted support for pregnant women, new mothers, and young children, focusing on nutritious foods, health screenings, and education. There are ongoing efforts to limit whether SNAP benefits can be used to purchase UPFs. Advocates argue that preventing the use of public funds for products high in sugar, fat, and artificial additives could reduce diet-related diseases and promote healthier eating. Critics, however, warn that such restrictions risk stigmatizing recipients and limiting choice without guaranteeing access to affordable, minimally processed alternatives. As research continues to highlight the health risks of UPFs, debates over the role of food aid in shaping diets underscore the need for broader strategies: improving food environments, supporting affordable fresh foods, and ensuring that nutrition policies do not disproportionately burden vulnerable groups. This national shift reflects a state-level trend toward limiting access. Nebraska, Indiana, and Iowa have all secured federal waivers from Secretary of Agriculture Brooke Rollins to exclude sugary drinks from SNAP.[76] Arizona has passed a law to request a federal waiver, but the governor vetoed the measure. In her veto letter dated April 15, 2025, Governor Hobbs acknowledged the bill's intent—to "improve the health outcomes of Arizonans"—but firmly rejected it, stating that it would "unnecessarily deprive SNAP participants of their purchasing power and relegate them to a new underclass of grocery shoppers."[77] She argued that it was preferable to focus on expanding access to fresh, nutritious foods rather than restricting purchases. The governor's response exemplifies the challenge faced by many Democrats, who are deciding whether to support or reject the trend of restricting access to UPFs via SNAP. Many see the benefit of improving health outcomes in low-income communities and the long-term economic benefits that will be achieved but fear that it might be a backdoor way to restrict SNAP and cut support for social welfare by slowly undermining and eroding the program.

Global Interventions

The WHO has tried to encourage national governments outside the United States to implement policies to encourage better diets, in line with their responsibilities to drive improvements in public health. The WHO is clear that there is growing evidence that taxes on unhealthy foods, like sodas, can lead to lower demand and consumption.[78] Inversely, subsidies for foods that contribute to a healthy diet, such as fruits and vegetables, make these foods more accessible and affordable. However, single interventions are unlikely to tackle the problem. Addressing the crisis will likely require a diverse set of policy interventions, reflecting the complexity and interdependence of the food system and the multiple factors that shape its outcomes. One method to tackle UPFs is the introduction of sugar taxes. These taxes introduce a sugar threshold over which companies need to pay an additional tax. This is a similar approach to the "sin" tax imposed on cigarettes and alcohol. After all, if these UPFs are designed to be addictive, it seems pragmatic to consider the same policy interventions that we have used to curb the use of more traditionally addictive substances. This approach tests the elasticity of demand and presumes that when the cost of the product increases, consumers will purchase less.

In the approximately seventy countries where taxes on sweetened drinks have been implemented, the biggest impacts were felt in poorer countries where consumers are more sensitive to price changes.[79] Mexico is one of the earliest and most studied cases of the introduction of a soda tax. In 2014, the country implemented a nationwide tax of 1 peso per liter (approximately a quart) on sugar-sweetened beverages.[80] This is nearly 10 percent of the total price. The results were swift and striking. Within the first year, purchases of taxed beverages fell by 6 percent, and by the second year, they dropped by 10 percent.[81] But the most notable change came from lower-income households, who reduced their sugary drink consumption by 17 percent in the same period.[82] These families, more sensitive to price changes, quickly adjusted their habits, often switching to untaxed options like water or homemade drinks. A similar pattern emerged in South Africa, where a Health Promotion Levy was introduced in 2018, taxing beverages based on the grams of sugar they contained. Within the first year, household purchases of sugary drinks fell by 28 percent, with the greatest decline seen in low-income communities.[83] They also found a 51 percent reduction in sugar, a 52 percent reduction in calories, and a 29 percent reduction in volume of beverages purchased per person per day following implementation of the tax.[84] These data demonstrate the effectiveness of such taxes in regions with price elasticity and encourage wider adoption. A study in India estimated that a 20 percent tax on sugar-sweetened beverages could prevent over 11 million cases of obesity and 400,000 cases of diabetes over a decade, with the lowest income groups benefiting the most.[85]

However, in most instances, the companies opt to reformulate their product to ensure that the sugar levels sit just below the threshold and therefore are exempt from the tax. For example, the UK Government announced their intention to tax sugar drinks in 2016, and it was implemented in 2018.[86] By the time the tax was implemented, half of the drinks covered by the levy had already been reformulated to lower their sugar content.[87] Coca-Cola, in particular, reduced the sugar content in its regular drinks, such as Coca-Cola Classic, to avoid the higher tax brackets. By 2020, the reformulation of drinks in response to the levy was so successful that almost 50 percent of sugary drinks sold in the UK were tax-free due to their reduced sugar content.[88] The primary goal is shifting public behavior, but a happy by-product is that these policies raise public funds to offset health and social costs. For example, in the UK, Professor Tim Jackson calculated that the total food-related cost of chronic disease as £268 billion ($362.86 USD), derived by combining health care (£67.5bn or $91.39bn USD), social care (£14.3bn or $19.36bn USD), welfare (£10.1bn or $13.67bn USD), productivity (£116.4bn or $157.78bn USD), and human cost (£60bn or $81.3bn USD) of chronic disease attributable to the current food ecosystem.[89] The sugar tax that taxes drinks with over 5g (0.176oz) of sugar at 18p ($0.23 USD) per liter (approximately a quart) and 24p ($0.32 USD) per liter for those over 8g (0.28 oz) accumulates approximately £350 million ($474 million USD) each year for the UK Treasury—less than the estimated revenue since most companies chose to reformulate their product to escape the levy.[90]

Another policy intervention that has been trialed is mandatory labeling to reduce UPF consumption. In Chile, the ready availability of convenient, packaged food has increased the consumption of UPF across all economic strata and in both rural and urban populations. Today, 67 percent of Chileans over fifteen years old are overweight or obese.[91] The World Economic Forum reported Chile as the world's largest per-capita consumer of sugary drinks, at 188 calories per person per day, ahead of Mexico (157 calories) and the United States (158 calories). To combat this crisis, Chile introduced a new nutrition label in 2017. All foods that exceed 275 calories, 400 milligrams of sodium, 10 grams of sugar, or 4 grams of saturated fats per 100 grams would now be labeled with black stop signs on packets. Products that feature these labels are not allowed to be advertised to children, cannot include toys and cannot be sold in or near schools.[92] While praised by health experts, countries including the United States, Canada, and EU members argued that Chile's labels could be trade-restrictive, stigmatize certain foods, and deviated from Codex Alimentarius standards, which serve as international food guidelines. They also claimed that the labels might act as nontariff barriers, affecting multinational food companies and therefore violating the World Trade Organization (WTO) rules. Despite pressure, Chile maintained the policy, defending it as a necessary response to its escalating obesity crisis. The Chilean government emphasized the measure's scientific basis, transparency, and its right under WTO agreements to protect public health. Chile argued that the labels were essential to inform consumers clearly and effectively—especially children and low-literacy populations. Though formally raised at WTO technical committees, no official dispute was filed. Ultimately, Chile's persistence paid off: the labels remained in place and inspired similar on-packaging labels to be introduced in countries like Peru, Mexico, and Israel. Today, Chile's approach is widely seen as a benchmark in using food policy to tackle noncommunicable diseases, illustrating how public health can prevail despite international trade tensions.

Case Study D: Pouched and Pressured—the Feminist Dilemma behind Baby Food Convenience

The growing prevalence of ultra-processed baby foods (such as pureed pouches) has sparked a heated debate, highlighting the intersection of children's health, corporate influence, and the societal pressures faced by parents, especially mothers. While Bee Wilson, a leading UK food writer and journalist, warns about the nutritional shortcomings of these convenient products, many parents feel caught between health ideals and the realities of daily life.[93] These products, often marketed as healthy and time-saving, are high in sugar and low in essential nutrients. Their overuse has been linked to problems such as tooth decay and speech delays in toddlers, who are increasingly favoring ultra-processed snacks over fresh, nutritious meals.[94] Bee Wilson criticizes the baby food industry for exploiting the time constraints of modern parenting by promoting these snacks as suitable options.[95] Guardian columnist and author Rhiannon Lucy Cosslett points out that many parents turn to these products out of necessity, not negligence.[96] In a 2021 multicountry study of Argentina, Australia, Brazil, Chile, Colombia, Mexico, the UK, and the United States, toddlers aged two to five in the UK and the United States derived 61 percent and 58 percent of their total energy intake, respectively, from ultra-processed foods, which were the highest figures among these countries for this age group.[97] The trend continues into later childhood and adulthood: individuals aged six to nineteen in the United States and the UK consume 66 percent to 68 percent of their energy from such foods, while the figure drops to just 19 percent in Colombia, which also shows a comparably low rate of 18 percent for children aged two to five.[98] The study shows that early reliance on ultra-processed foods often persists into adulthood, raising long-term health concerns.[99] Experts argue that lax regulation has allowed companies to frame ultra-processed snacks as suitable for babies, prioritizing corporate profit over public health.[100] In combination with the so-called maternal guilt industrial complex this debate frames reliance on ultra-processed baby foods as a sign of parental failure rather than systemic shortcomings. It also underscores a broader issue: the need for a public health approach that considers social support. Tackling the rise of ultra-processed baby foods means not only improving regulations and transparency but also providing families with realistic, affordable, and accessible alternatives. Only then can we create an environment where parents are supported (not shamed) in making the best choices.

Wellness and Diets

The growing attention on the health problems of poor diets has led to an effort to reverse the trend. Over the past few years, there has been a rise in weight-loss medications hitting the market. In 2024, the weight-loss drug market was estimated at $100 billion USD. Experts now forecast the market value at $150 billion by the early 2030s.[101] The most infamous of these is Ozempic, which has become a cultural phenomenon, but there are numerous others such as Wegovy and Zepbound. These drugs (classified as GLP-1 drugs) work by mimicking naturally occurring hormones that reduce feelings of hunger and signal fullness to the brain. In fact, those who used GLP-1 drugs over a two-year period saw a 15.2 percent decrease in body weight, while those on placebos only saw a 2.6 percent decrease.[102] Weight loss in itself is the goal of many of the people who take the medication, but there are numerous knock-on benefits. According to a recent study, weight loss can lead to reduced risk of forty-two conditions including heart disease, cancer, kidney disease, and even psychiatric conditions such as dementia and addiction.[103] Further, it has been shown to be a significant aid for those suffering from hormonal disorders such as polycystic ovary syndrome (PCOS). Common symptoms of PCOS are unexplained weight gain, hormonal irregularities, and insulin resistance, which can all be relieved by taking GLP-1 medications.[104] Yet, despite the many benefits, some have raised concerns. There are some notable physical side effects (such as muscle loss, pancreatitis, and kidney issues), and there might be some significant mental impacts since it could trigger eating disorders or overreliance on pharmacological measures to lose weight.[105] Finally, there are issues of equity of access. These medications are expensive, averaging around $1000/month without insurance, making it much easier for wealthy and privileged individuals to access them.[106] This is particularly problematic given that, as discussed, the obesity rate and the associated NCDs are much higher in low-income communities. The people who would most benefit from weight-loss drugs would not be able to readily access them since the cost is price-prohibitive, which will further contribute to health disparities across socioeconomic classes.

However, a focus on weight loss is not the only manifestation of the rising awareness of diet within the Global North. Wellness and "clean eating" have become a significant trend in recent years. While there is no clearly accepted definition of the term, the Harvard University Department of Public Health explains that clean eating refers to foods that are as close to their natural state as possible. With minimal use of additives and preservatives, a clean diet focuses on consuming whole fruits and vegetables while limiting highly processed foods with added sugar and salt. In summary, clean eating seemingly promotes health and wellness and tends to be associated with younger consumers. Plant-based and cultured meat alternatives have tapped into this trend, differentiating themselves from traditional meat, which is associated with poor health outcomes. A McKinsey poll across the UK, United States, France, and Germany in 2022 on conscious clean eating habits found that 40 percent of conscious consumers report eating fresher, healthier foods, especially fresh produce.[107] In total, 33 percent try to choose food without artificial ingredients.[108] A total of 50 percent of respondents identified healthy eating as a priority for them.[109] As a result, the global dietary supplements market has experienced rapid and sustained growth over the past decade as people try to "hack" their health. The market is projected to expand even further. Valued at approximately $177.5 billion in 2023, the market is expected to reach around $327.4 billion by 2030. Asia Pacific leads the market, accounting for 35 percent of global revenue in 2023, with North America and Europe following closely behind.[110]

This focus on wellness has led to increased public awareness of the role of the microbiome. More than half of the cells that the human body contains are not human.[111] They are "microbes"—a catch-all phrase to describe organisms too small for the unaided eye to see. This includes well-known microorganisms such as bacteria, viruses, and fungi but also others called archaea and protists.

These microbes collectively form the human microbiome, symbiotic microbial communities that live in and on us. They can be found at places in contact with the outside world, including the skin, lungs, mouth, and gut.[112] The gut microbiome plays an essential role in the digestive tract. The microbes in our gut allow us to get more out of the food we eat. They further break down compounds that we can't, helping us to access more energy from our food. Gut microbes also produce various metabolic compounds that are essential to (human) cellular functioning. For example, vitamin B12, folate, and short-chain fatty acids (SCFAs), which are utilized by epithelial cells to help regulate things like cell differentiation, proliferation, gene expression, and cell death. They also protect against pathogens, disease-causing organisms, as well as keep a strong mucosal barrier and epithelium.[113] The composition of the gut microbial community is heavily influenced by diet. Altering the community's composition can have far-reaching effects that have been associated with a variety of illnesses. The gut microbiome has been implicated in mediating various neurological and autoimmune diseases. For example, demonstrating the gut-skin axis, one study found that an eight-week probiotic treatment significantly improved symptom severity in psoriasis patients, along with changes in the gut community.[114]

Yet, even for those who are trying to maximize the amount of fresh produce in their diet, they face obstacles. According to a number of scientific studies, over the past several decades, the nutritional density of vegetables has declined.

A 2004 US study found important nutrients in some garden crops are up to 38 percent lower than in the middle of the twentieth century.[115] On average, across the forty-three vegetables analyzed calcium content declined 16 percent, iron by 15 percent, and phosphorus by 9 percent. The vitamins riboflavin and ascorbic acid both dropped significantly, while there were slight declines in protein levels. Similar decreases have been observed in the nutrients present in wheat, and one must eat three apples to receive the same iron content as a single apple in 1940.[116] Several factors are believed to contribute to this trend. Thought-leader Michael Pollan has addressed the decline in the nutritional density of vegetables in his article *Unhappy Meals* and links it to industrial agriculture's focus on yield and efficiency over nutritional quality.[117] In addition to crops being selectively bred to grow faster and larger, soil degradation and overuse of synthetic fertilizers have also reduced the availability of key minerals in the soil, further limiting what plants can absorb. With less nutrients in the food, it means that the body needs more fruit and vegetables to access the same amount of nutrients. In Chapter 3, we will discuss the rise of more industrialized forms of agriculture and the introduction of new breeding programs that focus on increasing yields above all else. In focusing exclusively on yields, the nutrient values of many foods have decreased over recent decades.

Given the interest and potential growth in the wellness sector, the future could see further innovation and expansion. There is investment into personalized nutrition programs. The global personalized nutrition market size accounted for $15.56 billion in 2024 and is anticipated to reach around $60.94 billion by 2034.[118] These are customized dietary plans that are designed to the individual's unique nutritional needs. At the moment, they are fairly expensive and only available to wealthy individuals, but as the costs come down, they could have a transformative impact on people's diets. One example of this is Zoe—a popular personalized nutrition program with tailored dietary advice, costing from £1500 to £2000 ($2000–2700 USD) for a year. There is room for this approach to become mainstream. As David Julian McClements acknowledged in his book *Future Foods*, "In the future, personalized nutrition programs may be supported by governments, companies or the health insurance industry if it can be shown that they provide health benefits outweighing the costs."[119] This could be seen as a preventative healthcare method to tackle dietary ill-health through government health programs. Similarly, this increasingly personalized approach to diet could utilize 3D printing technology to create supplements that are tailored to each individual's needs. Although this creates some challenge, as Chloe Rutzerveld acknowledged in her book *Food Futures*, "To print food, you need raw materials. These must be processed into a puree, powder or paste so that the printer can extrude or glue together the food. Especially in the case of fruit or vegetables, the processing causes a decrease in nutritional value and a loss of texture."[120]

Conclusion

Global diets are in crisis. In 2022, 2.5 billion adults were overweight and 890 million adults were obese.[121] As of 2023, over 60 percent of adults in OECD countries (primarily in the Global North) are overweight or obese, with one in four adults classified as obese.[122] But this crisis isn't limited to the Global North. The rise in obesity is happening around the world. The WHO reports that nearly one in three people in both South-East Asia and Africa are now overweight.[123] This is due to the rise in processed foods and the decline in traditional foods. It is driving ill-health, perpetuated by a lack of essential nutrients. The effect of this obesity crisis, and its predominance in low-income communities, is that it is creating cycles of ill-health and trapping people in poverty. This is all the more concerning in children who can see academic attainment decline with poor diets. But this is not a problem of the individual. This is a societal problem. The dynamics of our food system have allowed this situation to arise. The lack of early intervention by governments and the vast power of food businesses has resulted in obesity expanding unchecked, damaging our future. Our growing obsession with wellness and weight loss is another manifestation of this crisis. People respond differently to our broken food culture. The response is often driven by socioeconomic status. Consequently, while low-income communities have higher rates of UPF consumption and obesity, wealthier people tend to be more fixated on wellness. Both manifestations stem from the same root cause—our broken food system. Ultimately, the future of food will both define, and be defined by, the future of the public dietary crisis.

3

Social and Political Challenges

In 2019, leading agronomist and farming expert Robert Saik opened his book, *Food 5.0: How We Feed the World*, saying, "When you woke up this morning, did you worry about a plague? Did you worry about war?"[1] He continues by dismissing the immediate threat of either disrupting our food system—how wrong he was! The years since 2019 have been dominated by a global pandemic that exposed the fragility of our supply chains, while killing millions. There has been a marked rise in global violence, including war in Ukraine (the breadbasket of the world) and conflict in Gaza. Disparities in affluence have driven resentment and spurred migration, elevated by an exploding global population. These forces have driven (and exacerbated) the rise in populist politicians around the globe.[2] Part of the populist platform is championing tariffs and economic nationalism, which can have a dramatic and inflationary impact on food prices. The public attention that the price of eggs garnered in the 2024 US election is an example of how food can get caught in the crosshairs of political agendas. This chapter considers how domestic politics and international tensions directly impact and destabilize our global food system.

Population Growth

One of the key challenges in considering the future of food is ensuring that we have enough food to feed the world. In 2022, 735 million people were hungry out of a global population of 8 billion.[3] Yet, populations around the world are growing. By 2050, the global population is expected to reach 9.8 billion, with the majority of that increase coming from the Global South.[4] How can we ensure that we can feed all of those people, especially when we are facing the environmental crises that we discussed in Chapter 1? An ethical question arises from that question too—are we going to ensure that everyone has sufficient nutrients to thrive? If so, does that mean that others (primarily those in the Global North) will have to adjust their diets and have less? While the vast majority of people would want everyone to be fed, it is not quite such a popular idea to start restricting access to food so that it can be shared equally.

This is not a new question. In 1798, Thomas Malthus wrote *An Essay on the Principle of Population.*[5] He argued that the population will always expand to the limit of subsistence. Essentially, populations would continue to grow until they are halted by famine or other constraints, such as war. He saw this as a real and present danger as the nineteenth century dawned and believed that the demand for food from an increasing population would soon overtake the land available for crops and result in hunger and disease. This didn't happen. Instead, new agricultural technologies were invented to increase production such as improved plant breeding, horse-drawn hoes and seed drills, and better crop rotations. The Industrial Revolution also meant that food could be shipped from region to region more easily with the expansion of railroads and intercontinental shipping. Through the nineteenth century, achieving food security for the short term and guaranteeing it for the long term within Malthus's era was deeply connected with colonialism. Some European nations expanded into new territories, increasing their access to agricultural land and resources. These new territories kept the population of Europe fed: canned beef from Argentina, wheat from the Punjab, butter from New Zealand. Colonial powers used military force to ensure food security for themselves, often to the detriment of indigenous people. More than just establishing an extractive economic system to feed the Global North, they created a violent, abusive system of slavery to continue the supply of food. Sugar production from the Caribbean and Brazil relied on the enslavement of Africans. Over 1.8 million are estimated to have died as a result of the Slave Trade, although the true horror and pain can never fully be known.[6] Tea, coffee, and cocoa plantations used exploitative and cruel practices to meet the demand from the Global North.

Yet, despite the fact that the horrors that Malthus predicted did not come to pass, these fears over how to feed a growing population didn't disappear. In 1968, Paul Ehrlich gave voice to these fears once again in his book, *The Population Bomb.*[7] He argued that the planet's agricultural systems were nearing their breaking point and could not sustain the rapidly growing global population. He predicted that widespread famine was imminent and that thousands would soon die of starvation. In response, he promoted population control measures to ease pressure on food systems. Ehrlich's language was deeply patriarchal and alarmist—his doomsday rhetoric fueled aggressive campaigns to reduce birth rates in the Global South, leading to grave human rights violations, including coercive sterilizations. While Ehrlich's predictions captured global attention, they were ultimately proven overly simplistic. While it is true that devastating famines have occurred—particularly in the Global South— and thousands have died of hunger and malnutrition, on a global scale, his prediction of mass, inescapable starvation due to population growth did not materialize. Instead, agricultural innovation once again outpaced demographic pressures. Technological advances, particularly the Haber-Bosch method and the widespread adoption of agrochemicals, technologies, and other elements of the Green Revolution, dramatically increased crop yields (global food production has increased by 390 percent since the 1960s) and reshaped global food production.[8]

Today, voices warn that the food supply cannot keep pace with a growing population, while others dismiss such fears as misguided, pointing to the possibility that new technologies will boost yields and avert disaster—just as past predictions of collapse by Malthus and Ehrlich proved wrong. Neither Malthus nor Ehrlich could have anticipated the degree of global integration and connectivity enabled by the Digital Revolution, which has fused economies and cultures far beyond anything they imagined. Information moves around the world lightning fast, facilitating the exchange of goods by allowing companies to communicate more quickly and efficiently than ever before. Over the past century, our economies have globalized and today are dependent on each other. This has transformed the way food moves around the planet and permanently altered our diets. Whereas societies of the past were constrained by their levels of domestic production, nations can now import their food security. Our grocery stores are stocked with produce extracted and exchanged from farmers thousands of miles from where we live: lamb from New Zealand, tomatoes from Morocco, green beans from Kenya.

Rising Urbanization, Rising Incomes

Technology and globalization are not the only differences from the worlds of Malthus and Ehrlich. Populations are increasingly urban, a tension that we will return to in Chapter 9 when we discuss feeding cities. Secondly, people across the world are overall richer. Consequently, they demand better food, not just the bare minimum to survive. From 1820 to 2015, global average GDP per capita (adjusted for inflation) rose from $1,500 to $17,735, and poverty decreased from 90 percent to 10 percent.[9] This has happened particularly in China and India.

Daily income levels in China rose by 126 percent from 2001 to 2011 with the middle class expanding to roughly 707 million.[10] In India, income levels rose by 24 percent during the same time period.[11] Moreover, we have also seen this trend in other countries that have become economic powers, such as Indonesia. That is not to say that poverty, inequality, and resulting hunger are no longer issues in these places, but there are millions more people who expect to be able to purchase and eat the food they want and particularly food that they associate with wealth. A lot of this is imitating Global North diets, which heavily feature meat. In China, between 1961 and 2019, meat production increased by more than 3,000 percent, while per capita meat consumption went up fifteen-fold.[12] This is decidedly different from the context that Malthus and Ehrlich were writing in when the vast majority of the global population was impoverished and could not afford meat or other luxury goods. Part of this trend is the conspicuous consumption we discussed in Chapter 2, where people want to be seen consuming food that denotes wealth.

Globalized Instability

By the 1990s, the global order under which Malthus and Ehrlich were writing had changed considerably. The rise of globalization, coupled with the end of the Cold War and new technologies, pulled economies closer together. Academics in the 1990s suggested that the world might be moving into a period of stability and peace.[13] Yet, that dynamic was quickly proven false. The new century did not bring about an era of calm. Despite (or perhaps because of) this interconnectedness of economies and culture, the world has become less and less stable since the turn of the millennium. These geopolitical shifts have profound implications not only for security and governance but also for the future of global food supply. As climate change, resource scarcity, and population growth put increasing pressure on agricultural systems, control over fertile land, water resources, and food trade routes becomes a strategic asset. In the coming decades, the rivalry among powers is likely to play out in part through competition over sustainable food systems and control of global supply chains.

Moreover, the past decade has seen two powers rise to challenge American hegemony—China and Russia. Since 1978 when it introduced policies to open up its economy, China has seen extended and consistent growth—ranging from 9 percent to 13 percent annually.[14] It has become a manufacturing powerhouse, and per capita incomes have almost quadrupled.[15] Experts predict that China will become the largest economy (overtaking the United States) in the 2030s.[16] It can use its newly found economic power to influence and shape the future of geopolitics with a robust economy and a world-class military. It increasingly asserts itself, whether in the territorial disputes within its region (i.e., Spratly Islands and Taiwan) or by seeking to increase influence in intergovernmental institutions, such as the UN. As the United States did with the Marshall Plan in the 1950s, it utilizes foreign aid and investment to expand its sphere of influence.[17] China has invested heavily in ports, railways, and airports across the Global South to establish allegiances and strengthen its hand. China's investments in agriculture and infrastructure across Africa, Asia, and Latin America aim not just at influence but at securing long-term access to food and raw materials to sustain its population.

The expansion of the Chinese economy and the associated increase in global influence have been simultaneous with the reemergence of Russia as a counter to Western democracy. Following the fall of the Berlin Wall and the end of the Cold War, Russian influence in global politics declined. However, starting in 2000, President Putin's ascension to power has returned the country to a powerful force. He consolidated political control and leveraged the country's vast natural resources to boost the economy. Oil and gas have long been Russia's main tools of economic and diplomatic power, but it also uses its vast grain exports and fertile farmland in much the same way—shaping global markets, influencing food prices, and gaining leverage over import-dependent countries. He uses populism (coupled with intimidation, repression, and state violence) to maintain his support and, increasingly, far-right language to build his power base and present an image of himself as authoritarian.[18] Given his strongman tendencies, he had pursued a more assertive foreign policy, particularly in neighboring regions, and moved away from intergovernmental organizations, like the UN. This has led to rising tensions with nations around the world and a shifting alignment in alliances.

The world is also experiencing a rise in violence, instability and conflict. The war in Ukraine has caused huge amounts of destruction and displacement and constitutes the biggest threat to peace and security in Europe since the end of the Cold War. Russia invaded Ukraine in February 2022, sparking months of brutal conflict. A total of 12,456 people have died, and 8 million people have been internally displaced.[19] The invasion in 2022 followed the Russian annexation of Crimea by force in 2014. President Putin mounted a major military incursion into eastern Ukraine that locked the region in conflict for years. President Zelenskyy (with US and European support) continuously pushed back against

Russian forces and, despite Ukraine's smaller size and military prowess, created a stalemate in the war. In addition to the horrors of war, Russia's full-scale invasion severely disrupted Ukrainian agricultural production. In 2022, the agricultural sector's contribution dropped by 39 percent compared to 2021.[20] Prior to the conflict, Ukraine was one of the world's leading agricultural producers and exporters, particularly known for its vital role in supplying oilseeds and grains to global markets. In 2021, agricultural products made up 41 percent of Ukraine's total exports. Over 55 percent of the country's land is arable, and the sector provides employment for around 14 percent of the population. In the 2021–2 marketing year, Ukraine ranked first globally in sunflower production, second in sunflower oil and meal, fourth in barley, sixth in corn and rapeseed, and seventh in wheat.[21] Experts have described Russia's war in Ukraine as the greatest military-related disruption to global agricultural markets in at least a century.[22]

In areas near the front lines, Russian forces have started using drones to set fields on fire. According to Dmytro Yunusov, the head of agriculture in Kherson province in Ukraine, Russian drones can destroy 1,000 hectares of crops each month.[23] This is not limited to wheat. Both pork production and viticulture have also been targeted by Russian forces.[24] Russia has openly admitted to using food as a geopolitical tool. In April 2022, former President Dmitry Medvedev referred to food as Russia's "silent weapon," and by August 2023, President Vladimir Putin stated that Russia "has the capacity to replace Ukrainian grain both commercially and as free aid to needy country's [*sic*]."[25] Ukrainian Minister of Agriculture, Vitaly Koval, describes the attacks as "a big threat to us," saying they are intended to cut Ukraine's exportable supply while pushing up global prices to benefit Russian farmers.[26] Following the invasion, Russia increased its wheat exports to sub-Saharan Africa (in particular North Africa, where it is the top exporter) and has also increased its market share in Western Asia.[27] By deepening other countries' dependence on its agricultural exports, Russia has used food as a soft power tool to expand its geopolitical influence and weaken Ukraine's role in global food security. The conflict continues at the point of writing with no clear end to the conflict.

Escalating global instability and further demonstrating the new tactic of using food as a weapon is the ongoing conflict in Gaza. The conflict opened a new chapter with a surprise attack by Hamas on southern Israel on October 7, 2023. In that attack, militants breached the border, killed over 1,400 Israelis, and took more than 200 hostages back into Gaza.[28] Israel responded with a large-scale military operation aimed at dismantling Hamas's infrastructure, launching extensive airstrikes and ground incursions into Gaza. The Israeli campaign has resulted in widespread destruction and civilian casualties, with Palestinian health authorities reporting tens of thousands killed with more wounded, and severe damage to homes, hospitals, and basic infrastructure.[29] Efforts at ceasefire negotiations, mediated by Egypt, Qatar, and the United States, have repeatedly stalled amid disputes over hostages, prisoners, and the terms of a lasting truce.[30] The conflict has drawn international concern, with calls for restraint and increased humanitarian access, but no resolution has yet been reached as fighting continues in mid-2025.[31]

The ongoing conflict in Gaza has created a severe humanitarian crisis with hunger now at calamitous levels due to the weaponization of food.[32] Famine was declared in parts of the territory.[33] Israel's military strategy has included tight controls on the flow of goods, in particular food, into Gaza, citing the need to weaken Hamas by cutting off its resources.[34] Since the conflict began, Israel has restricted or periodically halted the entry of food, fuel, and humanitarian supplies through key border crossings, most notably Rafah and Kerem Shalom, though humanitarian organizations have urged that the collective punishment of civilians violates international law.[35] Israeli officials have described these measures as part of a broader effort to stop Hamas from diverting aid from civilians to meet their own needs.[36] Still, the blockade has also impacted the civilian population, the majority of whom are already dependent on aid.[37] The United Nations and international NGOs report that nearly half a million of Gaza's population now face starvation, with widespread malnutrition, particularly among children.[38] Aid deliveries have been hampered not only by Israeli restrictions but also by damaged infrastructure and attacks on convoys.[39] The use of food and access to aid as a tool of war has sparked global concern and calls for unfettered humanitarian access.

All of these conflicts, and others such as Sudan, Haiti, and Myanmar, have led to record levels of displacement.[40] This has been exacerbated by the pull of economic opportunities that many hope to find in the West. As of June 2024, there are now 43.7 million refugees in the world.[41] This is not to downplay the devastating death and destruction caused by wars and conflicts throughout history—violence has always left deep scars. But, in terms of food, the global consequences were limited. Wars in previous decades (and previous centuries) caused hunger and famine, but food supply was more localized. Consequently, the impact of conflict on food supply was contained, and the global disruption was, therefore, minimal. Today, with our highly globalized food system, that is no longer the case. Now, supply chains span oceans, and many countries' food security depends on imports. This means that wars and conflicts, even those happening halfway around the world, can disrupt food supply, drive up prices, and push hunger levels higher—not to mention the added challenge of feeding millions of displaced people. Consumer have also recognized the weaponization of food and have been using their pocketbooks to voice their rejection of certain actions. For example, following the invasion of Ukraine, many consumers boycotted Russian food products.[42] Similarly, there have been boycotts of Israeli products in opposition to their actions in Gaza.[43]

Case Study E: World Central Kitchen—Overcoming Geopolitical Barriers in Myanmar

A magnitude 7.7 earthquake struck Southeast Asia on March 28, 2025, with its epicenter in Myanmar. The disaster left more than 5,000 people dead and thousands more displaced as homes, roads, and railways were destroyed.[44] Getting food into the affected regions immediately became a daunting challenge. World Central Kitchen (WCK) mobilized within hours. Through its #ChefsForMyanmar initiative, the relief team partnered with local restaurants and community leaders to prepare and deliver over 1.5 million meals to survivors in hard-hit areas like Mandalay.[45] Family cooking kits, filled with locally sourced ingredients, were also distributed to displaced households. In the process, WCK helped neighbors of different faiths—Muslims, Hindus, Buddhists, and Christians—work together in the recovery effort. Myanmar, with more than 130 ethnic groups and a tapestry of religions, embodies both the promise and the difficulty of WCK's model. This diversity has long enriched the country, but it has also fueled conflict, most painfully in the persecution of the Rohingya.[46] Against this backdrop, WCK's insistence on neutrality and inclusivity was not only a matter of logistics but also of diplomacy. Delivering food meant navigating deep political divides, local mistrust, and a government wary of outside influence.

Since its founding in 2010 by chef José Andrés, WCK has transformed the way humanitarian food relief is delivered, proving that fresh, locally prepared meals can reach people faster and with greater dignity than packaged rations. But Myanmar illustrates that the obstacles are rarely just physical. In crisis zones, authoritarian controls, shifting alliances, and bureaucratic red tape often slow or restrict aid. Securing permits, negotiating with multiple political actors, or simply moving supplies across contested territories can be as challenging as repairing damaged infrastructure. WCK's experience in Myanmar underscores a broader truth: in disaster response, political realities are as decisive as logistical ones. By building trust with local communities, sourcing food within local markets, and navigating complex political environments with diplomacy, WCK shows how humanitarian organizations can adapt and still deliver impact. Their work highlights not only the urgency of feeding people in times of crisis but also the delicate balance between humanitarian action and political constraint.

International Trade and Tariffs

As mentioned earlier, our food system is now globalized. In 2023, the global food trade was valued at $2 trillion, nearly four times higher than in 2000, and is predicted to grow at a rate 2.5 percent to reach $6.5 trillion by 2050.[47] Global trade has increased the choice and availability of food for the world's population—from the poorest to the wealthiest. About one-quarter of the world's food is traded internationally with WorldWatch estimating that the average American dinner plate will have traveled a total of 1500 miles.[48] Much of this trade is dominated by commodity crops. In the 2023–4 period, 2.307 billion tons of grain were produced with the leading producers being China (638 million tons), the United States (439 million tons), and India (369 million tons).[49] Out of this grain, a vast proportion goes to China. China imports over 50 million tons of grain annually with 36 percent coming from the United States.[50]

This increased interdependence and connectivity is a result of the push in the twentieth century to liberalize economies and embrace free trade. One example of this is the establishment of the World Trade Organization (WTO) that governs the rules of trade between nations and aims to ensure that trade flows as smoothly, predictably, and freely as possible. It seeks to reduce protectionism and promote globalization. The WTO has key agreements within it, such as the "General Agreement on Tariffs and Trade (GATT)," which governs different aspects of trade. Another key body in the global trade of food is the Codex Alimentarius. Established by the FAO and WHO, it sets international food standards to ensure consumer safety and promote fair trade.[51] The Codex Alimentarius guidelines influence everything from pesticide residue limits to labeling of novel foods like lab-grown meat. With growing global reliance on UPF and globally traded foods, the impact of this standard strengthens the need to balance innovation with health and sustainability. However, critics warn it may favor corporate interests over smallholder farmers and localized food systems. Once a member of the WTO, a state can officially complain that another member state is in breach of a WTO agreement. This happens mainly when a nation has adopted "protectionist" measures to benefit a domestic industry or limit imports. While not a regulator, the WTO is a forum for settling disputes about such agreements. If there is a dispute, the WTO will encourage the states to reach an agreement with the goal that any protectionist measures are removed. If agreement cannot be reached, then the matter is referred to a panel of experts for a ruling. To date, 594 disputes have been brought to the WTO, with over 350 rulings being issued.[52] Compliance with the ruling rests exclusively on social norms and conventions, so states can choose to ignore them, but members comply with rulings in about 90 percent of cases.[53]

In addition to the WTO, there has been a rise in bilateral or multilateral free trade agreements since the 1990s. These are between two nations (or a group of nations) that come together to agree a set of terms for free (or at least preferential) trade. NAFTA is the leading example of this type of FTA. Once NAFTA was in place, the total value of agricultural trade between the United States, Mexico, and Canada grew 233 percent (accounting for inflation) between 1993 and 2013, demonstrating the power of FTAs to deliver growth and make food cheaper.[54] However, these agreements are incredibly complicated to establish. As Professor Tim Lang wrote, "Politicians talk of new trade deals as though they are flicking a switch, but have little understanding of actual food flows (or the labyrinthine nature of trade negotiations)."[55] Many fail to get off the ground—particularly when negotiations prove too challenging to get the deal across the line or face public opposition. For example, trade negotiations between Japan and South Korea collapsed in 2019 over historical tensions of wartime atrocities. These trade deals have increased the flow of foods but have also created a food trade gap. Global North countries can rely on imports to ensure their food security. By contrast, many Global South countries remain dependent on exporting their agricultural products while still struggling to feed their own populations. This imbalance deepens inequalities in the global food system and leaves vulnerable regions exposed to market shocks and rising prices.

However, despite this globalized model of trade liberalization, there has been a significant shift away from free trade toward protectionist policies in recent years. This has primarily been driven by the marked rise in authoritarian populism across the world.[56] This has happened in the United States with President Trump, who appealed to a huge swath of the country with an anti-establishment agenda and secured a second term in 2024. India and Turkey have experienced similar rises with Prime Minister Modi in India, whose party, Bharatiya Janata Party

(BJP), espouses right-wing populism, and Turkey's President Erdogan, who won a third Presidential term in 2023 (although the election was not considered free or fair) on a platform of division.[57] This populist rise around the world has overlapped with the rise in extremist political views. Europe has seen the rise of far-right political parties. For example, the election of right-wing Giorgia Meloni in 2022, when her party, Fratelli d'Italia, won; Rassemblement National (previously known as Le Front National) reaching the second round of the French Presidential election in 2022 against President Macron; and Germany's AfD making significant gains in the 2025 election to become the second-largest party in Parliament.[58] Outside of Europe, far-right views have gained significant traction in the United States in recent years. While far-right movements have often existed on the fringes of American politics, they have become more visible and influential, particularly since the mid-2010s, and events such as the 2017 Charlottesville rally and the January 6 Capitol attack highlight the growing confidence and organization of far-right groups. This wave of populism and political extremism has begun to shape global food systems in significant ways. Rising nationalism and skepticism toward globalization have led to more protectionist trade policies, restrictions on agricultural imports and exports, and disruptions to international supply chains. In some countries, populist leaders have used control over food prices and distribution as a political tool to shore up domestic support, often at the expense of long-term sustainability. Meanwhile, the polarization and instability associated with these movements threaten international cooperation on food security. As climate shocks, resource competition, and population growth strain food systems, the rise of insular and antagonistic politics risks undermining the coordinated global response needed to ensure a stable and equitable food supply for the future.

The rise in populist protectionism has had a negative impact on food, causing disruptions in flows and higher prices. This wider rejection of free trade is best seen in President Trump's economic policy of "America First." His plan to "Make America Great Again" rests on reducing the reliance on foreign imports and bolstering domestic production. In his first term (2016–20), President Trump focused on China and imposed tariffs. This had a significant impact on US food and agriculture since China is the biggest export market for American farmers and in particular for US grain with the majority of this grain going into Chinese pork production.[59] China produces over 57 million tons of pork annually, so the retaliatory tariffs caused economic losses in the American farming sector.[60] Exports of grain from the United Staes to China decreased by 81 percent from 2017 to 2019.[61] In order to offset this loss, the US Department of Agriculture (USDA) paid farmers tens of billions of dollars to compensate for the decline in income.[62] In response, President Xi has attempted to wean his country off food from the West. Amid rising geopolitical tensions, he sees self-sufficiency and diversity of supply as an increasingly urgent need. Strategically, China switched supply and began importing from Brazil. In 2018, China imported 187,000 metric tons of soy from South America.[63] This shift had a profound impact on trade flows since China was able to purchase a huge quantity of the South American grain, limiting access to those exports for other nations.

President Trump's objection to the trade deficit with China was a key priority from the first day of his second term. In early 2025, President Trump reignited his trade war with China, starting with a 10 percent tariff on all Chinese imports in February, later raised to 20 percent in March citing concerns over fentanyl trafficking.[64] China retaliated with tariffs on American goods and nontariff barriers.[65] On April 2, dubbed "Liberation Day," Trump announced sweeping tariffs on nations across the world with particular emphasis on China —raising US duties on Chinese goods to up to 145 percent—and China responded with tariffs of up to 125 percent.[66] Facing sharp market reactions, both sides agreed on a truce in May 2025 and US tariffs dropped to 30 percent, Chinese tariffs to 10 percent.[67] President Trump claimed that the tariffs "will be great" for American farmers—but, in a mention of candor admitted there "may be a little bit of an adjustment period."[68] The impact of these tariffs will take time to quantify but are likely to be disruptive to the future of food given China's dominance of the US grain export market.[69] In response to Trump's tariffs, the country announced it will reciprocate tariffs on American food and agricultural products including soybeans, red meat, and chicken.[70] This will likely have a similar scale of impact as it did in President Trump's first term, and a recent study, commissioned by the National Corn Growers Association and the American Soybean Association, found that in the event of a new trade war, American soybean exports to China could drop by 51.8 percent, and US corn exports to China could plummet by 84.3 percent.[71]

President Trump's rejection of free trade is not solely focused on China. Mexico and Canada are America's biggest trading partners but have come under intense scrutiny from the Trump administration.[72] Throughout his 2016 campaign, then-candidate Trump rejected the terms of NAFTA and, once in the White House, he renegotiated a new trade agreement for North America—USMCA. The USMCA updated policies and procedures to deliver a better deal for the United States. American farmers are gaining increased access to the Canadian dairy, poultry, and egg markets, in exchange for Canada's increasing access to American dairy, sugar, peanuts, and cotton markets. Many of these policies returned when President Trump returned to power in 2025. Framing it as a step toward "America first" production, President Trump announced sweeping tariffs on imports from Canada and Mexico. Signed into effect on February 1, the policy imposed 25 percent tariffs on most goods from both countries and 10 percent specifically on Canadian energy, citing national security concerns. After brief negotiations, a one-month delay was granted on February 3, but the tariffs officially took effect on March 4. Canada responded swiftly with reciprocal 25 percent tariffs on US goods, while Mexico took a more cautious approach, delaying its response.[73] This dispute remains ongoing at the time of writing.

The Trump doctrine of protectionism expanded globally with "Liberation Day." On April 2, 2025, Trump declared a national emergency via Executive Order 14257 and imposed a 10 percent baseline tariff on virtually all imports, alongside steeper reciprocal tariffs—up to 54 percent—on countries with large US trade deficits, including the EU, Japan, China, Mexico, and Canada.[74] Framed as America's "economic independence day," the move sought to revitalize US manufacturing, reduce deficits, and pressure trading partners to renegotiate more favorable terms. The impact on the US agriculture sector was marked, with ripple effects extending across global markets. American farmers, already operating on thin margins, are now facing higher input costs for equipment, fertilizers, and animal feed—many of which are imported and now subject to steep duties. At the same time, retaliatory tariffs from trading partners like China, Mexico, and the EU have sharply reduced export opportunities for key US commodities such as soybeans, corn, pork, and dairy. This double hit—rising costs and shrinking markets—has left many producers financially strained, with some warning of widespread farm closures if relief or policy adjustments are not introduced. The effects of these tariffs, however, extend beyond the farm gate. Economists have warned that the increased costs of production will inevitably be passed down the supply chain to consumers. As a result, food prices are expected to rise, with low income households feeling the brunt of this inflationary pressure. Although the full impact remains to be seen at the time of writing, the disruption highlights just how deeply interconnected global food systems are—and how trade policies, even when driven by domestic priorities, can destabilize markets and threaten food security around the world.

Shipping Routes

The global exchange of food is only possible because of the network of container ships crisscrossing the oceans. Our globalized society is reliant on shipping.

A total of 95 percent of all manufactured goods in the world are transported via container ships, equivalent to $4 trillion.[75] Over $1 trillion of this is agricultural goods—food, seed, and fertilizers.[76] These ships are massive. One ship alone would have the capacity to carry 740 million bananas—enough for one European each—and is 1320 feet long and 200 feet wide.[77] The critical nature of container shipping to our global food economy cannot be overstated. Industry feels confident of their position too, expanding their capacity in 2023.[78] Part of this confidence stems from the fact that there is little alternative. Flying goods is expensive and therefore decidedly unpopular. Only 0.16 percent of food is flown by air.[79] The small number of foods that are flown in need to command a high price to justify the expenditure. For example, seafood is often flown from South America to the United States to ensure freshness.

Case Study F: The Suez Canal—a Food Security Chokepoint

The Suez Canal opened in 1869 and provides the shortest maritime link between Europe and Asia. It carries roughly 12 percent of global trade and a significant share of the world's grain, vegetable oils, and fertilizers, making it central to ensuring global food security.[80] Because of its centrality, the canal has often been entangled in geopolitical struggles. This is special, but not unique. There are fourteen major maritime, coastal, and inland choke points that underpin global trade. These critical trade locations are known as "choke points" where any disruption can ripple across supply chains. Many lie in politically volatile regions, where even smallscale disruptions can cascade into wider crises and destabilize markets, driving up food prices.

The Suez Crisis of 1956 was the first major disruption of the modern era. The crisis demonstrated how control over Suez could be used as leverage, with direct consequences for food-importing nations. Egyptian President Gamal Abdel Nasser nationalized the canal, prompting Britain, France, and Israel to launch a military response. The conflict halted shipping and pushed vessels to reroute around the Cape of Good Hope, delaying grain deliveries to Europe and raising transport costs. Further conflicts reinforced this vulnerability. During the Six-Day War in 1967, Egypt closed the canal. It remained shut until 1975, until the Sinai Interim Agreement ended the conflict after the Yom Kippur War. For eight years, world shipping avoided the passage, significantly raising transport costs for grain shipments from the Black Sea and Persian Gulf. The closure particularly hurt Global South nations reliant on wheat imports, underscoring the dependency of food security on maritime stability.

More recently, the canal's strategic importance has been tested by regional instability. Piracy in the Gulf of Aden during the 2000s and, more recently, attacks by Houthi rebels in the Red Sea have once again made Suez vulnerable. In late 2023, escalating attacks forced nearly 90 percent of shipping traffic to divert around the Cape of Good Hope, adding two weeks to delivery times and raising freight costs dramatically.[81] The knock-on effects were immediate: in early 2024, Red Sea disruptions contributed to food price inflation of up to 0.7 percent in many economies.[82] The Suez Canal illustrates how geopolitical conflict can translate directly into higher food prices and greater insecurity. For import-dependent nations, especially in the Middle East and Africa, stability at this chokepoint is not a matter of convenience but of survival.

President Trump recently highlighted the vulnerability of the Panama Canal early in his second term. President Trump claimed that the Canal was falling under Chinese control.[83] The Canal is currently operated by the Panama Canal Authority, but the United States controlled the Canal from 1903 to 1979.[84] President Trump expressed a desire for the United States to regain control of the Canal, calling the 1999 handover from the United States to the Panamanians a "foolish gift."[85] He accused Panama of overcharging US vessels and allowing Chinese influence.[86] While controversial, his claims highlight the essential nature of the Panama Canal to US interests. Nearly 60.4 million tons of US goods go through the Panama Canal.[87] Following the president's statements, the US company BlackRock acquired a majority stake in the two port operators at either end of the Canal that had previously been held by the Hong Kong-based, CK Hutchison.[88] That said, the Canal might end up becoming redundant in coming years due to climate change. Falling water levels in the Gatún and Miraflores lakes on either side of the Canal continuing to drop could make the Canal inoperable.[89] If that were to happen, the Bering Straits in the Arctic Circle could pose an alternative to the Panama Canal for commercial traffic in the long term since the passage could be navigable for a number of months each summer with the global temperature rises. However, the passage through the Arctic would require significant investment for escort vessels (given the risky nature of the journey) and staging ports. The Canadian commercial marine transport industry does not anticipate that this route will be a viable alternative to the Panama Canal for at least another ten to twenty years.[90]

Conclusion

People have always been concerned over population growth and whether we would retain the ability to feed ourselves. Fearing famine, history has proposed drastic solutions, while governments across generations have used it to push for more expansionist policies. The fears that expanding populations will pressurize the food system are nothing new. However, what is different today is the urbanization and the globalization. Since the 1990s, food supply has become increasingly globalized as our world economy becomes intertwined. Our supply chains have become longer and longer, relying on technology to facilitate exchange. Over the last few decades, this has helped to reduce costs and reduce hunger—both in the Global South and the Global North. Yet this makes our food supply vulnerable. Since our food system is now so connected and interdependent, when there are periods of geopolitical turmoil and instability (such as there is now) the food supply becomes compromised. This is pertinent to our consideration of the future of food since we are entering into a period of instability. War and violence are on the rise around the world.[91] With connected economies, these conflicts wipe out harvests and the flow of food by closing off transport routes and destroying food infrastructure. They also displace people, including farmers and food producers, and create hunger in refugee communities in the immediate and more widely among the people who depended on the food grown in the conflict region. However, disruption is not always from war. It can come from trade disputes—these are on the rise with populist leaders advocating more isolationist policies—and from climate change with rising instances of extreme weather events. Market shocks can devastate food supply and impact nutrition halfway around the world. In particular, the vulnerabilities within the shipping industry are substantial, particularly given the overreliance on several key choke points. With a delicate balance of just-in-time readiness and the tight margins, disruption in one choke point causes a ripple effect, delaying exports around the world and impacting food security. With all these risks bumping along together, it seems increasingly likely that something soon will cause a dramatic disruption to our food supply. Consequently, we need to think how we can develop a more resilient future for food supply to ensure that we are better able to weather the shocks.

4

Farming in Crisis

The world's food system is under strain, and farmers are bearing the brunt—on the frontline of multiple crises, many are reaching breaking point. Yields are dropping because of changes in the climate with increased frequency of extreme weather events. Land is eroded with soil fertility declining. Despite this, since the 1950s, there has been a push to make food cheaper to buy to ensure that fewer citizens go hungry—an admirable goal. This has been achieved through the industrialization of food production and globalization of food trade. Food is produced where it is cheapest (i.e., the Global South, where land and labor are less expensive and production standards are lower) and then shipped around the world to meet demand. Yet, even though 9–10 billion tons of food are produced each year, farmers' incomes have been stagnant or declining for the last decade.[1] Consumers have come to expect readily accessible and inexpensive food. This expectation is making it hard for farmers to earn a living. The rising input costs (driven by agrochemical companies) and increased geopolitical uncertainty with increasing instances of protectionist tariffs mean that farming is no longer economically viable for many. Although 2022 was a record year for net farm income in the United States, 57 percent of farms still operated at a loss.[2] Approximately 30 percent of all farm businesses in England are running at a loss, and a further 25 percent are making a farm business income of less than £25,000 ($30,000 USD).[3] In response, smallscale and family farmers are exiting the industry. If the situation does not change, the future of agriculture might result in transnational agribusinesses running a patchwork of corporate farms.

Input Companies

Farming is capital-intensive, and nowhere is that clearer than in machinery: a commercial tractor can cost up to $250,000 USD. Smaller farms rarely have that kind of money on hand, often struggling even with midsized models priced around $50,000 USD—leaving many dependent on debt. Other key components of modern farming are expensive too. According to the USDA, US farmers spent $35.8 billion on fertilizers and other soil improvers and $21.6 billion on pesticides in 2023.[4] Farmers are overwhelmingly price-takers, meaning that they have little ability to influence the prices that they pay for fuel, fertilizer, and feed. They have to accept the prices set by retailers primarily since there is no other option. A few very large transnational companies dominate the agricultural sector, and they are so powerful that they can set prices. The top six firms—Bayer, Syngenta, Corteva Agriscience, BASF, FMC Corporation, and ADAMA Agricultural Solutions—now account for 78 percent of all sales of pesticides, herbicides, and other chemical products.[5] A similar consolidation has occurred within the fertilizer sector. Nutrien, Yara International, CF Industries, and The Mosaic Company are central to the production, transportation, and sale of fertilizers—whether phosphate- or nitrogen-based fertilizers. Collectively, they control roughly 50–60 percent of the global fertilizer market. The same is true for machinery. John Deere has approximately 28 percent of global market share, Mahindra & Mahindra has approximately 18 percent.[6]

This pattern repeats across seeds, where Bayer, Sygenta, Corteva, and BASF are also the major players and collectively control approximately 65–85 percent of the seed market.[7] Two of the other companies, Monsanto and DuPont Pioneer, have now been folded into Bayer and Corteva, respectively, increasing this control still further. The expansion of biotech seeds has exacerbated this problem, which we will discuss in Chapter 8. Yet, even aside from those new plant-breeding techniques, purchasing new high-quality hybrid seeds can be a huge outlay. Hybrid seeds (also known as F1) are varieties bred from the controlled cross-pollination of two different parent plants. They are typically chosen for desirable traits and are designed to produce plants with improved uniformity, vigor, and yield. But the seeds from these varieties cannot be saved, a traditional practice, since the next generation tends to be unpredictable and inconsistent. For farmers in the Global North, purchasing new seed can equate to $129 per acre for corn.[8] For farmers in the Global South that is even more. According to Timothy Wise's *Eating Tomorrow: Agribusiness, Family Farms and the Battle for the Future of Food*, the majority of farmers in Malawi could only afford hybrid seeds with government subsidies; they were far too expensive on their own.[9] However, the subsidy program has brought an end to the routine practice of seed saving, meaning that farmers are tied into purchasing hybrid seeds year after year (often pushing them into debt) since the breeding method sees productivity drop in year two.

With all this power in the hands of very few companies, farmers have little ability to shop around for a bargain. They have to accept the price set by these companies and often find themselves facing steep increases. With these rising costs, it is only the big businesses that can survive due to the economies of scale. Small- and medium-sized farms (often family owned and operated) are not able to survive. There has been a notable rise in corporate farms—something that the Biden Administration attempted to address through Executive Order that the Trump administration rescinded.[10] Corporate farms tend to cover thousands of acres and only focus on one product—soy, corn, pigs, and chickens, for example. These enterprises are often operated by corporations with multiple holdings worldwide, as opposed to individual farmers. Companies include Cargill, Tyson Foods, and JBS and have deep pockets and significant resources, so they are better able to weather shocks. After Russia's invasion of Ukraine, the cost of agricultural inputs (seeds, agrochemicals, etc.) soared, in particular in Europe. In the last chapter, we discussed the impact of the conflict on the global market for commodity crops. However, there was an energy impact too that was felt globally because of Russia's natural gas exports. Without access to Russian gas, the prices of European energy, fertilizer, animal feed and diesel collectively rose 38–50 percent in the five years to 2024 with an 85 percent spike in 2023.[11] According to IMF research, the UK was the worst hit due to its overreliance on gas.[12] This sudden rise caught many farms unawares. They had not budgeted for such an enormous climb in costs, and many smaller farms could not survive. The larger ones were better able to endure and weather the storm. Yet, the impact that the rise in Russian gas had on food prices was not limited to Europe. The disruption from the conflict caused fuel and fertilizer prices to rise globally, contributing to inflationary pressure and increasing food insecurity.

Supermarket Power

The retail sector has seen similar consolidation, akin to the agribusinesses just discussed. This gives supermarkets huge power to control the supply chain. This means that they have the dominant bargaining power and can set the terms and prices with the farmers that supply them—pushing for lower and lower prices. There are also allegations that supermarkets fail to honor the original agreed price or fail to pay in a timely manner, creating serious cash flow issues for farmers, and opt for short-term contracts so that farmers have little confidence over the price or demand for certain produce when they start sowing.[13] They have to gamble that they will be able to sell it at an acceptable price come harvest time. In the US groceries sector, Walmart controls 23.6 percent with Kroger and Costco making up another 19.3 percent.[14] In Germany, Aldi and Lidl together make up 20.1 percent—two supermarkets that are expanding their reach globally and emerging as global leaders.[15] In the UK, four big supermarkets—Tesco, Sainsbury's, Asda, and Morrisons—control around 70–80 percent of the market, and more than two-thirds of grocery sales pass through the tills of those four supermarkets.[16] The weak position of farmers within the supermarket system was highlighted in the UK in 2015 when milk prices dropped due to the removal of the EU's milk quota system. The EU milk quota system, introduced in 1984, was designed to curb overproduction by setting strict limits on how much milk each member state—and each farmer—could produce. Farmers who exceeded their quota had to pay a levy, which helped stabilize markets and keep prices from collapsing. Without the quota, there was a glut of milk. Dairy farmers were only paid 15–20 pence (0.20–0.27 USD) per quart. In response to the controversy and associated protests, the Grocery Code Adjudicator was tasked to ensure fairness and transparency within the food sector.

To challenge this control, there has been a rise in farmer cooperatives. Cooperative farms are groups of individual farms and farmers that work together to boost profits. It empowers individuals by fostering collective decision-making, resource-sharing, and risk mitigation, thereby enhancing the economic and social well-being of its members. The collective bargaining power afforded to them by virtue of functioning as a cooperative has also helped to protect them from exploitative contracts and unfair practices from larger companies. This has worked successfully in Italy since the 1990s. Italy's Law 381/91 on social cooperation was established in 1991 to support and encourage the establishment of cooperatives.[17] The legislation outlines various forms of support for agricultural cooperatives, including tax benefits and financial assistance for cooperative development. Ten of the fifty largest Italian agro-food enterprises are cooperatives, including Agricultura Nuova, which has been running since 1977 and now employs forty people over 300 hectares of land producing organic cereals and vegetables. Cooperative dairy has been successful in challenging the power of supermarkets mentioned earlier. Some of the biggest cooperatives around the world are in the dairy sector—Arla in Europe has a turnover of $8.77 billion USD, Fonterra in New Zealand has a turnover of $13.48 billion USD, and Land O'Lakes in the United States has a turnover of $16.8 billion.[18] It is worth noting that cooperatives have had a transformative impact on agriculture in the Global South, allowing smallscale producers to access lucrative international markets that they would never have otherwise been able to. This power can be seen in high-value commodity crops produced exclusively in the Global South, such as tea, coffee, and cocoa, among others. For example, Kuapa Kokoo Ltd. is a successful farmer cooperative that fosters long-term partnerships with around 30,000 smallscale cocoa farmers. As one of the largest cocoa trading companies in Ghana, it provides members with bonuses and other benefits.[19]

Case Study G: Farms of the Future—Educating a New Generation of Rural Brazil

As is true the world over, Brazilian agriculture is dominated by big businesses. The largest meat company in the world, JBS, is Brazilian and accounts for an annual revenue of $77.18 billion USD, while the Brazilian BRF S.A. is one of the world's largest poultry exporters and is present in over 140 international markets.[20] These large multinationals have made Brazil a major agricultural player in global markets, with total agribusiness exports reaching $164.37 billion USD in 2024.[21] However, despite the dominance of large-scale agriculture in Brazil, small- and medium-sized rural producers struggle to survive. From generational succession, low productivity, and lack of technical knowledge, many face economic instability. To address some of these challenges, Brazil has invested in rural education and technical training. One key initiative is the Technical and Managerial Assistance (ATeG) program, developed by the National Service for Rural Apprenticeship (SENAR). The program promotes sustainable practices, strategic planning, and technological adaptation, aiming to improve competitiveness and quality of life in the countryside. ATeG offers a free twenty-four-month training cycle that includes personalized diagnostics and ongoing support. Over 6,700 field technicians and service providers deliver support across a range of sectors, with key focus areas including dairy, fruit growing, and beef cattle.[22] The program has reached over 387,000 rural producers and operates across more than 14 million hectares, an area five times the size of Argentina.[23] Participants vary in both education and age: 42 percent have completed high school, 35 percent elementary school, and 12 percent hold a higher education degree, while the average participant is forty-eight years old, and 92 percent are smallscale farmers.[24] While ATeG can help address generational succession and encourage youth to remain in (or return to) agriculture, uncertainties remain. Brazilian agriculture faces scrutiny both domestically and internationally over issues such as environmental degradation, deforestation, and land rights disputes, particularly those affecting Indigenous communities. The challenge ahead is to combine smallscale farming competitiveness with responsibility.

Government Support

Most governments around the world provide financial support to their farms in recognition of the important role that they play in maintaining food security and delivering public goods. For example, farmers can get government grants for tree-planting. The aspiration is that these payments can smooth the volatility of the food market so that one poor harvest does not bankrupt a farm. However, subsidies and supports are often inconsistent and fluctuate depending on which party is in power at the time. This can make it difficult for farmers to plan financially. In the United States, the Farm Bill is the mechanism that ensures support for farmers. There is a small funding allocation for conservation and other environmental schemes, but the majority of funding primarily comes from the commodity programs for major crops. In recent years, US farmers have received $10 billion to $20 billion annually from these programs due to factors like fluctuating crop prices, market instability, and the impact of natural disasters.[25] In addition to this, there is the Federal Crop Insurance Program (FCIP), which allows farmers a financial mechanism to manage risk. It provides different levels of coverage that the farmer can select. Catastrophic Risk Protection (CAT) is the most basic and least expensive form of coverage. It provides up to 50 percent of the average yield at a price set by the government, and farmers typically pay a small administrative fee (usually $655 per crop). It is intended to cover extreme losses but may not cover typical crop failures. Multiple Peril Crop Insurance (MPCI) is a more comprehensive insurance policy that covers multiple types of risks, including weather, pests, and disease. Farmers can choose different levels of coverage, ranging from 50 percent to 85 percent of their expected crop yield or revenue. The higher the coverage level, the higher the premium. These support mechanisms have been fairly consistent for decades, yet due to the ongoing disorder within Congress, the Farm Bill's future might not be guaranteed. Despite the fact that a new Farm Bill should be approved by Congress every seven years, the 2018 Farm Bill had to be extended as part of the American Relief Act of 2025 since no agreement could be reached.[26]

This extension is due to end on September 30, 2025. The lack of certainty on the future of these programs makes it extremely difficult for farmers and ranchers to plan their coming seasons. Beyond the Farm Bill, the USDA offers support to farmers. Yet, the USDA programs are at the discretion of the administration and so are not guaranteed. For example, President Trump's Secretary of Agriculture, Brooke Rollins, cancelled the Local Food for Schools Cooperative Agreement Program and the Local Food Purchase Assistance Cooperative Agreement Program when she took office in 2025.[27] Together, these programs delivered $1 billion of support for farmers, primarily smaller growers, and without it, many farmers will lose a valuable market.[28]

Yet, the United States is not alone in having uncertainty over the future of its agricultural support programs. The Common Agricultural Policy (known as CAP) is the program of agricultural support in the EU. Under CAP, farmers receive direct payments to help stabilize their earnings and ensure a basic income level, particularly for those who might struggle with fluctuating market prices. Payments are often linked to the area of land they farm or to the number of livestock they raise. Over the years, the EU has amended their approach, and today, a portion of the direct payments is tied to environmental sustainability. Farmers who follow specific "green" practices, such as crop rotation, maintaining buffer strips, or preserving natural habitats, receive additional financial support. There is also a crisis fund that can be deployed in an emergency. CAP is revised and adjusted every seven years. In the last revision, more freedom was awarded to individual member states so that they can create their own National Strategic Plan based on the stated CAP's goals, adapting it to their own context. This has created a period of adjustment for farmers who will need to see how the changes affect them.

Following the UK's departure from the EU and its exit from CAP, each of the nations within the UK had to create their own farming support scheme. This created a huge amount of uncertainty and change for farmers in the region. Many felt confused and poorly informed as to what the schemes might look like. Given that farmers need to plan years into the future to select varieties and manage land, they need certainty to know what support they will receive. The most problematic of these changes happened in 2025 when the government suddenly and without warning announced the closure of the Sustainable Farming Incentive that provided farmers in England with public money to help deliver environmental benefits (such as planting trees or hedgerows). The immediate closure of the scheme shook the industry and left many farmers reeling over how they would fill the gap in their budget.

For farmers from minority backgrounds, the uncertainty of government support is compounded by entrenched inequality and systemic discrimination. Farming requires significant upfront investment—seeds, equipment, and inputs must often be purchased on credit months before a harvest generates returns. When access to loans or credit is restricted through bias in financial programs, the consequences can be devastating, undermining the viability of entire farm operations. In the United States, Black farmers have faced a long history of exclusion from federal credit programs, a pattern that has contributed to widespread land loss, reduced intergenerational wealth, and the erosion of farming livelihoods. These systemic barriers mean that government policies designed to provide stability often fail to deliver equitably, leaving minority farmers more vulnerable to economic shocks and less able to benefit from public support. As a result, throughout the twentieth century, vast stretches of farmland, once owned by Black farmers, were foreclosed upon.[29] Today, discrimination occurs along similar lines, with government loans from the USDA denied to Black farmers applicants at a much higher rate than to their white counterparts; furthermore, Black farmers earn on average just one-seventh of the income of white farmers.

Labor in Crisis

Uncertainty regarding government support is just one of many factors driving people away from the industry and contributing to a crisis in farm labor. Across the Global North, farmers have struggled to access the workers they need to bring in their harvests—a situation amplified by the disruptions of COVID-19. Much of this work depends on migrant labor, particularly for so-called low-skilled tasks such as fruit and vegetable picking. Yet the term "low-skilled" undervalues the reality of the job, which demands remarkable dexterity, speed, and stamina. Workers often endure long, physically taxing days in difficult conditions, racing against time during the narrow window when crops are ripe. When farms cannot secure enough labor to meet this critical window, they are often forced to plow crops back into the soil—resulting in the devastating loss of an entire harvest. This labor shortage extends beyond the fields. Food processing plants, which rely heavily on migrant workers to clean, cut, package, and ship agricultural products, have also been hit hard. These facilities face similar challenges: demanding, repetitive work that few domestic workers are willing to take on, compounded by high rates of absenteeism during the pandemic due to illness, quarantines, and unsafe working conditions. Bottlenecks in processing have led to spoilage, further waste, and financial losses across the supply chain. Together, the farm and processing labor crises threaten the stability of food production and highlight the dependence of modern agriculture on a vulnerable migrant workforce.

There are 2.4 million farm workers in the United States.[30] More than half of all these farmworkers are immigrants, doing the most difficult work like fresh fruit and vegetable harvesting or working in slaughterhouses. In California, the fruit basket of America, 79.3 percent of farm workers are foreign-born.[31] That said, this official number of workers may not account for undocumented workers. The higher pay tempts many across the border illegally. Estimates suggest that in the United States, 30.1 percent of all farm workers are undocumented.[32] This undocumented labor pool has become highly contentious in recent years due to President Trump's agenda to reduce illegal immigration. Beginning in his first term, the Trump administration progressively restricted legal immigration pathways, which has significantly affected the agricultural workforce. For years, many farms relied on seasonal foreign workers admitted on temporary H-2A visas. However, these visas became increasingly costly—averaging about $2500 per worker—and processing delays often meant that workers arrived too late with 72 percent of farmers reporting that labor arrived after crops had already been lost.[33] The COVID-19 pandemic exacerbated these challenges. Travel restrictions prevented many foreign workers from crossing international borders, particularly after the H-2A program was temporarily suspended. For those already in the country, absenteeism increased due to infections, quarantines, inadequate PPE, and fear of exposure, leaving many farms critically short-staffed.

When President Trump returned to office in 2025, his administration renewed its focus on curtailing immigration, further straining the agricultural labor market. H-2A processing times have grown longer, increasing the risk that workers will miss the narrow harvest windows.[34] The administration also suspended the Farm Labor Stabilization and Protection Program—a pilot initiative aimed at improving conditions for workers while helping farmers offset the housing costs required under the H-2A program—leaving farmers uncertain as to whether funding will resume. In addition, the termination of a USAID effort to expand H-2A eligibility to agricultural workers from El Salvador, Honduras, and Guatemala eliminated a pathway that could have increased the legal workforce and reduced incentives for unauthorized migration.[35] US Immigration and Customs Enforcement (ICE) raids and arrests on farms and food processing facilities resumed during Trump's second term in office. According to reports, their goal was to make a minimum of 3,000 arrests per day. Matt Teagarden, the head of the Kansas Livestock Association, stated, "These raids contribute to fear in these communities and lead to workers—fully legal workers—staying home out of fear." Together, these policy choices have worsened the labor shortage, leaving many farmers struggling to find enough workers to sustain their operations.

US farmers aren't alone in facing huge consequences for wider action on curtailing immigration. In the UK, the ongoing challenges with securing farm labor have been exacerbated by Brexit. UK agriculture relies on migrants for picking soft fruit and harvesting vegetables. Around 20 percent of all regular full-time employees in agriculture are foreign.[36] This is higher for seasonal workers.[37] Also, a number of food processing facilities are reliant on foreign labor, such as egg packing centers (50 percent of staff are EU migrants) or slaughterhouses.[38] The majority of these low-skilled workers come from Eastern Europe. However, the UK's decision to leave the EU meant that these workers were no longer afforded the "freedom of movement" that exists within EU member states. In August 2021, with COVID-19 raging, the UK food and drink sector had over 500,000 job vacancies, and a 33 percent shortage in seasonal workers.[39] The EU Energy and Environment Sub-Committee at the House of Lords observed that the "entire food supply chain will be adversely affected by any loss of access to that labor pool."[40] Consequently, competition to secure seasonal worker contracts will likely rise at a time when the industry is already struggling to find candidates. Eastern European economies have strengthened in recent years meaning that workers have more opportunities within their home country, while, simultaneously, the UK economy has declined. The weakening of GBP against the Euro has made other markets more attractive to migrant laborers. The perception that migrants are not welcome and rising xenophobia also factor in.

While media narratives often focus on undocumented immigrants and labor abuses in the Global North, unpaid, illegal, and undocumented labor is equally pervasive—and often more entrenched—in agricultural production across the Global South. In the Ethiopia–Sudan borderlands, for example, the lucrative sesame trade relies heavily on seasonal migrant workers whose cheap, flexible, and exploitable labor is made possible by broader forces of land grabs, dispossession, conflict, and climate change. In Tanzania, agricultural investments enrich local elites but leave vulnerable workers facing exploitative conditions, social insecurity, and cycles of debt. Gendered exploitation also marks agricultural workplaces, as research in Costa Rica's banana industry has shown, with male domination and sexual harassment embedded in the hierarchy of the shop floor. In South Africa, both domestic and foreign farm workers endure systematic violations of their legal rights—including unpaid minimum wages, denial of leave and overtime pay, unlawful deductions, and barriers to injury compensation—leaving them trapped in precarity despite the legal protections ostensibly afforded to them.[41] Together, these examples expose how global commodity chains in agriculture are sustained by deeply rooted systems of inequality and exploitation in the Global South.

Child exploitation within the agricultural sector remains widespread across the Global South, particularly on export-oriented commodity farms. In Brazil, for example, a fifteen-year-old boy was recruited to work on a coffee farm supplying Starbucks.[42] He received no pay, was denied proper protective equipment, and labored almost continuously from 5:30 a.m. to 6 p.m. A subsequent investigation described his experience as "child labor in hazardous conditions" and noted he had been "trafficked and subjected to slavery-like conditions." While some children work legally alongside their families on small family farms, much of child labor in commercial agriculture is both illegal and exploitative. The cocoa industry in West Africa illustrates the scale of the problem. Ghana and Côte d'Ivoire together produce nearly 60 percent of the world's cocoa, yet an estimated 1.56 million children work on cocoa farms in these two countries.[43] These children often perform dangerous tasks such as wielding machetes, carrying heavy loads, and handling toxic agrochemicals—typically without adequate safety training or equipment. Many work long hours under hazardous conditions, forgoing education and enduring physical and emotional harm. Although international conventions and national laws, such as the UN Convention on the Rights of the Child and Brazil's Statute of the Child and Adolescent, prohibit child labor and trafficking, enforcement is weak, and the complexity of global supply chains allows abuses to persist while shielding companies from accountability.[44] Ending these practices will require not only stronger regulation and monitoring but also addressing the economic pressures that drive families to send children into such exploitative work.

Mental Health in Farming

For those who persevere in farming, there is a huge mental health toll. This can be exacerbated by the fact that farming is an industry with one of the highest levels of neurodivergent personnel, many of whom are often undiagnosed.[45] There are several mental health risk factors associated with agriculture. Farmers work long hours, often alone, and live in isolated areas. They can be under significant financial pressure, often required to take on significant debt to purchase the land and equipment required to operate. And in most cases, a farmer's place of business is also his or her home, meaning there is no easy way to get away from the stress. These challenges often go unacknowledged, and, tragically, farmers (predominantly men) take their own lives at a higher rate than workers in any other industry.[46] Often, farm accidents can mask suicide. Ready access to firearms and agrochemicals might also contribute to the elevated rate of suicide. This isn't contained to the Global North. Elevated rates of suicide were particularly notable in India in the early years of the twenty-first century. While we will discuss the circumstances in greater detail in Chapter 8, an estimated 270,000 farmers in India committed suicide between 1999 and 2013 due to rising costs, a move away from traditional farming methods, and the inability to make ends meet.[47] It is only recently, with increased societal awareness of the importance of mental health, that campaigns have started to talk about this issue and offer support. Yellow Wellies is a campaign in the UK that focuses on raising awareness in the farming sector of the importance of mental health and uses technology to help farmers get the support that they need. The USDA provides a free mental health helpline for farmers in rural states with counselling services for family farmers being run by the Farm Family Wellness Alliance.

Case Study H: When Farms Become Destinations—Can Traditional Agriculture Keep Up?

A growing trend in agriculture is reshaping agritourism: as more farmers adopt an entrepreneurial mindset, they're creating sustainable and inviting farm experiences that appeal to environmentally conscious visitors. This shift reflects both changing consumer expectations and a broader move toward responsible farming. In 2019, the global agritourism market was valued at over USD 69 billion and is projected to exceed USD 197 billion by 2032.[48] That year, Europe led the market with a 47 percent share.[49] Agritourism, where visitors engage with working farms and rural life, often through hands-on experiences like tours, harvests, and farm stays, presents farmers with new opportunities to diversify income, reach broader audiences, and improve financial sustainability.

Located near Zürich in Switzerland, Jucker Farm combines farming with tourism, education, and sustainability. The farm initially struggled to maintain a stable income through traditional farming. Its turning point came with pumpkin production and the launch of an annual pumpkin exhibition. Since then, Jucker Farm has expanded into four "Erlebnisbauernhöfe" (experience farms), hosting over 500 events yearly and attracting thousands of visitors. These farms offer interactive experiences such as fruit picking, barefoot trails, animal feeding, and an apple tree labyrinth that educates visitors about apple cultivation and cider making. Jucker Farm emphasizes regenerative agriculture and promotes seasonal, regional products in its farm shops and restaurants. The company further engages with customers through its e-commerce store and direct-to-consumer delivery network. It leverages digital tools for event planning, logistics, and staff scheduling and maintains an active presence on social media. In 2015, Microsoft profiled Jucker Farm as a digitally led, modern, medium-sized business with multiple revenue streams, noting that its growth was driven by a natural flair for business.[50] Despite the success of tourism-focused agritourism, this model raises important concerns. Traditional farms without tourism components risk being left behind, and many farmers lack the entrepreneurial skills—like business planning and investor relations—needed to develop hospitality ventures. It's crucial to recognize that farming itself should be economically viable without forcing farmers to become hospitality operators. Additionally, challenges remain in balancing genuine farm experiences with entertainment, ensuring sustainability goes beyond marketing claims, and managing the environmental impact of increased visitor traffic.

Exiting the Industry

The combination of all these things, in addition to the pressures that we discussed in previous chapters, has resulted in many farmers in the Global North exiting the industry. In Australia, the number of farms has decreased from approximately 22,000 to 4,000 since 1980, while the United States saw a reduction from 648,000 to 24,470 farms between 1970 and 2022.[51] Many of these farmers have retired or left the industry without a successor. In past generations, the family farm would have been passed down to the next generation to take over. However, after seeing how difficult it is to make a living, fewer and fewer children want to take over the family farm. Instead, they want to pursue other professions. As a result, many farmers are forced to continue farming long after they might otherwise have handed over the business. Without a new generation taking over, the average age of US farmers is 58.1 years old, while the average age of European farmers is 57 years old.[52] Once those older farmers decide that they have to retire, there are few options for them other than to sell up. Europe has seen high levels of farm abandonment, particularly for small to medium-sized farms and those in less productive areas.[53] These farms often come to market and get consolidated into larger corporate farms. For example, Missouri had 23,000 independent pig farmers in 1985. Today it has just over 2,000 with independent cattle farms falling by 40 percent over the same period.[54] Alternatively, they get bought by developers to turn into housing. In the UK, research by CPRE shows that almost 14,500 hectares of the country's best agricultural land, which could grow at least 250,000 tons of vegetables a year based on typical yields, has been permanently lost to development since 2010.[55] While it is impossible to dispute the need for housing overall, these developments are often luxury or second homes—known as rural gentrification, which we will return to momentarily. Some campaigners are working to ensure that the best quality agricultural land is afforded additional protections under planning regulations to protect it from development.

The demise of family farms is not new. There was a particular rise in foreclosures of family farms in the 1980s, known as the Farm Crisis. This was due to the fact that many US farmers took out large loans in the 1970s when times were economically buoyant to fund their expansion. This expansionist approach was supported by the government. In the early 1970s, US agriculture secretary Earl Butz pushed the idea of large-scale farming with the mantra "get big or get out." However, in the early 1980s, global commodity prices started to drop, prompted by the 1979 Soviet Union grain embargo. With farming incomes dropping, many farmers were not able to manage the repayments. This was exacerbated by other nations rising up. Brazil started becoming a global powerhouse for grain exporting and started to challenge the US prowess. By 1985, more than 250,000 US farms (many family-run) had declared bankruptcy. In response, legendary musicians Willie Nelson, Neil Young, and John Mellencamp organized a concert in 1985 (known as Farm Aid) to raise awareness about the loss of family farms and to raise funds to keep farm families on the land. The Reagan Administration also stepped in to create support programs to help farmers survive.

Farming within the Rural Economy

The effect of the closure of family farms is a weakened rural economy. That is not to say that farming is the sole factor that contributes to the rural economy, but it is a core component and the lifeblood of many rural communities. As a result of the decline of farming incomes, there has been a rise in rural unemployment in the Global North. In 2023, several countries in the EU, including Bulgaria, Lithuania, Hungary, Poland, Romania, and Slovakia, observed that the highest unemployment rates were among people living in rural areas.[56] Furthermore, from 1990 to 2024, rural unemployment has consistently been higher than urban unemployment in West Virginia, North Carolina, South Carolina, Maryland, and Virginia.[57] With this backdrop, young people want to leave rural areas because they do not find them exciting or see any future prospects. This creates a brain drain of talented young people leaving for the city. This happens the world over. Young people want to move from the rural region of Yendi to the Ghanaian Capital, Accra, just like kids from Lake Okoboji in Iowa want to move to New York. Many rural communities have been hollowed out as farmers sell up and families leave—all that remains of vibrant villages are boarded up storefronts. In 2024, the poverty rate in the rural United States was 13.5 percent, while the countrywide rate was 11.1 percent. Furthermore, the median rural household income was $62,250, compared to the countrywide median of $80,610.[58] Limited employment opportunities and declining economies have contributed to the rise of opioid addiction in rural areas in the United States. This has been particularly acute in recent years in the United States, where rural areas face higher rates of addiction and overdose, and the poorer the region, the higher the addiction rates.[59] From 1999 to 2014, a 1 percent increase in the unemployment rate in rural counties led to a 7 percent increase in opioid overdose-related emergency room visits.[60] To make matters worse, around 56 percent of rural counties do not have a provider who can prescribe buprenorphine, a medication used to treat opioid use disorders.[61]

In response to the rural decline, some rural municipalities facing a decreased population and tax base have attempted to come up with some solutions to encourage repopulation of these areas. For example, some municipalities in Italy are selling off villas for 1 euro in the hope of encouraging young people to return to these communities and revitalize the area. Similarly, the "Heartland Visa" was proposed by the Economic Innovation Group (EIG), a bipartisan think tank focused on economic development, and endorsed by the US Conference of Mayors in June 2024.[62] The Heartland Visa, unlike traditional visas, focuses on specific geographic areas with labor shortages, addressing regional economic disparities, and is limited to high-skilled individuals. France enacted a program to help with land matching. This scheme allows farmers to connect with young people who might want to take over the farm so they can avoid selling it for development. The new entrants come and work together with the retiring farmer for a number of years to learn the land before taking over. Every year from 2015 to 2023, this scheme saw approximately 14,000 new entrants secure a holding of their own.[63]

In some regions, there is a different challenge—rural gentrification. Some rural areas are facing increased housing pressure with rents and house prices skyrocketing. Certain parts of the countryside become tourist destinations. The move of middle-class families from cities out to the countryside is linked to the rising prices of urban living and the fact that the COVID-19 pandemic turbocharged a shift to home-working. Those relocating often select rural locations that are still accessible to urban centers, either by train or highway. However, these "affluence arteries" push up prices and have a detrimental impact on the future of farming. Areas such as the Cotswolds in the UK, Hudson Valley in New York, or Umbria in Italy have become magnets for visitors looking to enjoy the landscapes and the slower pace of life. This has pushed up prices and transitioned the economy toward tourism, making it challenging for the communities that traditionally lived in these regions and worked the land to remain. With limited stock of available housing, rural rents in England rose by 27 percent, compared to the 17 percent increase nationwide.[64] For the people who do want to farm in these regions, many of them face challenges in accessing land. As the price of land increases, it becomes more and more challenging to purchase.

Conclusion

Farmers are facing a tough commercial environment. In addition to the challenges caused by a changing climate and the increased geopolitical uncertainty, the environment within their sector is increasingly hostile. The price of inputs is increasing, driven by wider economic inflation but also a demand for greater profit from agrochemical companies that have consolidated the market. This landscape is causing many to leave the industry since farming today is becoming economically unsustainable for smaller producers. It is only the large corporate farms that seem able to weather the storm. Any consideration of the sustainable future of food has to consider the sustainability of farmers' incomes and farming communities. Through the rest of the book, the importance of economic sustainability and the future of farming communities needs to be weighed into the broader question of the future of food.

Part 2

The Future of Protein

Introduction

This part will consider the biggest question in the future of food: what is the future of protein? Protein is essential for the human body to function. It helps to build and maintain muscle and is essential for bone health, lowering the risk of osteoporosis and fractures. The Dietary Guidelines for Americans 2020 to 2025 recommends that adult females consume at least 46 grams of protein per day and adult males 56 grams.[1] Yet there are issues with our current model of protein production—primarily livestock. Livestock has been correctly identified as a climate problem. It is a significant carbon emitter through carbon dioxide from land use change. It also causes methane release during the digestion of ruminant animals, such as cattle. Globally, the livestock sector contributes approximately 14.5 percent of all anthropogenic greenhouse gas emissions—comparable to the entire transportation sector—making it a major culprit in the climate crisis.[2] Moreover, increasing economic prosperity in the Global South has led to rising meat consumption. For example, India is predicted to grow its annual meat consumption by 17 percent between 2005 and 2025.[3] As global meat consumption increases, these cumulative emissions will rise too, causing further concern.

Emissions from livestock are both direct and indirect. Cattle emit methane as a direct byproduct of digestion (enteric fermentation) and manure decomposition, making livestock farming a significant source of harmful greenhouse gases. A single animal generates approximately 220 pounds of methane annually.[4] Methane has a relatively short lifespan in the atmosphere, lasting around seven to twelve years, but, despite its relatively brief atmospheric presence, methane is approximately twenty-eight times more potent than carbon dioxide in its climate impact.[5] Livestock systems are an indirect contributor to carbon dioxide emissions arising from the associated land-use changes, such as cutting down native forests to make grazing pastures. The prime example of LUC for meat production is seen in Brazil, where LUC accounts for approximately 66 percent of CO_2 emissions in Brazil.[6,7] The production of animal feed drives massive environmental damage, from deforestation in the Amazon for soy to nitrogen pollution from industrial corn, making it one of the hidden engines of biodiversity loss and climate change. For example, deforestation, driven by the double threat of cattle ranching and soybean farming that produces livestock feed, is responsible for 80 percent of current deforestation in the Amazon.[8] Beyond greenhouse gas emissions, meat production's environmental ramifications extend to water and air pollution, soil degradation, and biodiversity loss. These interconnected ecological damages underscore the fact that agriculture is a major contributor to the joint climate and biodiversity crises.

Most meat produced today comes from industrial systems, also known as factory farms or Concentrated Animal Feeding Operations (CAFO). Factory farms now account for 72 percent of poultry production, 55 percent of pork production, and 43 percent of egg production worldwide.[9] These systems focus on producing large amounts of meat quickly and cheaply by keeping thousands of animals in small, confined spaces. This causes issues with air and water quality in the surrounding areas. Animal advocates say that these conditions cause pain to the animals and create cruel and inhumane conditions for animals. They believe these animals should be given enough space in a natural environment to roam and graze.[10] For cattle, this means they are often housed in cramped enclosures that give rise to a myriad of health problems.[11] Pigs are often placed in such small enclosures that they can't even turn around with their tails docked. Hundreds of thousands of chickens are kept together in windowless sheds. They are often debeaked so the birds do not peck each other while being kept in such cramped conditions. The use of hormones to increase their rate of growth also raises welfare concerns. Cows, for example, are commonly treated with the growth hormone rBST. This, however, results in painful ailments such as mastitis (infection of the udder) and lameness.[12]

These intensive systems contribute to the global challenge of growing antibiotic resistance since antibiotics are routinely administered to livestock. The intensity of these CAFOs compounds the problem. The tight confinement inevitably causes stress, and evidence shows that animals are more susceptible to disease when stressed, since stress-released hormones reduce immunity. Furthermore, intensively reared animals are often bred for maximum yield, rather than for natural resilience and disease robustness. As a result, prophylactic use of antibiotics is commonly used to prevent infections.[13] About 75 percent of all antibiotics used globally are in livestock production.[14] This directly contributes to the world's antibiotic resistance problem since the higher the use of antibiotics, the higher the risk of a mutation that results in resistance. Antimicrobial resistance is expected to kill 10 million people a year by 2050, up from around 1.2 million in 2019.[15] Antibiotics have been used in livestock since the early 1950s, when they started being given to animals to boost their growth. According to the World Organization for Animal Health, 109 countries no longer allow growth promotion, including those in the EU, where it has been banned since 2006.[16] However, administering antibiotics for growth is still allowed in at least thirty-six countries, including the United States.[17] In Asia and Oceania, farm antibiotic sales per livestock unit are nearly four times higher than in Europe. Antibiotics are used worldwide to control disease outbreaks—livestock producers even use antibiotics classified by the World Health Organization as "highest-priority critically important" for maintaining human health. Some consider new breeding techniques to be a solution to the problem posed by the routine use of antibiotics in farming since they could be used to create disease-resistant animals, which could allow further intensification of production systems. Others see this as a false solution since it perpetuates a dangerous production system and argue for more higher welfare farms that have lower disease risk.[18]

Another risk arising from such a packed setting is zoonotic diseases. Prophylactic antibiotic use doesn't always work to stop deadly diseases from arising and spreading quickly from animal to animal or animal to human. For example, African Swine Fever, a deadly and highly contagious disease in hogs, broke out in China in 2018 and wreaked havoc in the global pork industry. The disease is devastating with a mortality rate of nearly 100 percent.[19] It spreads quickly between animals by direct contact with bodily fluids.[20] Infected pigs experience symptoms such as high fever, vomiting and diarrhea, coughing, difficulty breathing, skin lesions, decreased appetite, and weakness.[21] Within a year of the outbreak being identified, 40.5 percent of all Chinese pigs died or had to be culled. Since China is the largest pork producer in the world, the decline in stock spiked the global price of pork.[22,23] Foot and Mouth disease and bovine spongiform encephalopathy (known as BSE or "Mad Cow Disease") are other terrifying examples. CAFOs also increase the risk of virus mutations since the cramped conditions provide ideal breeding grounds, jumping from animals to humans and wreaking havoc in their new host species.[24] This is increasingly recognized as a high risk for a future pandemic.

Avian flu (H5N1) is already devastating both wild and farmed bird populations, causing mass mortality events among wild species and the large-scale culling of poultry flocks, often through inhumane methods such as mass asphyxiation. The virus has now been detected in a growing number of mammalian hosts, raising concern among public health experts that, with the right mutations, H5N1 could acquire efficient human-to-human transmissibility and trigger a major public health emergency. Worryingly, avian flu has now crossed into cattle. H5N1 is impacting dairy cows in the United States—possibly caused by infected wild birds, such as geese or ducks, coming into contact with grazing cattle. This development indicates the virus's ability to mutate to cross into new species—a trend that has continued since dairy and poultry workers who were exposed to diseased animals have also become sick. Globally since 2003, twenty-three countries have reported human infections due to bird flu viruses to the WHO with a total of 889 cases and 463 deaths.[25] The terrifying threat is if the virus mutates to move from human to human. To date, that hasn't happened, but the risk is not insubstantial. In recognition of this threat, the CDC is watching the situation carefully, monitoring infection rates through its flu surveillance systems.

As time goes on, the public has become more aware of the issues surrounding meat production—whether that be the ethics of factory farming or the climate impact. This has happened through public campaigns from organizations, such as PETA, and documentaries such as the Netflix series Cowspiracy. These media moments have often included sharing disturbing footage of the conditions that animals endure. Rising public awareness has led to a corresponding rise in dietary change. Additionally, many consumers are concerned over the health impact of consuming meat, in particular red meat. Consumption of red meat has been linked to colonic cancers, and high levels of processed meats have been associated with hypertension and heart disease. The WHO classifies red meat as "a probable carcinogen."[26] The shift away from red meat consumption, including a shift toward chicken, is directly attributable to the increased perception that it is bad for your health. For example, red meat consumption in the UK decreased by 26 percent across the last decade while chicken consumption increased 6.7 percent annually.[27,28] This shift in personal consumption patterns has been mirrored in public policy. There has been the conceptualization of Veganuary, a commitment to avoid meat during the first month of the year (akin to Dry January), and the campaign for Meatless Mondays. In the United States, 59 percent of Americans who eat less meat than a year ago are doing so mainly for health reasons; another 10 percent cite climate change as their primary motivation with 9 percent cutting back over animal welfare concerns. Among a younger demographic (eighteen–twenty-nine years), concern about the climate is higher (24 percent).[29] Recognizing the negative impacts associated with meat consumption, some countries have explored trying to reduce meat consumption by creating a meat tax. From 2030, farmers in Denmark will be taxed 300 kroner ($43 USD) per ton of methane emissions from cows, pigs, and other animals kept on their land.[30] This will be the first of its kind globally—only time will tell whether it might catch on in other countries.

Given the increased awareness of the damaging impacts of our current livestock model, the future of protein is a contentious and critical topic. How can meat production be altered in the coming years to make it healthier, more ethical, and more environmentally sustainable? All while simultaneously protecting the livelihoods of the farmers and communities that depend on it? There has been a strong emergence of alternative forms of traditional proteins. Most common is the use of protein powders from whey, which has transformed a waste product into a lucrative income stream for dairy farmers. Developments in aquaculture are transforming the fishing industry, while companies are looking at insect protein and legumes to see how they can be transformed into new products that might better resonate with the consumer.

Beyond just reimagining what already exists in new forms, entirely novel types of protein are being created in laboratories. Lab-grown meat ranges from plant-based meat and cell-cultivated meat to microbe protein and lab-grown dairy. Can they scale up production from a niche product to dominate the protein market? What impact will that have on our food system overall? Since it is an entirely new product, it is difficult to predict the consequences that might result from the shift, but there are questions around equity, corporate control, public health, and accessibility that need to be answered. Yet, surely substituting the worst quality and most environmentally damaging industrial meat with a cell-cultivated alternative that is more humane, less polluting are potentially more nutritious is a good thing?

However, while all this resource and effort are being put into innovating the protein sector, there is value in considering whether we might be reinventing the wheel. Part of the issue with our industrialized meat industry of today is that it has become disconnected from nature and lost its connection with the traditional forms of livestock management that are environmentally sustainable. While there is wide acceptance that we need to eat less meat overall, is there an opportunity to transition the meat that we do eat to be of better quality and produced in harmony with nature? New regenerative agricultural movements are trying to reclaim the traditional methods of producing high-quality, high-welfare meat on well-managed grasslands. If successful, they could turn livestock production from a climate problem into a climate solution by using mixed farming systems to sequester carbon back into the soil and improve biodiversity—all while improving farming incomes.

In all likelihood, it will need to be a combination of all of these to successfully deliver a sustainable future of protein. The scale of the challenge is so substantial that one solution alone is unlikely to make enough of a difference at a global scale. Each has its opportunities, and each has its drawbacks. In this part, we will dive more deeply into each potential future for protein.

5

Reconceptualizing Alternative Protein Sources

To avoid the worst environmental outcome, we need to reduce the amount of meat we eat. Our current consumption levels are unsustainable. The world now produces more than 350 million tons of meat each year, which has a significant impact on our planet and its future—something that we have discussed.[1] In order to reduce our levels of meat consumption, consumers are being encouraged to switch to alternative forms of protein. Many of these are familiar to us. There are traditional forms of protein that have been part of our diets for centuries and have smaller environmental footprints than livestock. These could help to provide a solution to our dietary future. New startups and innovative approaches are helping transition these foods from niche into viable solutions for feeding the world's population by getting consumers in the Global North to adopt them.

Farmed Fish versus Wild Stocks

As we discussed in the Introduction, in 2019, the leading thinkers on healthy diets came together to form the EAT Lancet Commission. They considered what future diets might look like and what changes need to be made to make our consumption patterns more sustainable.[2] Their report laid out their findings, and one of the leading recommendations was that people should increase the amount of fish in their diet, as it is a source of protein that is high in health value and recommended 28 grams of fish per day (one–two servings per week).[3] If we were all to shift to a healthier diet under the guidelines of EAT Lancet report that we discussed in the Introduction, it would result in an increased worldwide demand for fish.[4]

However, there are serious environmental questions about overfishing and the risk of depleting wild stocks. Certain species are preferred by consumers, like salmon, tuna, and cod, which has put pressure on these wild stocks.

Overfishing is one of the most serious threats to the oceans. Because pole-caught fishing is time-consuming and expensive, many fisheries use deep-sea trawling to maximize the number of fish they catch. These massive trawl nets are dragged along the sea floor, scooping up everything in their path with little regard for species or size. While intended for commercial fish, they also entangle nontarget species, including dolphins, turtles, sharks, and other marine life—a phenomenon known as "bycatch." Because dolphins hunt many of the same fish species sought by humans, they are particularly vulnerable, often drowning once trapped in the nets. The result is that countless marine animals that will never reach the dinner plate become unintended casualties of industrial fishing, contributing to the depletion of biodiversity and the collapse of fragile ecosystems. Beyond environmental costs, there are also serious human rights concerns: investigations have repeatedly documented forced labor and even modern slavery aboard deep-sea trawlers, with the problem especially acute in parts of Southeast Asia where migrant workers are subjected to brutal conditions, debt bondage, and physical abuse.

With the ocean quickly becoming depleted by overfishing, many producers are turning to farmed fish. Fish farming has a long history. In fact, the Romans built pools, known as assyrian vivarium, where freshwater fish could be kept alive until it was time to eat them. In more recent history, in the 1960s, farming fish was relatively niche, with an output of a few million tons per year.[5] However, as demand rose in the 1980s, annual production increased rapidly. In 1990, the world produced only 17 million tons. It now produces over 110 million tons.[6] Today more fish are produced through fish farming than are wild caught. While many consumers might choose farmed fish thinking that it is more sustainable than depleting wild stocks, there are some serious environmental concerns associated with fish farming, which, if unresolved, could pose an even greater threat to the environment than traditional fishing. Parasitic sea lice are rampant in fish farms, particularly those that farm salmon.[7] In the wild, it is difficult for sea lice to find hosts. However, because fish are packed together so tightly in fish farms, sea lice can spread very quickly, as it is easy to jump from fish to fish. When sea lice go unchecked, it can cause mass mortality. Further, fish that have a sea lice infection cannot be sold commercially due to lesions, which impact the quality of the meat. When farmed fish that are infected with lice escape the farms, they can spread sea lice to wild populations, putting them at risk. Given the problem posed by sea lice, producers eliminate sea lice from their stocks using harsh chemicals. These chemicals wreak havoc on the ecosystems that surround the fish farms. The chemicals also damage the fish themselves, causing them to lose their appetite and stopping them from growing. Besides, the margin between killing the lice and killing the fish is incredibly tiny, so often the chemicals end up wiping out the entire cohort of fish instead.[8] Some farmers prefer to use cleaner fish, such as wrasse, as a more environmentally friendly method of killing the sea lice. However, these fish are more effective in some areas than others, due to issues with hibernation and climate.

Beyond the issue of sea lice, fish farm waste can cause a great risk to the health and safety of the surrounding waterways and, consequently, to public health. This is particularly due to the fact that prophylactic antibiotics are regularly used in fish farms to promote growth and prevent infections. As discussed in Part 2 Introduction, the overabundance of these drugs promotes antibiotic resistance in humans.[9] The more that antibiotics are used, the more the bacteria adapt to develop resistance, making antibiotics less effective and bacteria more lethal. Antibiotic residue, combined with fish feces, chemicals, and feed waste, is a deadly combination when it spills into waters surrounding fish farms. This not only endangers other marine life but can also pollute waterways used for irrigation and drinking water.[10] Unless these forms of pollution are heavily regulated, fish farming may begin to cause a huge issue to public health as it becomes more and more widespread.

Farmed fish eat a carefully controlled diet designed to promote growth and health. Tiny crustaceans, known as krill, are the main feed for fish. The majority of commercially harvested krill comes from the Antarctic. While there is currently a cap on krill fishing in the Antarctic at 620,000 tons to protect the ecosystem, fisheries are seeking to double the amount of krill that they can catch.[11] Environmentalists are concerned that removing so much krill from the Antarctic will pose a threat to the diets of penguins, whales, and many other marine animals that rely on the tiny shrimp. Some farmers prefer to use fishmeal (ground fish meat) and fish oil instead of krill, but this is also unsustainable. Fishmeal and fish oil are highly resource-intensive because they are mostly made from fry and juvenile fish, like anchovies and herring, which has led to the wild population collapsing. Because these fish are at the bottom of the food chain, population collapses often have severe ramifications further up the food chain, destabilizing the ecosystem. The fishmeal and fish oil industry exists almost solely to feed fish on fish farms; 75 percent of fish oil and 69 percent of fishmeal go toward farmed fish.[12]

In recognition of these problems, innovative companies have been trying to create a model for fish farming that is environmentally sustainable and economically profitable. Hima Seafood is a Norwegian company that opened a land-based fish farm to produce trout using recirculated groundwater.[13] They use Recirculating

Aquaculture System (RAS) technology. Despite the global recognition and demand for Norwegian salmon, Hima Seafood chose to produce trout since they are less territorial and rely only on freshwater, which is easier to recycle. Because the water is so clean and the environment is sealed off from pollutants, the fish don't require antibiotics or other vaccinations. Even though the fish are unvaccinated, the farm still experiences low mortality rates (the industry's mortality rate is around 16 percent).[14] The founders of Hima credit this to the amount of space the fish have, which allows them to remain in constant movement. Mindful of their footprint, Hima Seafood has publicly committed to finding ways to lower their carbon footprint. They recently partnered with a Norwegian colocation company, Green Mountain Data Center, to reuse the heat waste generated by data storing. The pipeline, built in 2023 between the data center and the farm, regulates water temperature to maximize the effectiveness of the RAS technology. Sten Falkum, the CEO, said that "Green Mountain's waste heat represents a significant cost savings in our production, and we are thrilled that our heating requirements can help reduce the environmental footprint of Green Mountain and help cool the data center in return. This truly is a win-win solution for both parties."

Being a very new technology, RAS still has a long way to go—the Conservation Fund is researching ways to improve its efficiency and enhance fish welfare.[15] They hope to help farms using RAS to transition from manual data collection to remote data collection using sensors and other data analytics. By monitoring fish mortality using sensors and AI, RAS can be made more efficient and effective. Instead of manually checking a few times a day, Internet of Things (IoT) technologies would do round-the-clock monitoring and would alert the farmers if there were higher mortality rates than usual. This would increase yields, as farmers would more quickly be able to stop the spread of disease and decrease mortality rates. This would also improve the health and safety of the fish themselves. Companies are also looking at different varieties of fish too. While salmon are relatively easy to keep in pens, tuna, another consumer favorite, have migration patterns that are much more difficult to simulate in captivity. Next Tuna, a German startup, has successfully replicated the reproductive cycle of Atlantic bluefin in facilities on land. This is a mammoth task since tuna can grow to 2000lbs and can swim across the entire Atlantic. The end goal would be to use RAS systems that can mimic this journey by adapting the light, oxygen levels, temperature, and so on to fool the fish into thinking that they have swum across the Atlantic without having ever left their tank in Germany. The company aims to complete research and development phases by 2028, but already lots of logistical issues have come up.[16]

Case Study I: Reviving Tradition through Sea Cucumber—Aquaculture in Fiji

In Fiji, the collapse of wild sea cucumber stocks was felt not only as an economic loss but also as a rupture in cultural ties to the sea and local diets. For generations, coastal villages had depended on the bêche-de-mer trade as a modest but important source of income, while also valuing sea cucumbers as a nutrient-rich food, prized for their high protein content and health benefits. Ecologically, these creatures played an equally vital role: by cycling nutrients through sandy and seagrass habitats, they sustained the health of reefs that fed and sheltered entire communities. But global demand pushed sea cucumbers to near-extinction through overfishing, threatening the collapse of a careful balance of food, health, ecology, and culture.

The path to recovery for sea cucumbers in Fiji came from a revitalization of local traditions. Fijian coastal communities have long used "qoliqolis" (designated fishing areas) and "tabu" (fishing grounds temporarily closed to allow marine life to replenish).[17] In recent years, with support from the Pacific Community (SPC) and the Ministry of Fisheries, these traditional methods have been revived through a network of 400 communities known as the Fiji Locally Managed Marine Areas (FLMMA), using customary management techniques that have often been ignored for decades.[18]

Traditionally, a reef would be marked with sticks and a coconut leaf tied on it by the community elders. That action would make the area sacred, and any fishing would be forbidden. The success of the closure depends not on fences or external enforcement but on the authority of village elders and cultural duty. When the tabu is lifted and fishing is permitted, it is done carefully, ensuring that breeding populations remain to sustain the next cycle.

This blend of culture and aquaculture has delivered multiple benefits. The ecological role of sea cucumbers in maintaining reef health is being restored, strengthening the productivity of coastal ecosystems. Households gain a steady income but on terms that reinforce collective stewardship rather than undermine it. Perhaps most importantly, the practice reaffirms cultural identity: acknowledging the critical role of ancestral wisdom in maintaining ecological health. Fiji's sea cucumber farms show that community aquaculture is most powerful when it does not replace tradition but channels it. Here, the tabu is more than a conservation tool—it is a living cultural practice adapted for new challenges, sustaining both people and the sea that defines their way of life.

Sustainable Shellfish

Given the challenges of fish, it's possible that bivalves (clams and oysters) could provide a better option for marine protein. Shellfish are both a sustainable source of protein and improve their surrounding environment, making them ideal for farming and environmental programs. They are low in calories, high in protein, and contain essential vitamins and nutrients like zinc, vitamin B12, and iron. Shellfish are low to no input, meaning that they require minimal care. The best shellfish to farm are clams, cockles, mussels, oysters, and scallops.[19] Part of the reason that bivalves are so good for the environment is that they don't require krill. Shellfish make their own food by filtering algae from the water, and therefore they don't require regular feedings. Since they consume the algae in the water, shellfish also prevent hypoxic blooms and absorb excess nitrogen from the water, which they use to fortify their shells. Oysters are arguably the most sustainable seafood because they increase the health of the ecosystem. According to the National Oceanic and Atmospheric Administration (NOAA), "shellfish and seaweed farms removed an average of about 575 lbs of nitrogen per acre. That's roughly the weight of a grizzly bear per acre every year."[20] The value of that ecosystem service is estimated at $1,321–7,739 USD per acre each year.[21]

Various environmental groups have used shellfish in an attempt to restore habitats. In New York, the Billion Oyster Project is regrowing the oyster reefs in the New York Harbor. Before European settlement, New York City harbor was home to half of the world's oysters. Oysters are crucial to the ecosystem of a harbor, as they filter nitrogen from the water and prevent "dead zones," places where there's little oxygen, from being created. Oyster reefs are like coral reefs in that they provide a 3D structure for hundreds of organisms, spurring biodiversity, which is crucial to a healthy ecosystem. While the oysters in New York harbor are not yet safe to eat due to the current levels of pollution, sewage, and other water quality issues, the Billion Oyster Project hopes to kickstart a circular economy whereby restaurants regularly donate their used oyster shells and receive a batch of fresh oysters in exchange. The Chesapeake Bay Foundation has a similar project to revive the bay's oyster reefs with the help of local fishermen who sign up to be oyster gardeners. These projects are a glimpse into what the future could be: a circular economy that provides the community with sustainable, fresh food that improves its surrounding environment.

Seaweed and Algae

While many readers might have eaten mussels or clams before, few will be as familiar with seeing algae on their dinner plate—beyond perhaps some nori on your spicy tuna roll. Algae, particularly certain types of microalgae, such as spirulina and chlorella, are considered a sustainable, environmentally friendly alternative to animal-derived protein—and without some of the environmental problems discussed in relation to fish. Seaweeds are healthy. Some algae have protein levels that are akin to legumes or nuts. Red algae have a protein content between 19 percent and 47 percent while green and brown algae have around 10–20 percent in dry weight.[22] In comparison, both legumes and nuts have between 20–30 percent.[23] But microalgae have even greater potential. Microalgae can achieve a protein content higher than 70 percent.[24] These microscopic algae are extremely rich in protein with a good blend of vitamins, amino acids, and other nutrients. Moreover, they require no land to grow on and have a high growth rate (without the need for agrochemicals or irrigation). As a sector, it is taking off with high levels of investment. The global market for microalgae for human consumption was worth $590 million USD in 2018 and is expected to hit $970 million USD in 2025, says Japanese company Sun Chlorella.[25] However, despite the significant potential, the process of extraction and processing remains problematic for producers, and there have been regulatory challenges in getting products approved for sale, particularly within the EU. So far, approval has been limited to protein powders that can be added to smoothies. Perhaps future innovation will allow for these challenges to be overcome, and we will see meat alternatives featuring microalgae filling our supermarket shelves. Yet, for the time being, there is a focus on using microalgae as a more sustainable alternative to fishmeal. Instead of making fishmeal by grinding up smaller fish to give farmed fish the oils they need, they could use algae, which is far more sustainable. UK supermarket Tesco plans to source more fish that are fed on algae-based feed, while Thai Union declared an intention to become 'zero fishmeal' by switching over to algae-based feed for farmed shrimp.

Insect Protein

One of the most commonly cited examples that people think of when they consider the future of protein alternatives is insects. This makes sense since they tend to have high nutritional value, reproduce rapidly, and don't take up much space at all. In terms of sustainability, it's an easy win. The emissions of carbon dioxide, methane, and nitrous oxide resulting from the farming of edible insects have been found to be lower by a factor of 100 per kg of weight in comparison with cattle.[26] And they are more efficient to produce. According to the FAO, insects can convert 2 kg of feed into 1 kg of insect mass, whereas cattle require 8 kg of feed to produce 1 kg of body weight gain.[27] The nutritional value is also positive when stood against livestock. In terms of protein levels, found dried crickets contain more protein per gram than beef, chicken or pork—with 61 grams of protein per 100 grams, compared to 31 grams of protein in beef.[28] Dragonflies and damselflies have 55 percent protein dry basis, while cockroaches and termites have the lowest at 35 percent dry basis. So perhaps there is no need to be eating cockroaches any time soon! Insects could prove useful in the fight against hunger and malnutrition. Academics at Harvard and the University of Wisconsin-Madison conducted research that found that if consumers in Africa and Asia added 5 grams of insect food to their daily diets, 67 million fewer people would be at risk of protein deficiency, with 166 million fewer people at risk of zinc deficiency and 251 million fewer people at risk of vitamin B12 deficiency.[29] Anemia would also fall considerably. Yet, while delivering against those statistics would be transformative for global health, it might not achieve the financial success that these companies are after.

Insects have long been part of traditional diets. In Ghana, winged termites are commonly fried or roasted. In Thailand, mealworms are a common snack. Japanese bee larvae are cooked with soy and sugar. And, even if consumers in the Global North are more reticent about eating bugs, the fact is that we already do—and lots of them. They are already in some of your favorite foods. Cochineal bugs have been used to produce bright red dye for centuries. Females and their eggs produce high levels of carminic acid that can be harvested and turned into dye. Originally from Mexico, it was a highly prized pigment in the Pre-Columbian Era and gained global fame when the Spanish took it back to Europe to be used in expensive fabrics and paintings. Today the red pigment is approved for use in many foods to enhance the color. Skittles, M&Ms, and TicTacs use the additive. It may come back into favor if efforts to restrict the use of the synthetic dye Red No.40 in the United States are successful.[30]

And lots of insects make their way into our food without notice. The US Food and Drug Administration allows for up to 20 maggots per 100g of mushrooms, 35 fruit fly eggs per 230g of golden raisins, 4 larvae per 500 grams of Brussels sprouts and 475 insect "fragments" per 50 grams of ground black pepper, among others.[31] Given the complexity of food production, this is a reasonable legal provision, but it still might feel a bit icky to some consumers. Yet, fragments of bugs hidden in our food without us noticing are very different from marketing a product as insect-based. For many consumers in the Global North, the idea remains unappealing and unpalatable. Yet, some people are willing to take the risk on the potential of insects. In fact, none other than Mark Cuban was willing to bet $100,000 USD investment when Chirps Chips went on US television show Shark Tank and pitched cricket-based chips.[32] Companies across the Global North are trying to tap into the bug market in recent years. Eat Grub launched crispy crickets in the UK supermarket Sainsbury's, suggesting that they could be added to salads. Yum Bug launched a bean and bug burger that you could throw straight onto the grill. Bitty slow roasts crickets before grinding them into flour that can be used as high-protein baked goods.

These insect products coming onto market have been helped by the regulatory changes that have worked to expand the potential market. For instance, in 2023, the EU approved a fourth insect (the lesser mealworm) for human consumption.[33] This puts insect protein on course to increase from a global market of $0.8 billion USD in 2022 to an estimated $3.3 billion by 2027.[34] However, that could be in jeopardy if one of the key consumer groups turns away from the products. Vegetarians and vegans are a big market for all alternative-protein companies and are a target for many companies aiming at expanding insect consumption. Yet, some NGOs (in particular Rethink Priorities) are raising welfare concerns over insect production and raising questions over whether insects are sentient and can experience pain—something that might turn away many vegetarians from trying insect protein. Such an argument is not unprecedented since vegetarians complained about the use of cochineal in Starbucks drinks. Ultimately, this led to the reformulation of a number of beverages to use a natural tomato-based alternative. If insect consumption is not considered to be vegetarian due to the welfare concerns, that might damage its ability to gain traction in the marketplace. Although entomologist Dr. Sarah Beynon, who runs a UK-based insect farm, disagrees about the ethical concerns, saying, "With insects, we can farm them intensively without compromising their welfare. They're actually happier when they're close to many other insects of the same species."[35]

There is one consumer who seems very happy to eat insects—your pet. And pet owners are happy to spend a lot of money to keep their fur babies happy with the market growing by 2.6 percent annually.[36] There are now a number of pet food brands that are using insect protein. Of the 8.9 million tons of rendered protein meals produced annually in the United States, pet food accounted for 16.85 percent so shifting production away from livestock would be a big win.[37] Similarly, there is potential to feed livestock on insects, reducing the dependency on soy and corn. We have already discussed the huge environmental damage that is caused in the production of grain to support intensive livestock systems. Switching feed over to insect-based protein, or even just covering a portion of it, would be transformative and alleviate a lot of the land pressure. Such a shift wouldn't just be beneficial for land-based agriculture either. Fish too could be fed insects, reducing the need for krill and other small fry. Importantly, when farmed and processed under proper standards, insect protein carries little zoonotic disease risk, making it a safe and sustainable alternative.

Case Study J: Too New for Europe—the Regulatory Roadblocks to Edible Insects

Insects have the potential to revolutionize protein consumption. However, bureaucracy might hinder their adoption. In the European Union (EU), Regulation No. 2015/2283 states that any food not widely consumed before May 15, 1997, and thus not considered part of the "typical" European diet, should be classified as "novel" and, therefore, subject to "Novel Food" regulations.[38] As a result, all insect-based foods require a safety assessment by the European Food Safety Authority (EFSA) before they can be legally sold for human consumption. To commercialize insect products across the EU, companies must obtain authorization from the European Commission, based on the EFSA's risk assessment, followed by endorsement from the EU member states. This takes time. The key principles for approval are that novel foods must be safe for consumers, properly labeled with allergy information, and, if intended to replace another food, must not be nutritionally disadvantageous to the consumer. While this approval process is meant to keep consumers safe, it can also slow the introduction of insect-based products to the EU market—a type of food that has been a staple in many diets around the world for centuries. For instance, Mexico alone reports the consumption of around 100 insect species, and Thailand is home to an estimated 20,000 edible insect farms, highlighting a strong entomophagy culture.[39] Given that people around the world have safely eaten them for centuries, a date of 1997 in Europe seems a bit arbitrary and fails to acknowledge the customs and traditions of other cultures. It also denies people of non-European heritage access to culturally appropriate foods when living inside the EU. This not only limits dietary diversity but also overlooks the importance of food in preserving cultural identity and community well-being. That said, traditional foods from non-EU countries, including insects, with a proven history of safe use for at least twenty-five years, can undergo a simplified EU approval procedure.[40]

Peas and Beans

Many of these alternative protein foods are couched in language of newness to make them appeal to the consumer. However, there is value in considering the other, more traditional forms of protein that we have historically eaten that might not seem as glamorous and exciting. No one gets that excited about boring old kidney beans, but peas and beans are amazing. They have been a staple of diets for millennia, and for good reason. They have high levels of protein and other beneficial nutrients. One cup of garbanzo beans, also known as chickpeas, has 14.5 grams of protein and is high in folate and manganese. One cup of cooked lentils has 17.9 grams of protein, while one cup of black beans has 15.2 grams of protein.[41] They are also great for the environment. Legumes hold the power to restore soil health, which, as discussed in Chapter 10, is hugely depleted. Adding legumes into sustainably designed crop rotations would increase the soil's natural nitrogen levels and help to reduce the use of synthetic fertilizers and the environmental damage associated with them. This is why beans are considered one of the three sisters of land cultivation for the indigenous people of Central and North America, with corn and squash being the other two. Encouraging the public to eat more peas and beans, instead of meat, would create a positive change. A company that recognized the value in expanding the market for sustainably produced beans is Hodmedod's. Hodmedod's has been working to establish a better market for UK-produced pulses and grains to try to encourage the wider domestic production of legumes. Hodmedod's has been particularly supportive of farms that cultivate rarer native varieties (such as Red Fox Carlin and Kabuki Marrowfat) that have fallen out of favor with consumers. This type of innovative leadership needs to be expanded—something that the UN recognized when it launched the International Year of the Pulses in 2016.

That said, one legume has dominated our food system in recent years—and not for the better. Soy is a legume and has the ability to fix nitrogen. It is produced in vast quantities across the globe, and its production for livestock feed has caused significant environmental damage. Certainly, the world doesn't need to produce more soy. In fact, we need to grow much less. But instead of growing vast amounts of soy for livestock, we could grow a more environmentally appropriate amount for human consumption. Currently only 6 percent of the soybeans grown worldwide are for human consumption—as tofu or fermented tempeh.[42] Encouraging more of those soybeans to be fed to humans, while growing less overall, would be a big environmental win. It might also provide a softer transition for farmers who could see markets disappear if soy for livestock is phased out. Tofu is produced from condensed soy milk and has played an important role in Asian diets for generations. A 100 grams portion of firm tofu has 17 grams of protein, plus beneficial levels of calcium, manganese, copper, and selenium. The tofu market is growing and is expected to reach $2.7 billion USD by 2032 (from $1.5 billion USD in 2023).[43] Tempeh, made of fermented soybeans that have been soaked and drained, has a protein content of 19 grams. When considering the future of protein, there is an important role for tofu and tempeh as alternatives.

Conclusion

Many consumers are beginning to scale back their protein intake, driven by concerns over health, sustainability, and the environmental footprint of high-protein diets. As the nontraditional protein market grows, there are a number of options available to them. Many are traditional forms of protein that have been repackaged to appear more exciting and innovative. Fish has been a traditional source of protein for centuries. While fish farming has caused significant environmental damage, alternative models of production that are more sustainable are gaining traction. This includes using technology to innovate land-based fish farms, but it also includes shellfish and bivalves that are more environmentally friendly since they help with ecosystem rehabilitation. Similarly, algae has been recognized for its high protein levels and is seeing increased investment in new formulations. Just as fish have been eaten for thousands of years, insects have long been part of our diets. There is increased interest to see if consumers in the Global North could be encouraged to try them since they provide a high protein source with low land-use and high efficiency. However, it is more likely that they will provide an environmentally sustainable option for pet food and fishmeal (at least in the short term), since the majority of affluent consumers still remain hesitant. But it is also valuable to remember how previous generations used alternative proteins and not to forget traditional protein in a search for new solutions. We will return to this discussion in Chapter 7 when we discuss sustainable models of livestock management. Peas and beans are good sources of protein and also help to restore soil health since they can naturally fix nitrogen in the soil, reducing the need for artificial fertilizers.

6

Lab-Grown "Meat"

Traditional livestock production generates almost 15 percent of the world's GHG emissions.[1] As a result, many people are focused on how to make meat production more sustainable. The last chapter considered some of the more traditional alternative forms of protein. However, new technology and innovations are creating entirely new forms of protein. Lab-grown meat embraces technology to create new solutions to this problem that this chapter will discuss in turn: plant-based meat, cultivated meat, and microbe protein. Beyond that, companies are exploring the viability of lab-grown dairy. Venture capital numerous obstacles to scaling up production. Whether they can overcome these challenges is still to be seen. However, there are much deeper societal questions about the future of lab-grown meat. Will consumers purchase it? How will it be labeled? What are the ethical ramifications? These foods are classified as ultra-processed and geared towards a mass market, which raises questions over equity and public health that this chapter will consider. Given that this is a new technology, these are live discussions that will need to be resolved if lab-grown meat is to become the future of protein.

Plant-Based Burgers

As we discussed in Chapter 5, alternatives to meat (like tofu or insects) have long been a staple part of diets around the world. However, it was not until the 1990s that the veggie burger that readers would recognize became mainstream. The veggie burger made its commercial debut in 1982 when Gregory Sams featured it on the menu of his London macrobiotic restaurant, SEED. There had been recipes for vegetable burgers before that (such as in Jimmie Joan Thurmon's *Adventures in Vegetarian Cooking*), but these had never been sold commercially. Sams made his patty with seitan, tamari, aduki beans, and oat flakes. Inspired by the popularity of the dish, he decided to create a similar option that could be sold in supermarkets. In the United States, it took a little longer for the veggie burger to arrive. In 1985, Gardenburger was founded by Paul Wenner in Portland, Oregon, and commercially released its burger in 1992 with Boca Burger arriving the following year. While these vegetarian products were often used in place of meat, they were not trying to mimic meat since they were primarily smashed beans with some flavoring. These products grew in popularity as the public became increasingly aware of environmental concerns and personal health.

The arrival of plant-based meat technology is transforming the veggie burger. These are not like your bean-based patty of the past. The goal of these burgers is to use ingredients from various plants to mimic the taste and appearance of "real" meat via textured vegetable protein (TVP), which has a structure (and texture) similar to fibers.[2] TVP was invented in the 1960s and was initially used to bulk out recipes and make meat go further but has now been reconceptualized as a full substitute for meat.[3] The process by which plant protein is converted into TVP is known as extrusion. During extrusion, a mixture with low moisture protein like soy flour is passed through a barrel machine and squeezed out of a small opening called a die.[4] During this process, the mixture is exposed to high temperature, high pressure, and shearing.[5] Proteins denature (unfold) due to the high temperatures, and as the substance is pushed out of the die, the amino acid strands extend and align, creating a layered texture.[6] Food additives, such as binding and coloring agents, are then mixed with the TVP to create a final product.[7] A key additive is heme, which is used to replicate the odor and taste of meat since it contains iron. Heme also acts as a catalyst in many chemical reactions and molecular interactions. For their heme, Impossible Foods bioengineered yeast to produce leghemoglobin from legumes.[8] The use of GE could present additional challenges for the product given that there are regulatory restrictions on GE products around the world, which would limit the markets in which the product could be sold.

Cultivated Meat

Simultaneous to the emergence of plant-based meat has been the arrival of cultivated meat, grown from animal-derived cell cultures in vitro. Developing cell-cultivated meat involves five steps: (1) taking a biopsy of animal cells, (2) cell banking, (3) cell growth, (4) harvesting, and (5) food processing. Cells are first sourced from animals through tissue biopsy. Tissue samples tend to be collected from the animal's muscles under anesthesia, as well as collecting embryos in the uterus before they implant.[9] However, it is worth noting that these biopsies could equally be done post-slaughter. A variety of different individual cell types can be used, but stem cells are favored because they are undifferentiated—meaning that they have the ability to turn into a multitude of other cell types.[10] Moreover, stem cell lines can be immortalized, meaning that the cells won't stop dividing or generating new cells. As a result, creating immortal cell lines would mean that very few tissue samples would be needed.[11] The types of stem cells that are the most promising include:

- Embryonic: known for their ability to differentiate and proliferate (rapidly grow and accumulate). However, a challenge is that it is difficult to stimulate differentiation exclusively into cells that will produce skeletal muscle cells.[12]
- Myosatellite: skeletal muscle stem cells that regenerate muscle. Besides being a rare cell type, the downside is that these cells are limited in their ability to proliferate.[13]
- Mesenchymal: they are multipotent, meaning that they can differentiate into a limited number of cell types, including myogenic cells (which produce skeletal muscle cells). Mesenchymal stem cells (MSCs) face similar proliferation and differentiation limitations, however.[14]

Multiple cell types need to be cultivated together to successfully advance to the next stage. For example, myoblasts (which form muscle tissue) must be grown with fibroblasts, which are cells that generate extracellular matrix (ECM). ECM connects the cells together and aids in cellular signaling and regulation, as well as a physical structure for cells to scaffold onto.[15] ECM also contributes to the elasticity of tissue.[16] The cell cultures are placed in a medium with vital nutrients for growth that provides amino acids, sugars, salts, and vitamins.[17] Animal serum is often added to provide many of the other biological necessities such as growth factors and hormones.[18] Serum is essentially the liquid part of blood after removal of clotting factors, cells, and fibrin.[19] Fetal bovine serum (FBS) is the most common option. Controversially, FBS is extracted from the blood of dead fetuses from pregnant cows that are slaughtered—something that animal rights activists find abhorrent and will be discussed in greater detail momentarily. The serum consists of a myriad of elements that help the cells to grow including hormones, fats, proteins, carbohydrates, minerals, vitamins, and inorganic compounds.[20]

In larger-scale production, the cell culture is placed in a bioreactor once the muscle culture has matured.[21] In the bioreactor, the cells themselves are implanted into a tissue scaffold designed to give structure as well as guide cell differentiation and growth into proper alignment that imitates how muscle cells would grow in the actual live animal.[22] Bioreactors are big machines that can control the culture's environmental factors like temperature and pH, as well as control diffusion of nutrients (mixing or stimulation).[23] This is carefully managed and coordinated by technicians since the required combinations of nutrients and growth hormones change throughout the course of cell growth.[24] When the tissue comes out of the bioreactor, the growth is complete, and the cultivated meat can be processed into a food product, whether that is nuggets or a burger patty.[25] From start to finish, the process takes a few weeks, rather than the months or years required for traditional meat.

Regulatory Process

Given that this technology is entirely new, the regulatory framework to assess its safety has to be conceptualized from scratch—it does not neatly slot into any existing regulatory models. After a lengthy Congressional debate, the FDA and USDA announced that they had formally agreed to oversee cell-cultivated meat jointly.[26] This division is because the FDA and the USDA have different regulatory responsibilities. The FDA is responsible for regulating "food," which includes "articles used for food" and "articles used for components of any such articles."[27] The USDA, on the other hand, is solely responsible for the inspection of meat and poultry. Cultivated meat production sits in both their jurisdictions. In practice, this means that the FDA governs the collection of cells, cell differentiation, and cultures up to maturation (under the Food, Drug and Cosmetic Act and the Good Manufacturing Practices), and then the USDA takes over regulatory control at harvest time, overseeing aspects such as sanitary conditions, hazard analysis, product testing, and labeling.[28] Companies harvesting cells for human food are subject to FSIS regulations on sanitation, Hazard Analysis and Critical Control Point verification, and testing to ensure that the products are unadulterated, and properly labeled under the Federal Meat Inspection Act and the Poultry Products Inspection Act. Navigating that process is not easy, but, in June of 2023, both Upside Foods and Eat Just passed the final approval step and were cleared to start selling their products in the United States.[29]

Two countries outside of the United States have approved the sale of cultivated meat—Singapore and Israel. In 2020, Singapore became the first country in the world to approve the sale of cultivated meat, while Israel approved cultivated beef for human consumption in 2024. Both of these countries face high threats to their food security. As a small island nation of only 284 square miles, Singapore faces huge restrictions on their domestic food production. There is little room for traditional agriculture, and with a growing population, land is more and more restricted. Currently, only 1 percent of Singapore's land is used for farming. As a nation, they have therefore adopted the 30 × 30 policy to ensure food security, which seeks to build up the nation's agri-food industry's capability and capacity to sustainably produce 30 percent of nutritional needs by 2030.[30] Given the limited access to land and other resource constraints, Singapore has embraced technology to increase food production and has supported the sector with grants and R&D funding. Cultivated meat is one way that they are using technology to produce more food at home and reduce the dependency on imports. In line with the government's desire to develop food security through technology, the Singapore Food Agency (SFA) has streamlined the application process for novel foods, such as cell-cultivated meat, to speed up approval. This streamlining was likely aided by the fact that there isn't a strong livestock lobby to challenge or build pressure—unlike in Europe and the United States. As a result, Singaporean Eat Just was approved to sell their "chicken" nuggets in 2020, with further chicken products being approved in 2021. Vow's quails were also approved for sale in restaurants in 2024, and there is an expectation that SFA will approve cultivated chicken and pork from French firm Vital Meat and Dutch company Meatable, respectively. In Israel, on the other hand, only 29.7 percent of land is suitable for traditional agriculture with desert accounting for 50 percent of the rest of the country. Climate change is further shrinking arable land, heightening food security pressures that are compounded by ongoing conflict in the region. The war in Gaza has underscored how deeply food systems are entangled with geopolitics, disrupting supply chains and amplifying regional instability. In response to these risks, Israel—like Singapore—has increasingly turned to food innovation, investing heavily in technologies that can boost domestic production and reduce reliance on fragile imports. Aleph Farms developed a petit steak using stem cells from a Black Angus cow from California (nicknamed Lucy). The company is now working to develop lab-grown collagen. Interestingly, unlike other companies, Aleph Farms focuses on marketing that the product originated from a cow and therefore doesn't present itself as a vegetarian alternative. As an entryway into the EU, Aleph Farms has applied for approval to sell their steak in the UK and in Switzerland.

Case Study K: Startup to Scaleup—Singapore's Recipe for Food Tech Innovation

The Singapore Government has committed to expanding domestic food production in order to meet the growing demand. This commitment requires advancing new technologies and innovation since the highly populated nation lacks the necessary land for conventional farming. Consequently, the Singaporean government has invested millions into agrifood research.[31] Part of this investment into the future of food is the rollout of agribusiness incubators to support startups in the food space to scale. Singapore, currently the global leader in government-sponsored initiatives for alternative proteins and cultured meat, had around sixty alternative protein startups employing over 700 people in early 2024.[32] One example is ScaleUp Bio. Established in 2022, ScaleUp Bio aims to help startups focused on microbial fermentation succeed. It lends expertise in technology transfer and downstream process optimization so that these startups can grow from an idea into a fully fledged food business. In 2024, ScaleUp Bio secured a food manufacturing license from the Singapore Food Agency (SFA) for its Commercial Pilot Production facility in Tuas, western Singapore.[33] By providing lab facilities and expertise to new companies, ScaleUp Bio helps make it easier to develop and test products, as the cost of establishing a lab can be prohibitive for newcomers. The facility specializes in submerged microbial and precision fermentation at scales of up to 10,000 liters (approximately 2,641.72 US gallons). Submerged microbial fermentation involves cultivating microorganisms such as bacteria, yeast, or fungi in liquid nutrient solutions within bioreactors, allowing for the controlled growth of proteins and other compounds that can be processed into food products. Precision fermentation refers to the use of genetically engineered microorganisms to produce specific functional ingredients, such as proteins, enzymes, or fats. Once the startups have developed a viable product and proposal, ScaleUp Bio connects them to a network of partners, from research to market launch, including the Singapore Institute of Food and Biotechnology Innovation (SIFBI).

Milk Alternatives

The labs that developed cultivated meat aren't stopping there—they want to explore creating cultivated versions of other animal products, including milk. Most readers will be familiar with plant-based milks from their morning oat milk latte. The global market is already massive with dairy alternatives worth $27 billion USD in 2023 and projected to be worth $43.6 billion USD by 2028.[34] These plant-based milks are made by soaking the base ingredient (oats, for example) and then blending and filtering them before the mixture is sterilized. However, new technologies similar to those that are used for cultivated meat are being applied to the dairy industry. Many of these technologies aim to reproduce milk components, like whey and casein, by inserting the genetic code for various milk proteins into bacteria. The bacteria are fermented and churn out these proteins that can then be turned into dairy products. One Silicon Valley startup, Perfect Day, is using this approach to produce milk, yogurt, cream cheese, and ice cream. Other companies are taking things one step further. The Israel-based company Wilk, backed by dairy giant Danone, is producing cultivated milk by using milk-producing cells from mammals. In addition to disrupting the milk sector, Wilk hopes to produce a breast milk alternative that might challenge the infant formula sector. It is an enticing sector to get a slice of—revenue in the baby milk and infant formula sector equaled $53.57 billion USD in 2024, and the market is expected to grow annually by 5.61 percent with particular emphasis in China.[35] However, just as with the arrival of plant-based milks, this might cause a backlash from the dairy sector. In Europe, plant-based milks cannot refer to themselves using dairy-related terms, while in the United States, there is guidance from the USDA that encourages the use of a voluntary statement to identify that they are "nutritionally inferior" to dairy.[36]

Microbe Protein

In addition to plant-based and cultivated meat, laboratories might be able to grow meat from microbes. Readers in the UK might be familiar with one of the leading brands of alternative protein from microbes—Quorn. Quorn mycoprotein is derived from *Fusarium Venenatum*, an ascomycete, which is a type of fungus that naturally occurs in the soil. The fungus is grown on high-grade carbohydrates—derived from corn and wheat. In the first step of production, a few spores of the fungus are grown in a small conical flask with a little glucose and salt solution. Then it is introduced into the main fermenter, which has been sterilized and filled with water containing glucose and various essential salts. These fermenters are huge, measuring 130 ft high. To avoid the fermenter overflowing, the fermented broth is continuously extracted at exactly the same rate as the feeds are added, ensuring that the fermenter can produce continuously for weeks at a time. The harvested broth is gently heated, and the fungal mycelium formed in the fermenter vessel is harvested by centrifugation to give the Quorn's mycoprotein paste, or "dough." The mycoprotein is then seasoned and mixed with a little free-range egg, or plant protein in vegan products, to help bind the mix. It is then steam cooked for about thirty minutes, and chilled, before being shaped for a variety of different products. Quorn products have graced the shelves of UK supermarkets for over thirty-five years and continue to be considered a staple of vegetarian diets in the UK.

New actors in the United States are using a similar process to Quorn to produce meat alternatives with the hope to market these new products to millennial consumers who want to reduce their meat consumption. In 2018, Kozubal launched the company Nature's Fynd in order to bring Fy-based food products to market.[37] Former NASA researcher Mark Kozubal discovered *Fusarium str. Yellowstonensis* (or Fy) in Yellowstone in 2009—a microbe with unique properties that may make it highly suitable as a sustainable source of protein. Fed on a diet of water, sugar, and salt, a tray of Fy could provide 3,500 grams of protein, as much as twenty-five chickens. A 2024 study from the University of California, Davis about the potential for food products like Fy found that they could be ten times as efficient as beef in terms of land and water use, as well as three times as water efficient and two times as land efficient as chicken.[38] From their 3,000 square meter factory in Chicago, Nature's Fynd ships Fy products across America, including to major grocery chains like Whole Foods, offering yogurt, cream cheese, and breakfast patties. People seem to like the taste, even those with the most decerning palates. Michelin-starred chef Eric Ripert recognized Fy's potential and produced a squash blossom cheesecake containing the ingredient, saying he was "very impressed by the possibilities and versatility of what the product could offer."[39] This type of microbe protein alternative isn't limited to the United States either. Solar Foods, based in Finland, uses single-cell organisms that its cofounder Pasi Vainikka discovered on the Baltic Sea coast to produce a yellowish powder called solein.[40] The cells are fed with hydrogen, a mineral mixture, and CO_2. Then, the organisms are killed and dried with hot air, producing a yellow powder high in protein that can be used in the production of meat and dairy alternatives. Solein's Helsinki factory produces 160 tons annually. So far, solein has only been approved for sale in Singapore, where it has taken the form of a dairy-free gelato and chocolate bars. The company hopes to pass regulatory requirements in the United States and EU by the end of 2026.

Consumer Uptake

So the big question is whether people eat lab-grown meat? When Dutch scientist Mark Post unveiled the first cultivated product in 2013, his lab took time to consider what medium to shape the meat into to make it most appealing. The lab considered making a sausage, but sausages in Europe are too closely linked with artisanal production (think salami or chorizo). In the end, the international appeal of the hamburger won the day. As Benjamin Wurgaft wrote in *Meat Planet: Artificial Flesh and the Future of Food*, "The hamburger is an appropriate avatar for modern meat, associated not only with abundance, but also with industrial production, uniformity, speed, flexibility and frequently with automobiles."[41] So far, indications are that consumers are open to trying cell-cultivated meat. In the United States, the market reached $15.1 million USD in 2023, up from $3.3 million in 2022.[42]Recent studies tracking American consumer attitudes toward cell-cultivated meat have found a relatively positive outlook for a product yet to be available for mass consumption.[43] A YouGov survey conducted in May 2024 revealed a 50–50 split of the participants between those who preferred to eat animal meat over lab-grown meat—although 14 percent responded that they wouldn't eat either.[44] This 50 percent preference for animal meat was consistent across various categories—region, gender, politics, age, and race—with only slight variations. The most significant divergence was political: 68 percent of Republicans preferred animal meat over lab-grown meat. In Singapore, a survey funded by the Ministry of Education found that among local consumers, plant-based meat has the highest acceptance rate, at 29 percent, followed by cultivated meat at 25 percent and insect-based products at 7.2 percent.[45] However, in countries where cultivated meat has not yet reached supermarket shelves, there is greater hesitation. A 2024 YouGov survey found that while 74 percent of Britons were aware of cultivated meat prior to asking, only a quarter (26 percent) said that they would probably eat cultivated meat if they could purchase it at the supermarket.[46] In France, only 9 percent said that they would try it.[47] This reticence to try novel foods, even when those foods have passed food safety standards, is known as food neophobia. While more common in children when they experience new tastes, it is increasingly common in adults as new food products come into the market.

The More Sustainable Option?

One primary reason given for advancing lab-grown meat is animal welfare. A National Sanitation Foundation study found that almost 70 percent of consumers in the United States were influenced by animal welfare concerns in their purchases.[48] As discussed in the introduction to this part, the conditions that livestock experience in industrial farms are inhumane with animals packed in tightly without access to fresh air or even the ability to turn around. These cramped conditions create the perfect conditions for pathogens to arise and quickly spread. These diseases cause suffering to the animals and put human health at risk if the pathogens can transfer to humans. A switch to lab-grown meat would immediately nullify the risk of zoonotic diseases arising. It does not need massive numbers of animal sources to generate cell lines and can be done without slaughter. As a result, lab-grown chicken nuggets would not require factory farming and the disease-ridden conditions that come along with it. It would also reduce the amount of antibiotics used in agriculture since they would not need to be used routinely to control outbreaks. Lab-grown meat also addresses animal welfare concerns since it removes the need to keep animals in such intensive housing and eliminates the need for slaughter. Some animal rights activists may object to conducting tissue biopsies on animals; however, it is arguably much less invasive and can be done under anesthesia and without causing great harm to the animal.

Investment in lab-grown meat is also driven by environmental concerns. A global study done by the Organization for Economic Co-operation and Development found that, on average, about 80 percent of survey respondents in every country thought that their country should fight against climate change.[49] Globally, traditional livestock production generates almost 15 percent of the world's GHG emissions.[50] Lab-grown meat, then, would appear as an incredibly appealing opportunity that would not release nearly as many GHGs (or use that much land or water) as raising mass amounts of livestock. While it is understandable how the raising of only a couple of animals for cultivation would drastically reduce water and land use, there is some debate as to whether cultivated meat production at scale would actually lower GHG emissions compared to "real" meat production. Current cultivated meat production is still small, and for it to be a viable alternative, it would need to be scaled up significantly. However, cultivated meat production currently relies on natural gas, generating CO_2 emissions. A group of scientists at the University of California, Davis performed a Life Cycle Assessment (LCA) measuring the environmental impact of cultivated meat and found that the global warming potential (measured in kg of CO_2 equivalent) of cultivated meat, produced with various formulations, was higher across the board compared to traditional beef.[51] This analysis, however, was performed assuming that bioreactors and processes producing necessary resources were run on natural gas. If cultivated meat production were to be powered by renewable energy sources, it would produce much lower carbon emissions than "real" meat production.[52] Nonetheless, cultured meat will certainly reduce the land and water footprint. However, to see any climate benefits of the switch to cultured meat, sustainable technology will need to be developed for such use at a large scale.

Religious and Ethical Choice

Many people choose not to eat meat. This personal decision can be motivated by many reasons: religious, ethical, philosophical, health, and environmental. Does cultivated meat count as vegetarian or vegan? There is no hard and fast answer. Part of the challenge of answering that question is that it's a very personal choice. Each consumer has their own ethical standards, and the technology is still emerging and advancing. For the most part, the use of fetal bovine serum (FBS) as the growth medium for the animal cells means that cultivated meat cannot be considered vegetarian. FBS comes from the blood of a dead (fetal) calf.[53] However, a number of cultivated meat producers are experimenting with alternatives to FBS, such as plant-based medium. For example, Aleph Foods uses bovine FGF-2, a growth factor alternative.[54] By switching away from FBS, the cultivated meat could be considered vegetarian. Animal-free serums are being explored using fungi and plants.[55] However, cost might present a challenge. According to Benjamin Wurgaft, "while many commercially available growth media are free of the decidedly non-vegan fetal bovine serum, they have all been judged to be too expensive for use at industrial scale."[56] So it could become vegetarian, but being considered vegan might be more challenging. For some vegans, cultivated meat may cross an ethical line because it still relies on animal cells and byproducts. So, even if the animals aren't killed in the process of producing the cultivated burger, the process of extracting animal cells via biopsy is still exploitative. The animals from which the cells are being extracted are sentient beings that did not consent to the biopsy. The production is commodifying the animal. Therefore, any product that uses cells extracted from animals (either via slaughter or via biopsy) could not be considered vegan.

But what if the cells are extracted from another species that can give consent? Some people have joked that the future might allow a burger to be produced from the stem cells of celebrities, making them more appealing and potentially overcoming the hurdle to being classified as vegan. Perhaps people could be tempted to bite into a burger made from a Kardashian stem cell? This could, in theory, remedy some of the concerns expressed by vegans over consent and exploitation. While a cow might not be able to consent to a biopsy for the removal of its stem cells, a human could legally consent and be compensated accordingly. If exploitation and cruelty to animals are at the core of why some vegans don't eat meat, would a nugget grown from the cells of a consenting human be less problematic? Or does this come too close to cannibalism to be viable? However, more realistically, there might be value in considering the stem cells from other mammals. Just because a rib eye has historically come from a cow, a cultivated steak doesn't need to be grown from bovine stem cells. As a technologist and author of "Food Futures," Chloe Rutzerveld, asked, "Stem cells of mice, for example, grow much faster and more efficiently into muscle tissue than cows' cells. So why do scientists not use mice cells instead, in theory saving time, energy and costs?"[57] But would consumers be willing to eat a nugget grown from the stem cells of mice? Or would it be unpalatable? As we discussed in Chapter5, the "ick" factor for future protein sources is a big challenge.

What are the religious implications of cultivated meat? Singapore's Islamic Religious Council declared cultivated meat to be halal.[58] In the fatwa that the Council issued, they stated that they consider cultivated meat to be halal provided that (1) the cell source is taken from animals that are halal to consume, (2) every ingredient that makes up texture and composition of cultivated meat is halal, and (3) the product is nontoxic and clean. They commented that their ruling was "underpinned by the Islamic principles that it serves to (i) preserve human life; and (ii) protect the environment" but also acknowledged that "actual Muslim consumer acceptance of cultivated meat will also depend on other considerations like personal dietary preferences, taste, and cost."[59] The case for certifying cultivated meat as kosher is slightly more complicated. Jewish law permits the consumption of certain animal species (such as cows, sheep, goats or chickens) and prohibits others (such as pigs or camels). Flesh removed from any live animal, including kosher species, is not permitted. The key consideration for whether a cell-cultured product can be certified as kosher is where the starter cells originated from. Whether cultivated meat can be considered kosher depends largely on the source of its starter cells. If cells come from a live animal, the product may be deemed non-kosher, since Jewish law prohibits eating flesh taken from a living creature. Likewise, if the animal was not slaughtered in a kosher manner, the end product would not qualify. Cells taken from a kosher-slaughtered animal, however, could allow for certification. Complications arise around the prohibition against consuming blood: while this is less relevant when cells are drawn from fertilized eggs (as in the case of Aleph Farms), it is an issue when a version of FBS derived from blood is used. As a result, some rabbinic authorities insist that kosher slaughter requirements must still apply, others argue certain starter cells should be avoided, and still others suggest cultivated meat could, under specific conditions, be considered not only kosher but even pareve. What is clear is that opinion remains divided, and consensus is unlikely in the near term.

Case Study L: Feather to Fork—Turning Chicken Waste into Edible Protein

Kera, originally a startup from Thailand and now based in the UK, is tapping into the growing chicken production trend by creating a way to make feathers edible. Chicken feathers are composed of approximately 91 percent protein (keratin), 8 percent water, and 1 percent lipids. Kera's technology extracts keratin from raw feathers using a biological, water-based solution that breaks down the rigid, strongly cross-linked keratin structure in the feathers, then filters the extracted solution to obtain the desired powder form for consumption. From the feathers of a single chicken, up to 190 grams of hydrolyzed protein can be produced with an extraction efficiency of over 86 percent.[60] The amino acid profile of chicken feathers is similar to that of chicken breast and also contains a high level of antioxidants, comparable to those found in berries, says the startup.

The idea for Kera originated in 2018, and the business was established in the UK in 2023. With the vision of becoming a sustainable waste management solution for chicken feathers, Kera has the potential to achieve two key co-benefits. First, by turning feathers into an edible protein powder, more people could be fed by one chicken. Second, by making this waste product profitable, it creates an opportunity for chicken farms to generate additional income. This has huge potential, since chickens have been the most slaughtered animals for meat production worldwide for decades, reaching a yearly record of over 76 billion in 2023.[61] Its future success remains uncertain, with several critical issues still unresolved. Kera's product must overcome deep-seated consumer aversion to waste-derived ingredients. If vegetarians reject protein made from chicken feathers, its viability will depend entirely on attracting enough flexitarians. The approach hinges on the continued industrial slaughter of chickens, raising serious ethical concerns. And without curbing the global surge in chicken consumption, its potential to drive meaningful change will be limited. While consumers' reactions to this new protein alternative remain uncertain, Kera's concept is undeniably innovative. By leveraging technology to maximize the output of a single chicken, Kera has the potential to address pressing modern challenges like food waste, global hunger, and creating additional income streams for farmers, especially when adopted on a larger scale.

Increasing the Market

Given this confusion over whether cultivated meat is or is not considered vegetarian or vegan, many cultivated meat labs have marketed their products toward flexitarians instead. Although there is no official definition, "flexitarianism" refers to a style of eating that endorses mostly plant-based foods while enjoying some meat and other animal foods in moderation. The motivations for reducing meat consumption tend to be either planetary or public health. For example, flexitarians account for 9.2 percent of the UK population and have grown in size by 21 percent since 2018.[62] That number is only set to grow, with predictions that half of UK consumers will identify as flexitarian by 2025.[63] Given that flexitarians don't completely exclude meat products from their diet, the inclusion of FBS or bovine stem cells would not be a deal-breaker for those consumers. Demand for flexitarian products, including plant-based meat, rose from 2019 in both Europe and the United States. Plant-based meat, dairy, and egg sales in the United States were worth $3.9 billion USD in 2017 and worth $8.1 billion USD by 2023.[64]

This rise coincided with the COVID-19 pandemic. These two things are not unconnected and offer insight into how plant-based meat might position itself to increase its market share in the future. Firstly, there was increased fear of disease and illness during the pandemic—not irrational at a time when the world was facing regional lockdowns. At a time when there was rising awareness and fear over zoonotic disease, plant-based meat alternatives were perceived as safer than traditional meat since they were seen as less likely to be contaminated or cause food-borne illness than traditional meat.[65] Although it is possible to grow the meat without them, antibiotics are often used to avoid contamination in lab-grown meat—something that perhaps runs counter to public claims that these novel foods provide an alternative to the high levels of prophylactic antibiotic use in industrial livestock production that we have discussed.[66] Secondly, the meat supply chain was significantly disrupted during COVID-19. While there was disruption to all supply chains across the entire globe, plant-based products did not suffer as much as traditional meat products. First, because people could more easily stockpile plant-based alternatives since many are shelf-stable and didn't need to be refrigerated or frozen. Second, there were significant issues in maintaining labor levels in slaughterhouses. By May 2020, approximately 4,200 meatpackers at 115 processing plants in the United States had contracted COVID-19 with restriction protocols put in place at the remaining plants to prevent the virus from spreading further.[67] The restriction slowed the production line, resulting in 45 percent less meat being processed each day.[68] This was less of an issue for the production of plant-based products since they do not require the same level of labor, nor do they have the same regulatory requirements. As a result, many people tried plant-based meat products for the first time and discovered that they liked them.

Both the perception of plant-based meat as "safer" and the lower labor requirement provide opportunities for the lab-grown meat sector to grow further. First, the perception that plant-based meat is safer (whether accurate or not) and healthier than traditional meat could allow lab-grown meat companies to tap into the clean eating movement. We discussed the clean eating movement in Chapter 2 and highlighted its focus on reducing unhealthy foods (including meat) and opting for unprocessed foods with minimal additives and preservatives. This interest in "clean" eating is notably increasing in China, where 43 percent of consumers described themselves as "conscious consumers" in a recent McKinsey survey. This interest in "safe" and "clean" food is linked to the history of food crises (e.g., cooking oil in 2024 and milk contamination in 2008) in China with the most recent case being that of swine fever that saw 27.9 million metric tons of pork destroyed between late 2018 and early 2021.[69] This could provide an inroad into the Chinese market for lab-grown meat—a market that currently has some reservations about cultivated protein. According to a 2018 YouGov poll, only 26 percent of Chinese consumers said they would eat lab-grown meat, among the lowest rate of acceptance compared to other countries in Asia.[70] In comparison, 34 percent of Thai consumers and 52 percent of Vietnamese consumers said they would eat lab-grown meat. Engaging with Chinese consumers would be a huge success for meat alternatives since revenue in the Chinese meat sector is expected to reach $87.75 billion USD in 2023, and China's meat market is projected to grow annually by 20 percent between 2023 and 2027.[71] Labor dynamics is the second insight that the rise in plant-based products during the COVID-19 pandemic highlights. The lower labor requirement that lab-grown meat has in comparison to the traditional livestock sector could be a significant asset for lab-grown products to increase market share in the future. As we discussed in Chapter 4, farming is facing a labor crisis in the Global North. This is even greater in the livestock sector, which faces additional labor requirements. Few people aspire to work in a slaughterhouse, so there is a high level of turnover.[72] Additionally, due to the nature of the job, slaughterhouse jobs often require additional training, including having a qualified vet on site, which restricts recruitment. Working on the factory floor of a lab that produces lab-grown meat is likely an easier sell than convincing someone to work in a slaughterhouse. Beyond low-skill labor, there is also the fact that the highly technical and innovative nature of lab-grown meat might appeal to young people and encourage them to enter the food and farming sector when they would have otherwise been less interested in traditional farming jobs.

The consumers with the greatest openness to cultivated meat are in India with 20 percent—somewhat surprising since 38 percent of Indians are vegetarian, and 81 percent of citizens follow some dietary restrictions.[73] This is primarily due to religious and cultural reasons. Hinduism considers cattle to be sacred, while pork is eaten only by a few, and Jainism is lacto-vegetarian. There is also a Muslim population in India that does not eat pork. However, as incomes across India rise, traditional vegetarian diets are being abandoned, and more people are adopting omnivore diets. According to the Statista Global Consumer Survey, 30 percent of Indians said they were vegetarian in 2024, down from 33 percent in 2018.[74] Part of this is explained by declining religious identity and desire to emulate Western diets. However, it is also due to rising incomes. Many vegetarians in India do not eat meat because they cannot afford it, not because of ethical or religious concerns. Many members of the lower castes, Dalits (formerly known as untouchables) and tribespeople, identify as meat eaters but are required to have a vegetarian diet since they cannot afford to buy meat. Yet, India's economy is growing, and poverty is declining. According to Brookings, India has experienced a significant decline in both urban and rural inequality from 2011 to 2023. The Gini coefficient measures income inequality. India's urban Gini declined from 36.7 to 31.9 while the rural Gini declined from 28.7 to 27.0.[75] As incomes rise, more individuals who want to eat meat will be able to afford it. Cultivated meat could be well placed to meet this increased demand.

Lab-grown beef could be an appealing alternative since significant cultural challenges remain in creating a domestic cattle industry in a predominantly Hindu country.

The Economic Landscape

There are a lot of people willing to bet big money on cultivated meat being the future of protein. The Good Food Institute estimated that from 2010 to 2022, approximately $3 billion USD in private capital was invested in cell-cultivated meat and seafood companies worldwide.[76] Beyond venture capital, some of the world's largest food and meat companies are buying into cell-cultivated meat companies to future-proof their income. For example, Tyson Foods invested in UPSIDE Foods in 2018 and Believer Meats in 2021, while Cargill invested in UPSIDE Foods in 2017 and 2020 and invested in Aleph Farms in 2019.[77] However, growing meat in a lab is one thing. Doing it at scale is another. If cultivated meat is going to be the food of the future, it needs to be able to be scaled up dramatically to feed a growing population. To replace just 10 percent of the world's current meat supply, 4,000 new cultivated meat facilities would need to be built. This might be a challenge. Each facility would cost $430 million USD to build, requiring approximately 600 bioreactors running simultaneously to produce 22 million pounds of meat per year.[78] However, researchers have raised concerns about the potential of current bioreactors to achieve this level of bulk cell growth. The cost of these machines and the technical skill required to maintain their functionality is one of the biggest costs to producers—approximately 80 percent.[79] To be able to economically scale the production of cultivated meat, bioreactors need to be better designed for mass production, plus there needs to be less expensive starter cells and growth medium. In the short term, the price consumers pay at the supermarket for cultivated meat could be reduced by creating a hybrid offering. These hybrid options could use small amounts of cultivated animal product to "boost" the taste and quality of plant-based meat alternatives, enabling it to scale.[80] An economic analysis by Open Philanthropy concluded that cultivated meat faced intractable technical challenges at scale, including barriers in thermodynamics, cell metabolism, bioreactor design, ingredient costs, and phytosanitary standards to produce cultivated meat cheaply enough to displace traditional meat.[81]

However, on the provision that innovation can solve the roadblocks to cultivated meat scaling up, the cost of production (and therefore the consumer price) should come down. Currently, plant-based meat and cultivated meat are expensive. The first cultivated burger in Mark Post's lab cost $330,000 USD to produce.[82] While costs have declined, the wholesale price today is just over $28 USD per pound of cell-cultivated meat.[83] To claw back a profit, this cost would be passed onto the consumers with $45 USD per pound charged in a supermarket or $100 USD per pound in a restaurant.[84] Similarly, a six-pack of Nature's Fynd breakfast patties will run you just over $5 USD, about twice as much as patties made from pork. At that price, cultivated meat is only accessible to middle-class consumers with higher disposable incomes. The fact that lab-grown meat is a luxury good could not be better demonstrated by the fact that Fortnum and Mason, a luxury department store in London, was the first place in the UK to market lab-grown meat when it marketed a cultivated scotch egg. While they couldn't sell the product directly to consumers due to the fact that cultivated meat is not approved for sale, the decision to place the product in one of the most elitist spots in London reflects cultivated meat's current place in the market. Over time as production costs come down, the price of cultivated meat is likely to decline too. The Good Food Institute—the lobbying body of the alternative protein sector—says that lab-grown meat will be cost-competitive with traditional meat by 2030.[85] This could ultimately mean that cultivated meat could become a cheaper option than traditional meat within a decade. While this might seem unlikely today when looking at a luxury cultivated quail egg, this trajectory mimics the history of sugar. When sugar first came to the tables of Western Europe in the 1800s, it was the ultimate example of conspicuous consumption—it was expensive and denoted your high social status. However, as time went on and production increased (due in large part to the enslavement of Africans), the price came down, and it became accessible to all. This same accessibility journey could happen with cultivated meat. If lab-grown meat does follow that trajectory, it could ultimately become less expensive than traditional meat. That could be further compounded if livestock emission taxes (such as those to be implemented in Denmark in 2030) are imposed, raising the price of traditional meat further.[86] As a result, in a switch from the current dynamic, the future economy could mean that only wealthy consumers would be able to afford traditional meat, while low-income communities might only be able to afford the lab-grown meat alternatives. This compounds a two-tier food system where wealthy consumers are able to choose the best quality food while low-income communities aren't able to make choices for themselves and their families since they are confined by price.

Ultra-Processed?

So what does lab-grown meat mean for health? Arguments for the consumption of lab-grown meat tend to focus on the environmental and animal welfare benefits of these novel foods, but what about the human health implications? Numerous food regulators around the world with stringent assessment processes have deemed them safe for human consumption. However, there is a deeper question about whether they are beneficial to public health more broadly. There are certainly problems with high consumption levels of traditional meat—it is associated with an increased risk of heart disease and diabetes.[87] Red and processed meat, in particular, are linked to higher rates of bowel cancer.[88] Cultivated and plant-based meat alternatives claim to lower these risks. Cultivated and plant-based meat can be designed to be heart-healthy—the saturated fats that are problematic in traditional red meat can be replaced with healthier fats, such as omega-3 fatty acids. Similarly, L-carnitine, a substance found naturally in red meat and associated with heart disease, could be removed from cultivated meat. Given the public health crisis that we are facing with rising NCDs across the world (see Chapter 2), this might seem like a godsend. It could offer a way to reduce the damaging impacts of meat consumption without having to give up your game night burger or Friday night steak.

Yet, as with everything in food, it is not as straightforward a fix as it might seem. As we discussed in Chapter 2, the NOVA definition of UPFs is industrial food and drink products that contain multiple ingredients and undergo complex processing (such as extrusion, hydrogenation, or molding), with the goal of creating highly palatable, convenient products.[89] Lab-grown meat products would certainly fit within this definition. As outlined in Chapter 2, there is now a global movement to lower the consumption of UPFs to help improve public health since there is a link between UPFs and NCDs. The commitment to lowering UPF consumption rates raises questions about the future of lab-grown meat. If we accept, as a society, that UPFs are bad for public health, and lab-grown meat is a UPF, surely that would logically follow that lab-grown meat is bad for public health? If we agree that the consumption of UPFs needs to be reduced, can lab-grown meat truly be a positive solution for the future of food? It might well be preferable to industrially produced meat, but that is a pretty low bar and does not equate to it being a net positive. This becomes more problematic if the future consumption of lab-grown meat is concentrated in low-income communities. If cell-cultivated meat does become cheaper than conventional meat, as some forecasts suggest, it is likely to be positioned as a direct substitute for industrial meat in these markets. While this could reduce the environmental footprint of mass meat production, it also raises questions about equity and health. Low-income groups may become the primary consumers of a product still under debate regarding its nutritional profile, long-term safety, and cultural acceptance, while higher-income consumers retain access to more diverse and minimally processed food options. Such a divide risks reinforcing a two-tier food system, where affordability dictates not just quantity but also the quality and type of protein available to different social groups. This consumption distribution would have the potential to further entrench the disproportionally high levels of UPF consumption (and associated higher levels of NCDs) within these communities.[90]

Conclusion

There is huge potential for lab-grown meat to have a transformative impact on the future of food. Companies are testing different formulations to create a wide range of lab-grown meat alternatives from cultivated meat to fungi. Each product comes with its own opportunities and challenges. Innovation in the cultivated meat sector could create a route for heart-healthy meat that would be accessible to millions of people without the associated environmental and welfare implications of traditional meat. However, given that it is still so new, there are some significant hurdles to overcome. The biggest roadblock to the future of lab-grown meat will be scaling up production to reach the mass market. Without doing so, it is unlikely that lab-grown meat will ever occupy a significant market share. This includes developing bigger facilities with better technologies suited to mass production. If production can be scaled up, another significant challenge remains—whether consumers will buy it. Apprehension to novel foods can doom products from gaining market share. There are positive signs that consumers are willing to try lab-grown meat alternatives, especially if price comes down. However, outside of India, the consumer interest in lab-grown meat alternatives in the Global South hasn't been widely tested. Despite much of the forecast 73 percent rise in demand for meat by 2050 coming from developing countries, there is a significant lack of research on whether consumers outside of the Global North would accept cultured meat, especially in regions where livestock management is deeply linked to the culture.[91] To encourage that wider adoption, ethical questions need to be resolved. Is it vegetarian? Is it kosher? Is it halal? Religious scholars and animal rights groups debate these tensions, but it will likely be up to the individual citizens to decide whether they are comfortable eating these novel foods. Given all these concerns, it is critical that the labeling of these products is regulated and standardized globally to give consumers confidence in the product that they are buying. Worldwide standards and regulations for those products have not been established. There urgently needs to be voluntary international food standards, guidelines, and codes of practice for cell-cultivated meat products under the UN's Codex Alimentarius. Ultimately, lab-grown meat is at a pivotal time, and the next few years will determine whether it becomes a viable future for protein or remains a niche product for a select few.

7

Sustainably Raised Meat

While it's commonly accepted that meat is bad for the planet and for our health, there needs to be a distinction between industrially produced meat and sustainably produced meat. This essentially means distinguishing between meat that is part of the problem and meat that is part of the solution. Industrial meat refers to meat produced through intensive livestock farming systems designed to maximize output while minimizing costs. These systems typically involve high stocking densities and often the animals are kept exclusively indoors. The systems rely on grain to feed the animals, as opposed to grass or silage, and the animals are often given hormones to promote rapid growth and antibiotics to manage health. Industrial meat production is defined by its focus on economies of scale, often at the expense of animal welfare, environmental sustainability, and human health. Clearly this approach to meat production has negative impacts both for the environment and public diets. The need to reduce meat consumption (resulting in an associated drop in production too) is generally accepted. If a general and marked decline in meat production and consumption were to happen, it would help with climate concerns and improve public health outcomes. However, while accepting the need to reduce consumption, the entire planet is not going to become vegan overnight. Humans are omnivores, and meat has long played a key role in our culture. Moreover, meat consumption is a status symbol in many regions, and rising income levels are increasing the demand for meat. Recognizing that meat will likely continue to form a significant part of human diets, it is worth considering that there is a more sustainable model of meat production.

Sustainable Livestock Techniques

The cow itself is not a climate problem—after all, humans have been eating beef and other ruminant livestock for centuries. It is the method of intensive production that is environmentally damaging, and the vast quantity of meat that is being consumed globally as income levels rise. The mantra of these more environmentally sensitive ranchers is, "it is not the cow, but the how," highlighting the need to transition away from industrial production to more sustainable methods. But what are these more sustainable methods? According to the International Center for Tropical Agriculture, "sustainable livestock farming integrates environmental, social, and economic factors into animal production systems, thus minimizing negative impacts on ecosystems, conserving natural resources and increasing animal welfare."[1] But even within the sustainable livestock category, there are different approaches—some of which are more sustainable than others. For example, extensive grazing means that the animals are kept outside—it's certainly a more sustainable approach but does not adopt more progressive, ecological practices, such as ensuring diverse species-rich pasture.

"Regenerative agriculture" has been gaining traction and has become a buzzword for sustainable production. Many definitions exist and are much debated. For this chapter, we will use the term as defined by the state of California, which is the first legal definition of the term to exist: "An integrated approach to farming and ranching rooted in principles of soil health, biodiversity and ecosystem resiliency leading to improved targeted outcomes. Regenerative agriculture is not an endpoint, but a continuous implementation of practices that over time minimize inputs and environmental impacts[,] and further enhances the ecosystem while maintaining or improving productivity, economic contributions and community benefits."[2] Going one step beyond this, "regenerative livestock" denotes a more sustainable approach to livestock management that is pasture-based and high-welfare, where the animals function within the natural environment, primarily as part of a mixed-farming rotation. Some do not like the term, arguing that "regenerative" is too permissive in its tolerance for agrichemicals to be truly sustainable. Certainly, agrochemicals are widely accepted within the "regenerative" movement, which might run counter to the wider aspiration of a more sustainable farming future. Many regenerative farmers improve soil health through no-till agriculture techniques. No-till agriculture is a farming method where crops are planted without plowing, leaving soil undisturbed and covered with previous crop residues, which is positive for the ecosystem but relies on herbicides to kill off cover crops, hurting biodiversity.

Instead of "regenerative," some call for the adoption of "agroecology." The FAO defines agroecology as "an integrated approach that simultaneously applies ecological and social concepts and principles to the design and management of food and agricultural systems," adding that "it seeks to optimize the interactions between plants, animals, humans and the environment while taking into consideration the social aspects that need to be addressed for a sustainable and fair food system."[3] Since the terms "agroecology" and "regenerative" are both fairly fluid, some sustainability experts have advocated for "organic" since there are more stringent regulations that need to be met if something is to be classified as "organic" and these are set down within law. Whether "regenerative," "agroecological," or "organic," the key feature of sustainable livestock management is rotational grazing, which is when animals are regularly moved to allow the pasture longer periods of rest and regrowth. Within this management approach is mob grazing, which is when a high density of animals is moved from plot to plot with high frequency (one–two days). The pasture is then given a long period to recover. This approach is designed to mimic the natural grazing of herds in the wild. Ultimately, while the distinctions between these terms are important—and significantly affect both their sustainability and the ethical concerns they address—the mantra of sustainable meat, "less meat, but better," can serve as a useful catch-all umbrella.

Farmers and ranchers across the Global North are starting to advocate for this method of livestock management. Leading voices, such as the US farmers Gabe Brown and Nicolette Hahn Niman, have tried to promote a more sustainable vision for meat production across the United States while groups like the UK's Pasture for Life are advocating for a new certification scheme that recognizes regenerative management. Holistic grazing that mimics natural grazing patterns is also being championed in the Global South. These mob grazing techniques help restore landscapes while supporting viable livelihoods for local communities, many of which have long traditions of herding. Nomadic pastoralists like the Tuaregs and the Toubou have been practicing such techniques for generations. However, these practices have been lost through the colonization of Africa and the enclosure of land that had previously been managed as common. After years of mismanagement and overgrazing, coupled with the decline in wildlife, grasslands have been destroyed, and soil has eroded. Traditional grazing practices have been resurrected in recent years. By reintroducing animals to the land and managing them through sustainable mob grazing patterns, fertility can be brought back to the land, and the landscape can be restored. Allan Savory founded the Africa Centre for Holistic Management to restore grasslands in Zimbabwe in 2013, which has subsequently expanded into a global network. The Africa Centre has trained a new generation of landscape managers to help restore grasslands across the region so that they can once again function in harmony with nature. This benefits communities by ensuring that the environment is more resilient to climate challenges of the future and that smallscale ranchers earn a good livelihood through sustainable meat production and selling other byproducts such as leather or wool. So successful have these centers been that the term "holistic grazing" refers directly to Savory's approach.

Qualifying as truly sustainable also requires consideration of the grassland type. The most sustainable grazing landscapes are termed "seminatural" because they have been shaped over thousands of years by traditional farming practices—particularly livestock grazing—but haven't fallen prey to the industrial techniques of livestock management. As a result, they remain species-rich and healthy. Broadly speaking, seminatural habitats have low stocking rates, with grazing carried out at times of the year that best suit the habitat. Permanent pasture refers to any grassland that has been in place for more than five years. Prior to that, they might have been woodland, arable, or even brownfield sites. But grassland can be "improved" to make it more productive, damaging the soil and associated biodiversity. This includes techniques such as draining, plowing, reseeding with monoculture grasses (rye swards, in particular) and the application of synthetic fertilizer, or slurry. This trend to "improve" grasslands started happening in the 1950s as the Green Revolution took hold globally. When this type of intervention happens only once, it's referred to as "semi-improved." However, even when only done once, these techniques alter the landscape, nullifying some (but not all) of the benefits associated with external grazing that we have discussed. Grasslands can also be temporary, defined as land that has been under grass for less than five consecutive years. They are generally part of mixed farming systems where grass is grown within a rotation between other (usually arable) crops. These are often referred to as "leys." These can be managed in a sustainable way as part of an agroecology rotation or unsustainably, using the same "improvement" methods. A recent trend that delivers greater environmental benefit is the development of "herbal leys," which are much more diverse and include a variety of grasses, deep-rooting herbs (such as chicory or plantain), and forage legumes, including clovers and lucerne.

Case Study M: The Bovaer Case—Misinformation and Quick Fixes for Sustainability

Bovaer, a methane-reducing feed additive developed by DSM-Firmenich, is seen as a major breakthrough in sustainable agriculture. It has the potential to cut methane emissions by about 30 percent in dairy cows and up to 45 percent in beef cattle.[4] The additive works by inhibiting an enzyme in the cow's rumen that is responsible for methane production. Approved for use in multiple regions and sixty-eight countries, including the EU, UK, Canada, Australia, and Brazil, Bovaer has been rigorously tested for more than fifteen years and deemed safe for both cattle and consumers.[5] By early 2025, more than 200,000 cows had received the additive.[6] In Europe, Bovaer had already been commercially available for more than two years.[7] However, a recent controversy in the UK exposes the challenges of the additive's real-world adoption and highlights how misinformation, fueled by social media, can shape consumer perceptions.[8]

The issue gained particular traction in November 2024 when several British retailers partnered with Arla Foods, a Danish-Swedish multinational dairy cooperative, to launch a ninety-day trial of Bovaer on thirty British farms.[9] A key focus of the project was to unite the UK food industry to better understand Bovaer's use, its impact on costs and daily farm operations, and to prepare for scaling up the feed additive. However, just days after the announcement, misinformation spread rapidly on social media, garnering millions of views on platforms like TikTok.[10] Despite research proving otherwise, posts claimed that Bovaer contaminates milk and poses health risks, including cancer.[11] Others went even further, linking the additive to conspiracy theories about global elites and depopulation.[12] The backlash led to calls for boycotts of Arla products.[13] The crisis underscores the need for clear communication to bridge the gap between scientific innovation and public acceptance. In response to the controversy, regulatory bodies and scientific experts reaffirmed Bovaer's safety. The UK's Food Standards Agency (FSA) issued a statement clarifying that the additive does not transfer into milk and poses no risk to human health, animals, or the environment.[14] Nevertheless, while Bovaer is promoted as a way to reduce a farm's carbon footprint, some question whether it is just a quick technological fix that helps justify and expand industrial agriculture. Industrial farming often comes with other environmental problems, beyond carbon emissions, such as biodiversity loss and air pollution. This raises doubts about whether Bovaer is truly a sustainable solution that addresses broader environmental issues and supports animal welfare.

Livestock as a Climate Solution

As laid out in this book, modern industrial methods of agricultural production are key contributors to climate change and environmental degradation. Greenhouse gas emissions from agriculture account for about 30 percent of global warming and livestock, and fisheries are responsible for the majority, comprising 31 percent of total emissions from agriculture.[15] Yet, these emissions are from industrial livestock production, like those where cattle are fed soy from deforested lands, and they are kept in concentration lots. This approach to livestock is a climate problem—there is no argument over that—but managing our grasslands more sustainably can help reduce emissions. Temperate grasslands hold more carbon than any terrestrial pool other than wetlands and boreal forests. Globally, grasslands contain a third of all terrestrial carbon (90 percent in the soil).[16] These carbon sinks need to be protected. Where they have already been degraded, they need to be restored. The FAO estimates that 33 percent of global soils are degraded and therefore have the potential to increase their carbon levels.[17] Regenerative grazing can help to restore carbon into the soil, drawing it down from the atmosphere and locking it away. By allowing for natural nutrient cycling, cattle (and other ruminants) can help increase soil organic matter through their waste. The cattle feed on grasses and other plants that send their roots deep into the ground, furthering organic matter and allowing soil health to thrive. These sources of soil organic matter increase the soil carbon levels. This can be pushed even higher if you incorporate hedgerows and trees into the landscape. Globally, studies suggest that carbon levels are 15–20 percent higher in the soils under species-rich pastures compared with those that are species-poor.[18] Incorporating a wider range of plants into the grasslands can help too. For example, it has been estimated that introducing forage legumes into pastures has the potential to sequester 203 million tons of CO_2 per year globally.[19]

Through this process, livestock can be a nature-based solution to climate change. According to the IUCN, nature-based solutions are "actions to protect, sustainably manage and restore natural or modified ecosystems that address societal challenges effectively and adaptively, simultaneously providing human well-being and biodiversity benefits."[20] Carbon trading is entering the agricultural sector as the ability of soil to sequester carbon is increasingly recognized. Carbon trading allows companies to purchase offsets to meet their climate and ESG commitments. Carbon trading has tended to focus on forestry, but there is a potential for soil carbon credits to be bought and sold. However, the soil carbon levels would need to be better recorded with high levels of accuracy for the market to have trust in its validity. Currently, this is done through carbon calculators, such as the Cool Farm Tool, which uses an empirical model that is based on metadata from academic studies of soil and carbon at a global level. However, soil sensors (that we discuss in more depth in Chapter 10) might provide much more accurate and scientific readings that better reflect the farm management itself and could allow farmers to be paid precisely based on the percentage increase in soil carbon on their holding.

However, some have raised the risk that this carbon is not locked away and can easily be released if the grasslands are plowed up. Certainly, that is a risk. However, no more so than a forest being cut down and should not be a justification for not supporting regenerative methods. What it does mean is that grasslands should have some level of protection under law, akin to peatlands or woodlands. In the UK, there are efforts to create a classification of "ancient grasslands" (over fifteen years) that would restrict development and mirror the protections given to "ancient woodlands." There are also arguments against investment in soil carbon sequestration since soils can reach a saturation point, limited by soil minerals. However, this tension (while correct) is rather academic. Given the length of time that it would take for degraded soils to become fully saturated with carbon, and the overwhelmingly poor condition of agricultural soils globally, it might not be worth worrying about. Across Europe, for instance, it has been estimated that 80 percent of grassland topsoils are below their carbon capacity.[21] So, while saturation might be a theoretical problem, it would not be likely to happen for at least two or three decades—presuming that all those soils were managed in the most sustainable way possible, which is something that seems unlikely.

Beyond advocating that livestock should be managed in a more environmentally sensitive manner, some argue over the metrics that are used to calculate the GHGs that are attributed to livestock and used to condemn them as a climate problem. The primary GHG associated with livestock is methane, which cows belch out. At present, the way climate research measures methane emissions is via the GWP100 equivalence metric. Some climate researchers argue that this approach is flawed. In short, the GWP100 treats methane as a "stock" greenhouse gas that accumulates in the atmosphere and functions like other GHGs like CO_2 or N_2O, meaning that any methane emission contributes to global warming. However, a new understanding conceives of methane as a "flow" gas that has a short-lived nature and whose impact on climate depends largely on whether the rate of emissions is rising (leading to rapid warming), stable (leading to broadly stable temperatures), or falling (which causes a cooling effect). In recognition of this problem, a new version of GWP100, termed GWP*, has been developed that accounts for any changes to the rate of emissions. This new model more accurately reflects the impact of methane on the climate in any given emissions scenario and allows for a more accurate assessment of agricultural emissions. This new model doesn't excuse livestock's impact but frames it as an opportunity to build a system of sustainable management that contributes to climate solutions.

The potential climate benefit of regenerative livestock is substantial, but we should also consider the environmental impact beyond carbon. As discussed in Chapter 1, GHGs are only one part of the picture when considering the environmental impact of food production. If we only consider the carbon footprint of food, then intensive poultry would be the most sustainable option.[22] However, when you consider the wider Planetary Boundaries, then you can see the destruction to biodiversity and air pollution, not to mention the welfare damage.[23] Holistic grazed livestock can help ensure ecosystem health, allowing biodiversity to thrive. For example, meadows in Europe provide an opportunity to store and sequester carbon within healthy functioning ecosystems. Meadows include critical and endangered plant species such as snake's head fritillary and bird's foot trefoil. They also provide nutrients for pollinators such as bees and butterflies. These landscapes need to be actively managed through grazing to function as nature intended, and so the widespread adoption of holistic grazing techniques could help ecosystems to function more healthily. Healthy meadows can also enable water management. Functioning as floodplains, they allow the banks of rivers to expand and contract with water flow, protecting communities downstream. Using regenerative grazing can also reduce the risk of fertilizer run-off. This is primarily because the soil is healthier and therefore better able to hold water and, second, because it uses natural fertilizer from the cattle, as opposed to synthetic fertilizer. As a result, regenerative grazing can benefit waterways near fields by reducing the risk of eutrophication and improving marine biodiversity. In the grasslands of the Global South, regenerative grazing has also allowed degraded landscapes to recover. Areas that experienced significant desertification in sub-Saharan Africa have seen remarkable transformations as more sustainable management techniques have been introduced and communities have worked to ensure that the ecosystem returns to full health.

Animal Welfare and Regenerative Farming

Holistic grazing practices don't just improve landscapes, they also significantly improve animal welfare—something that the industrialized approach to meat production fails to do. Sustainable grazing practices allow livestock to express natural behaviors in diverse environments. Animals have constant access to fresh forage that not only meets the nutritional needs of livestock but also reduces stress associated with overcrowding and limited access to resources. These species-rich grasslands do more than just feed the animals. They give them the freedom to eat what they naturally want. Instinct lets animals seek out natural remedies, such as yarrow and plantain, when sick, reducing the use of antibiotics. These are known medicine fields and can help decrease the routine administration of antibiotics while maintaining animal health. Holistically grazed animals have the freedom to roam through the fields and demonstrate their natural instincts and practices. They can scratch in the dirt, roll in the grass, nuzzle, and sniff more freely.

However, delivering higher welfare through regenerative livestock management doesn't end at the farm gate. Sustainable livestock systems require wider supply chain infrastructure. It is not enough if you raise the animals sustainably if they cannot be slaughtered and processed to a similar standard. Smallscale local slaughterhouses, endorsed by organizations like the Sustainable Food Trust, improve welfare.[24] Firstly, by locating slaughterhouses within the immediate vicinity, it reduces the stress experienced by the animal in a tight and noisy truck. Secondly, it reduces emissions from transportation. A 2008 study revealed that transportation accounts for roughly 11 percent of food-related emissions.[25] Most importantly, smaller, less-industrialized slaughterhouses improve welfare standards. These smaller locations reduce the stress on the animal since there aren't hundreds of anxious animals being processed through an intensive system. There has been growing awareness of the welfare considerations of slaughter over the last few years, spurred by Temple Grandin's groundbreaking work on animal psychology.[26] Despite improvements on large-scale slaughterhouses, there are still huge welfare benefits to these smaller sites. Smaller slaughterhouses often process fewer animals at a time, which reduces crowding and noise and creates a calmer environment. With less pressure to move animals rapidly through the line, staff are able to handle them more carefully and monitor their condition more closely. The slower pace also allows for greater attention to stunning and slaughter procedures, lowering the risk of mistakes that can cause unnecessary suffering. These factors combine to create significant welfare benefits compared to the highly mechanized, high-throughput systems of large-scale plants.

Case Study N: Chickens Raised Right—the Question of a Good Life

Animal welfare is a central concern in the future of poultry production, shaping both ethical debates and consumer expectations. Defining and measuring welfare is complex, as behavioral and physiological indicators are difficult to interpret. Traditionally, welfare was understood as the prevention of suffering, focusing on reducing disease, pain, and poor housing. Increasingly, however, it is recognized that welfare also requires providing positive experiences. For poultry, this includes access to environments that support natural behaviors such as dust bathing, foraging, and perching, as well as exposure to natural light and fresh air. Low-stress handling further reduces fear. As understanding evolves, welfare in poultry production is no longer seen as simply avoiding harm but as ensuring conditions that enable a good quality of life.

Pasturebird, a farm in Georgia (United States), aims to give chickens a better quality of life by getting the birds back outside. They have created a mobile coop so that the chickens can be moved daily to fresh, green patches of pasture across the farm. This happens with the help of an automated system. The birds experience constant access to fresh forage, insects, and space to express natural behaviors like dust bathing and scratching. Unlike "free-range" systems, which only require birds to have some outdoor access, Pasturebird's chickens are raised outdoors with a minimum of 108 square feet per bird and spend the majority of their life on pasture—a standard that the USDA set in its updated definition of "pasture-raised."[27] By pairing this welfare-focused standard with regenerative land practices, Pasturebird aims to show how poultry farming can be scaled while giving greater consideration to animals and ecosystems. Yet questions remain. Even at 108 square feet per bird, critics point out that the density of animals may still feel industrial in nature, raising concerns about whether large-scale production can ever fully align with the ideals of pasture-based farming.[28] Still, compared to the cramped and barren sheds in which most chickens are raised, continuous access to pasture represents a significant improvement. It may not resolve every tension between scale, welfare, and sustainability, but it marks an important step in shifting the industry toward more humane and environmentally responsible systems.

Livestock and Livelihoods

Part of any discussion around the future of protein must involve the people who rely on livestock for their livelihoods. Communities all over the world depend on livestock production and have been for generations. Humans evolved to be omnivores, and livestock management has been woven through our cultural traditions for thousands of years. Raising livestock, in particular ruminant livestock, is deeply entrenched in our societies, and these deep cultural connections cannot easily be replaced merely by creating protein alternatives. This is true in the Global North and in the Global South. Ranchers in the

Global North (such as in the Western United States, Argentina, or Australia) have had a marked impact on our Global Culture with the powerful character of the cowboy transcending cultural boundaries. The cowboy is a figure that has endured long after the decline of cattle drives and wrangling in the open territories of the United States. The popularity endures today with television shows (such as Yellowstone) continuing to draw huge audiences from around the world, just as John Wayne's movies dominated box offices over seventy years ago. While less cinematic, sheep farming has a powerful economic influence on rural communities, particularly in the hilly landscapes that cannot support cattle or crop production. For example, in Wales, New Zealand, and Greece, sheep farming continues to define the rural economy. Without demand for lamb, the viability of these communities would decline drastically, and local economies would likely fail. This is particularly concerning when overlapped with areas already facing high poverty levels and unemployment. The traditional land management techniques (such as stone walling) would also likely fade. In the Global South, livestock ownership is deeply rooted in culture. Owning livestock has long been a method of securing family wealth and protecting yourself from future uncertainty. With rising climate uncertainty and increasing geopolitical tensions, it makes sense for these communities to remain loyal to their livestock traditions. Moreover, certain nomadic tribes (such as the West African Fulani) are cattle herders with unique cultures and traditions that are closely connected to the animals that they manage. These traditions and cultures can only continue through the management of livestock.

On the Plate

As we have discussed previously, meat consumption is rising worldwide with growing incomes, but many in the Global North are cutting back on red meat over health concerns. However, there is increasing evidence of the benefits of eating meat and dairy products that have been produced in a sustainable manner, in comparison to those from grain-fed animals. A *British Journal of Nutrition* study concluded that eating moderate amounts of grass-fed meat for four weeks would give consumers healthier levels of polyunsaturated fatty acids (PUFA)—in particular omega-3 and omega-6.[29] These PUFAs are often cited as being beneficial to human health because they help lower cholesterol levels and reduce the risk of heart disease. Humans cannot synthesize alpha-linolenic acid (ALA), which is an omega-3 fatty acid, and linoleic acid (LA), an omega-6 fatty acid, so they need to be incorporated through diet. ALA is formed in the chloroplasts of green leaves, so when cattle are taken off grass and fed a grain-rich diet, they lose access to a valuable store of ALA. Furthermore, longer chain PUFAs are also critical to health and are synthesized from ALA or LA. People who have sufficient levels of omega-3 are less likely to have high blood pressure or an irregular heartbeat. Studies have shown that meat from grass-fed animals has two to four times more omega-3 fatty acids than meat from grain-fed animals. However, it is the ratio between omega-3 and omega-6 that is one of the most important factors for human health. Grass-fed meat tends to have a lower, more balanced (and healthier) ratio of omega-6:omega-3 fatty acids of 4:1.

Pasture-fed beef can also be a good source of another type of PUFA—Conjugated Linoleic Acid (CLA). Grass-fed beef has higher CLA than grain-based beef because the grain reduces the pH of the digestive system in ruminant animals, which inhibits the growth of the bacteria that produce CLA. When ruminants are raised on fresh pasture alone, their meat and milk contain three to five times more CLA than similar products from animals fed conventional grain-based diets.[30] Higher levels of CLA may offer defense against heart disease, diabetes, and cancer since it's an antioxidant. Secondarily, pasture-fed meat also has higher levels of vaccenic acid, which can be turned into CLA. Beyond PUFAs, milk and meat from grass-fed animals also have higher vitamin levels, particularly in vitamin A (β-carotene). Carotenoids, such as beta-carotene, are precursors to vitamin A, which supports healthy vision, immune function, cell growth, and reproduction, while also helping maintain skin and tissue health. Grain-fed beef does not contain appreciable levels of carotenoids, for the simple reason that grains do not contain them. However, cows that eat carotenoid-rich grass and forage incorporate significant amounts of these compounds into their tissues. These carotenoids make the fat from grass-fed beef more yellow than the fat from grain-fed beef, so fat color can be a good indicator of how nutrient-rich the meat is. Grass-fed beef is also higher in Vitamin E (α-tocopherol) and some studies also show higher calcium, magnesium, zinc, and potassium levels.

We also need to consider whether we should be making fuller use of the animals we kill, eating a wider range of cuts, and incorporating more offal into our diets. If we value the sustainability of our livestock, it is also about making the meat go further. Consumers have become increasingly picky about the cuts of meat that they eat, opting for popular choices. This has driven producers to intensively breed poultry to meet this consumer preference to the detriment of the animal. Chicken breast is the most popular, but what about the other parts of the bird? Chicken livers are a gourmet delicacy but might not be enjoyed by many. Chicken feet are traditional foods in Asian cuisine but are not as widely consumed in other world cuisines. Just as insects are deemed too "yucky" by consumers, offal doesn't appeal to the mass market. Some chefs are trying to challenge this by advancing the concept of nose-to-tail dining that uses up all of the animal. For example, Fergus Henderson in the UK and Chris Cosentino in the United States are introducing menus that feature whole-animal eating. A German study concluded that increasing offal consumption by 50 percent by 2050 could reduce emissions from the German meat supply chain by 14 percent, by enabling the same amount of food to be produced from fewer animals.[31] This shift would also benefit diet health. Offal is a good source of nutrients and protein otherwise overlooked. For example, studies have found that beef liver is richer in iron, zinc, magnesium, and calcium than prime muscular tissue.[32]

Transparency

If the future of meat is "less but better," consumers need to be confident about what meat they are eating and that it truly has been raised to higher standards. When supply chains are short and consumers have personal relationships with their farmer, an individual might feel more comfortable purchasing uncertified meat since they feel that they can trust the farmer's word that the meat has been raised in a sustainable, humane way. However, when supply chains are longer, certification is required for a consumer to have confidence in what they have bought. Certification schemes ensure transparency and integrity. Yet, if sustainable meat is to play a significant role in the future of protein, there needs to be improvements in our certification and on-package labeling. There needs to be tighter regulation of the use of terms. Currently, much of the terminology of meat production is not legally defined, and therefore it can be open to "greenwashing" and tricky for consumers to truly know what they are buying. For example, the term "grass-fed" can be applied if the animal has been finished on grass or fed silage, meaning that the vast majority of their life can have seen them grain-fed and living indoors. Terms like "nature-friendly" or "sustainable" can be applied without any stipulation. While "non-GMO" or "organic" are better regulated, the terms still confuse the general public, who don't necessarily understand what those terms actually mean. Campaigns are advocating for more stringent regulations around on-package terms related to sustainability to match the existing rules for health claims. Campaigners in Europe are attempting to get terms like "humanely raised" or "raised with care" that denote higher welfare to be more tightly controlled. They are also aiming to better regulate the use of imagery on packaging. Many producers create "fake farms" that sound idyllic and bucolic and present pastoral images of grazing cattle on their products to create the illusion that they have a more sustainable approach to their livestock management. There are campaigns to stop this practice. Perhaps the future of labeling will allow on-farm data to be conveyed to consumers, providing consumers with a true picture of how their meat was produced. The complexity of this type of transparent data-driven labeling is tricky. It requires huge amounts of data to be collected and recorded, utilizing blockchain technology. Yet, the aspiration is there, and some of the new farming technologies (discussed in Chapter 9) can help to deliver it.

Conclusion

The reality is that eating meat is deeply entrenched in many cultures. North American cultural links to livestock might be the most cinematic, but they are not alone. Generations of sub-Saharan African communities have relied on livestock ownership to guarantee their safety in turbulent times. The connection between humanity (read masculinity) and meat consumption goes back thousands of years. It will be difficult to break that connection. Just creating cultured alternatives will not change hearts and minds. Therefore, it is worthwhile to try to create a model of sustainable meat production that celebrates our connection to meat-eating, instead of demonizing it. This includes using new technology and innovation to advance a more sustainable vision of protein. This could include utilizing big data to plan mob-grazing movements to maximize soil health outcomes, sensors to monitor animal health, or mobile abattoirs to avoid lengthy, overcrowded journeys often experienced in industrial systems. Communicating an approach of less but better will help to ensure a food system that values and respects both the animals and our connection to them. There is a model of livestock management that functions in harmony with nature. Livestock can be managed properly to be part of a climate and biodiversity solution. Ultimately, it is not the cow, it's the how.

Part 3

Agricultural Innovation

Introduction

If we are to tackle the enormous challenges that face the future of food, we need all the tools in the toolkit. This means embracing innovation and leveraging new technologies to tackle the problem of how to feed the world in an equitable, healthy, and climate-sensitive future. The technological revolution has provided new ways to meet this challenge. This trajectory of using technology to make farming more efficient began in the 1950s with the widespread adoption of agrochemicals, such as synthetic fertilizers and pesticides. It continued with the development of genetically modified crops in the 1990s and through to today with the innovation of genetic editing, which adds the potential to streamline breeding programs and ensure climate-resilient varieties are readily accessible.

Out of the GM crops that are currently cultivated, one of the most prevalent is glyphosate-resistant crops (particularly soy and corn). These resistant crops account for $174.72 billion USD of the global biotech crop value and make up over 80 percent of the market.[1] Their popularity stems from the promise of simplified weed control and reduced labor costs, making them especially attractive to large-scale, industrial producers, confident that they will kill the weeds but not the crop. Glyphosate works by blocking the shikimic acid pathway that the plants need to make amino acids and proteins. When the enzyme is blocked, plants die within a few days. Glyphosate-resistant crops have been modified using a gene isolated from Agrobacterium strain CP4 that proved resistant to the herbicide.[2] The most common formulation of glyphosate is Roundup, and the GM crops that are engineered to resist the herbicide are known as Roundup Ready. The popularity of this strain of GMO has concentrated biotech R&D onto a narrow set of traits, raising concerns about sidelining other potentially beneficial applications of biotechnology, such as nutrient enhancement or climate resilience, in favor of profitability.

Yet, improved breeding techniques are not the only ways in which technology can be leveraged to improve farming outcomes. There are hundreds of new technologies that aim at improving farming efficiency while protecting the environment. One of them will not be a panacea. With such a huge challenge, it will require a combination of technologies to be utilized together to reach a sustainable future for food. Solutions are still needed and are still coming forward. These innovations need to be embraced, but we also need to look to our past and learn lessons to ensure that we don't repeat old mistakes.

The Internet of Things (IoT)—"a network of devices, vehicles, appliances and other physical objects that are embedded with sensors, software and network connectivity"—is being implemented on farms with success and, if adopted widely, has the power to transform the future of food.[3] Soil sensors and drones present huge opportunities to monitor fields and stock remotely. AI will move this further forward by managing and automating interventions without the need for humans. This has the potential to eliminate human biases and errors, increasing yields and reducing waste. However, the more that our food system moves online, the greater the risk of data breach and the weaponization of our food system. In a period of growing political instability, the risk is increasing, and network intrusion has become a more powerful weapon for state and non-state actors alike.

If we double down on the belief that technology is the solution to the future of food, might we perpetuate a model of high-input, high-investment farming that has created so much environmental and public health damage? For example, are there lessons from the introduction of GM that we should consider before we wholeheartedly embrace GE? Creating guardrails against unintended consequences or damage caused by new technologies is not necessarily anti-tech, but rather a prudent calculation. Some argue that we need to reconsider the methods of the past. Humans are dazzled by what's shiny and new. Oftentimes, tried and true methods of the past are more effective, cheaper, and quicker to implement at scale. More importantly, they come with fewer risks. Yet, millennia-old technology isn't alluring and doesn't capture the imagination. This is all the more important to consider when considering agriculture in a development setting in the Global South. Well-meaning aid organizations, NGOs, and philanthropists push Global South countries to adopt new technologies to drive up yields, with the noble goal of reducing hunger. As Timothy Wise says in his aforementioned book, *Eating Tomorrow*, "Coming from a high-tech giant like Bill Gates, the answer to Africa's agricultural underdevelopment was, not surprisingly, technology."[4]

However, it might be more effective and pragmatic to help strengthen more traditional methods and work within a culturally appropriate framework. This is the position of groups such as Regeneration International or Sustainable Food Trust. These groups advocate for the reestablishment of seed saving and companion cropping while retaining traditional farming communities and providing stable incomes. Others disagree and say that supporting a Luddite adoption is akin to dooming communities to an inescapable future of subsistence farming and poverty. Robert Saik expresses this strongly in his book, *Food 5.0*, saying, "Deeply rooted in colonialism, elitism continues to be prevalent in political policy that tries to keep Africa down by not promoting modern farming practices. Condemning people to the era of muscle food production is only good for those that don't have to do it." He continues, "Why should these developing-nation farmers have to go through all the iterations experienced by other parts of the world? Can we help them leapfrog to ensure we move from the age of muscle farming to the age of abundance agriculture?"[5] This tension between preserving cultural practices and embracing technological modernization highlights the complex trade-offs facing policymakers that need to be actively weighed. The debate is about choosing one future over the other and considering the ramifications to resilience, equity, environment, and health (in the long and short term).

What do these innovations mean for rural livelihoods and economies? The trend toward automation has reduced the need for workers on farms and in food processing.[6] This has already reduced the number of jobs in rural economies and could lead to a further decline in the viability of rural communities. Since 2007, more people around the world live in urban environments than in rural ones.[7] This ratio is expected to increase to two-thirds urban by 2050, making urban food supply a major concern.[8] This concern is global. Of the thirty-seven megacities (a population over 10 million), twenty-four are in the Global South.[9] Given the growth of cities, it is critical to consider urban agriculture and food security in any consideration of the future of food. The expansion of cities has always been linked to the advancement of food technology. Cities cannot grow if there isn't a means to feed the population that lives there. The expansion of cities in the eighteenth century was linked to the invention of railways to bring food from the countryside into the heart of the urban environment. The next step will be to see how to reconnect urban communities with food. Vertical farming and the reinvention of community gardening and marketplaces offer new avenues for city dwellers to access local food. The affordability of this food, however, remains an important question since the share of poverty in urban centers is growing.[10]

But is technology inherently linked to an exploitative system of export-based commodity agriculture? Some argue that the economic model of capitalism and globalization that encourages the adoption of these new technologies is inherently wrong. They argue that we need to rethink the entire system and revert to a more localized food system that values environmental sustainability, regional economies, and good health outcomes. Vandana Shiva, a leading activist voice in Indian farming, said, "Everywhere I looked, local people were getting poorer and landscapes were being devastated as soon as this development or new technology came in."[11] This agroecological movement points to the switch that Cuban farmers made to agroecology when they could no longer access agrochemicals. Since 1997, more than 100,000 families—over a third of all Cuban farmers—have transformed their production systems to sustainable methods and become more climate-resilient to extreme weather events.[12] Yet, it would be tough to find many farmers in Cuba who would not consider switching from the model of low-input agroecology in Cuba to high-input commodity agriculture in capitalist Florida if offered the chance, especially when a 2023 Cuban Observatory for Human Rights (OCDH) report revealed that 78 percent of Cubans have stopped eating some meals, and nine out of ten Cubans barely earn "enough to survive."[13]

However, it seems mistaken to think that there is no room for the huge leaps that humankind has made in technological advancement in the last fifty (or even hundred) years to help increase agroecological yields to make it more competitive against conventional farming. Surely, there is an opportunity for farmers to adopt some of these innovative products while still maintaining their principles. Some farmers believe that these new technologies are compatible with a model of agriculture that is environmentally sensitive and socially just. The Soil Association has worked to create guidance for the adoption of agroecological innovation, stating a guiding principle of "Have we got the 'problem definition' right? Are we clear there isn't a mindset barrier or a farming or food system barrier before we assume that the barrier—and the 'silver bullet'—is tech?"[14] Given the urgency that humanity faces to ensure a safe, consistent, and environmentally sensitive supply of food for the next generation, it is worthwhile to consider how farmers and food producers can adopt and embrace new technologies to help meet that challenge.

8

The Future of Breeding

Whether through patient selection or precision editing, plant breeding is one of the most powerful—and controversial—tools we have to feed the world. In the 1990s, genetic modification (GM) transformed agriculture by creating plants that were higher-yielding and resistant to disease and pests. Genetically modified foods were first approved for human consumption in the United States in 1994. The Flavr Savr Tomato hit the market after securing FDA approval with a new gene that prevented the breakdown of the tomato's cell walls, thus extending its shelf life. Adoption of biotech rapidly expanded, particularly in commodity crops like corn and soy. In the last twenty-five years, GM crop production increased over 100-fold, and by 2021, farmers cultivated approximately 190 million hectares of biotech crops globally, which is approximately equivalent to the entire surface area of Mexico.[1] The recent emergence of gene editing has continued this trajectory with the creation of new varieties that can be genetically designed to address the challenges that farmers are increasingly facing. Advocates of these new plant-breeding techniques point to the potential to feed the world and ensure that yields continue despite changing climates (such as increased temperatures or inconsistent rainfall). However, they are not without their concerns. Biotech crops raise questions about corporate control of the food system and whether companies can "own" genetic materials. Those who advocate for agroecological methods point to the potential of landrace or heritage varieties and the need to widen the diversity of our crops, rather than to focus exclusively on yield. Biotech crops also raise questions about the global nature of our food system and how domestic regulation and phytosanitary standards interact with the principles of free trade. This chapter explores the opportunities and the risks within the future of breeding for agricultural crops. Will GM continue to dominate the commodity crop sector? Or will it fade as new plant breeding technologies take over? What insights can the trajectory of GM teach us about the future of plant breeding and the new techniques that have emerged in recent years?

Gene Modification and Gene Editing

Genetic modification is the process by which scientists change the traits of plants and animals by directly manipulating their DNA in the laboratory. Established techniques involve inserting whole genes that can be sourced from the same species or from a completely unrelated organism. There are two different types of genetic modification; cisgenic and transgenic. Transgenic is when genes are introduced from two completely different species. For example, GM canola includes genes taken from the California bay tree to alter its fatty acid composition. These new varieties could not naturally arise since the two parent species cannot reproduce outside of a laboratory. Cisgenic, on the other hand, uses genetic material from within the same species. In theory, these new varieties could arise naturally since the parent species are the same, even if the varietal is different. For example, GM potatoes utilize genes from wild varieties that are resistant to blight. Cisgenic breeding has further advanced in recent years with the development of gene editing (GE). GE focuses on making precise changes in the DNA sequence of a living organism, often by deleting or rearranging parts of the organism's existing genome. Gene editing uses enzymes, particularly nucleases, to break apart the target organism's DNA strands where it finds a particular sequence of nucleotides. A range of techniques is then used to encourage the cells to repair the DNA break in a way that achieves the desired result. This may involve the removal of existing DNA and the insertion of replacement DNA. Central to the advancement of gene editing is the CRISPR-Cas9 tool, discovered in 2012, and its further advancement in 2015 with the use of Cpf-1, rather than Cas9, as the nuclease paired with CRISPR. CRISPR uses the process that naturally occurs in bacteria as an immune defense. When infected with viruses, bacteria capture small pieces of the viruses' DNA and insert them into their own DNA. These segments are known as CRISPR arrays and allow the bacteria to "remember" the viruses (or closely related ones). If the viruses attack again, the bacteria produce RNA segments from the CRISPR arrays that recognize and attach to specific regions of the viruses' DNA. The bacteria then use Cas9 (or a similar) enzyme to cut the viruses' DNA apart and disable it. The gene editing process adapts this immune defense system to intentionally edit DNA. Labs create a small piece of RNA with a short "guide" sequence that binds to a specific target sequence in a cell's DNA, much like the RNA segments bacteria produce from the CRISPR array. This guide RNA also attaches to the Cas9 enzyme. When introduced into cells, the guide RNA recognizes the intended DNA sequence, and the Cas9 enzyme cuts the DNA at the targeted location, mirroring the process that occurs in bacteria. Once the DNA is cut, researchers use the cell's own DNA repair machinery to add or delete pieces of genetic material or to make changes to the DNA by replacing an existing segment with a customized DNA sequence. This is markedly different from the earlier processes used for genetic modification in the 1990s that was considerably less precise. That said, some commentators regard any differentiation between gene editing and genetic modification as redundant since in both cases science is manipulating the genes of a living entity to serve societal goals, and the negative impacts are the same.

Regulation around the World

In the United States, GMOs have been warmly adopted since the early 1990s. This stemmed from a 1986 document titled "Coordinated Framework for Regulation of Biotechnology" released by the White House Office of Science and Technology Policy. This document made clear that GMOs are equivalent to normal food products and therefore should be regulated using the same processes.[2] There are no additional hurdles for biotech products to pass than for any other food, although the National Bioengineered Food Disclosure Standard, which became law in 2016, requires that bioengineered food is labeled as such.[3] The assumption that GMOs can be assessed in the same way as other food means that GMOs in the United States are approved by a combination of the FDA and the USDA, the two governmental associations that are in charge of food regulation. The FDA makes sure that GM foods are regulated at every stage along the supply chain. The USDA Animal and Plant Health Inspection Service (APHIS) sets regulations to make sure that GMO plants are not harmful to other plants. Food manufacturers can determine if their product can be considered to be "General Recognized as Safe" (GRAS)—namely, if the ingredient and its proposed usage are generally considered unharmful by experts. GMOs would fall into this category since there have not been any scientifically proven adverse effects of consuming GM foods. It is worth noting that the Secretary of Health and Human Services, Robert F Kennedy Jr., has instructed the FDA to explore ending the self-affirming GRAS rule and increasing FDA oversight.[4]

This is decidedly different than the European approach. In the EU, the European Food Safety Authority (EFSA) sets all food regulations and standards, while individual countries carry out these regulations and enforcement.[5] The EU follows the guiding philosophy of the "precautionary principle." This principle states that if a new product poses a possible risk to human health or the environment, then it is prudent to restrict or prevent the introduction or use of that product or technology, even if the magnitude or nature of the risk is uncertain. Essentially, until something can be proved to be safe and have no negative impacts, it cannot be approved. As a result, the EFSA has much stricter standards on what it allows compared to the United States. For example, a lot of food additives such as titanium dioxide, potassium bromate, and brominated vegetable oil are allowed in the United States but banned in the EU.[6] GMOs are no different. The EU has approved far less GM products than the United States and has stricter requirements, such as mandatory labeling and traceability.[7] As a result, the only GMOs authorized in the EU are seven genetically modified crops (three corn varieties, two soybeans, one oilseed rape, and one cotton).[8] This position is not without controversy and opposition. That said, the EU has a more tolerant position on genetic editing, which they have termed "new genomic techniques" (NGT). In February 2024, the European Parliament relaxed the rules around the use of NGT while still maintaining a harder line on GMOs.[9]

This difference in regulatory approach over biotech products creates a tension within international trade between the United States and the EU. Under the WTO regulations, all barriers to trade must be eliminated to allow for free trade to occur. Since the United States deems GMOs to be safe, they consider the restrictions to the import of GMOs to be a nontariff barrier to trade that is unacceptable under the WTO. However, the EU deems GMOs to be of scientific concern. They therefore use the provision within the Agreement on the Application of Sanitary and Phyto-sanitary Measures (SPS) to stop their import—something that they are allowed to do if they deem the product to be unsafe or damaging.[10] This tension continues to be played out in international trade negotiations—most recently in the UK's attempt to secure a bilateral free trade agreement with the United States following their departure from the EU, where there was much debate as to whether US GMOs would be allowed to be sold within the UK borders.[11]

Ultimately, this means that the United States forms one pole and the EU forms the other. The United States presumes that GMOs require no further scrutiny because they are "GRAS" while the EU refuse to approve them while they haven't been proven to have no negative impacts under the precautionary principle. Countries around the world fall somewhere along that continuum, and so the Cartagena Protocol on Biosafety created the original framework that allows each country to determine its own relationship with biotech.[12] The Cartagena Protocol, signed in 2000, is supplementary to the Convention on Biological Diversity and seeks to mitigate the potential risks to biodiversity posed by biotech products—although it refers to them as LMOs, living modified organisms. The Biosafety Protocol allows Global South countries to make their own determination on whether to approve biotech products, based on clear criteria. It allows for the banning of GM imports if they feel that there is not enough scientific evidence to determine that the product is safe under the precautionary principle. It requires notification and advanced informed agreement. It has 173 signatories including the EU and China, but not the United States.

However, a lot has changed in the international arena since the Cartagena Protocol was signed. The fear over the rising challenges caused by climate change as harvests across the world start to decline and fail has increased. With less consistent production and rising populations (particularly in the Global South), many nations are considering whether they want to relax their stance on biotech products. This has been aided by the invention of gene editing and the slow-but-steady acceptance of the technology by the consumer. Additionally, the growing instability in geopolitics is causing nations to rethink their reservations over biotech and embrace new technologies—such as the EU's decision to ease their GE restrictions following the disruption to the flow of Ukrainian grain due to the war. In terms of shifting the global biotech market, the changes in both China and India will have huge impacts. Both nations have substantial populations, but also increasing affluence, which means a higher demand for food.

China took steps in 2024 to approve the use of GM crops. In late December 2023, the People's Republic of China's (PRC's) Ministry of Agriculture and Rural Affairs (MARA) approved fourteen GM soybean varieties and thirty-seven GM corn varieties.[13] These varieties were bred for stronger herbicide and insect resistance and produce higher yields. This was motivated by the desire to increase domestic food security to meet the demands of a sizable population while reducing the dependence on US imports. Prior to this recent development, only GM cotton and papaya were approved for cultivation inside China. China allowing the cultivation of GM soy inside its borders could profoundly alter the future of food production globally since China accounts for more than 60 percent of the global soy market but imports 88 percent of its soy, primarily from the United States.[14] Due to the ongoing uncertainly over President Trump's trade restrictions, China is looking to reduce their dependence on imports of American soy. However, there is some hesitation from the public over biotech food, and it is yet to be seen whether the Chinese public will accept GM soy as food products (e.g., in tofu).[15] As discussed in Chapter 6, there is a legacy of contamination incidents in China that have spark public outrage and generated distrust of domestic food standards within the Chinese consumer market. This legacy raises questions over whether safety assessments can be trusted.

Similar to China, India has been hesitant to back biotech crops. Nearly half a billion Indians are part of smallholder farming households, and, further, India is the fifth largest market for seeds in the world, making the GM debate here critical to the fortunes of global seed and agrochemical companies.[16] The only GM seed currently available for sale in India is cotton. No food crop has been approved for sale, although Indian scientists have developed GM rice, wheat, tomato, eggplant, and mustard in research settings. For crops to be approved under India's agricultural system, they must be sown over three seasons in different agro-climatic zones and prove themselves to be consistently better than their existing (non-GM) comparators. GM brinjal and mustard achieved this benchmark, but their approval has subsequently stalled. Approval for GM mustard is currently going through the Indian courts.[17] If approved, GM mustard will become India's first GM food crop. Insect-resistant (Bt) brinjal (also known as eggplant) previously came closest to approval. It was cleared by government regulators in 2009 but has been under an indefinite moratorium since 2010, ordered by the then environment minister, Jairam Ramesh, due to the wider social and environmental concerns that we will now discuss.

Aiding Adaptation to Climate Change

Farmers are facing a huge level of uncertainty due to climate change. The increasing temperatures are making certain locations less hospitable to crops that used to thrive there. Increasing instances of extreme weather mean that harvests are lost to sudden freezes. Lack of rainfall means crops can wither in drought conditions, or heavy precipitation can mean waterlogged soils and rotten roots. Biotech crops can offer a lifeline to farmers. GM crops can be designed to be more tolerant of the climate crisis—better able to withstand raised temperatures or longer periods without water. They can be engineered to have roots that are better able to cope with more saline soils or have stems that can withstand higher wind forces. This means that, as the climate changes and farmers face greater pressure on their livelihoods, the crops can be adapted to provide greater resilience. Embracing biotech crops can help provide a more consistent income to farmers who can have greater confidence in their yields. The troughs of production (such as in a drought year) can cripple growers, especially in the Global South, where margins are minimal. By opting for bioengineered crops, annual harvests would become more reliable, which would mean that low-income farmers would be able to better plan for the future and avoid a boom and bust cycle that could wipe out their business. Allowing rural communities to continue to thrive with consistent harvests and viable incomes reduces one of the key drivers for migration from rural areas into the cities. A side benefit that furthers the economic viability for farmers of GM crops is the reduction in labor and fuel costs. With GM crops, in particular herbicide-resistant crops, there is a lower need for labor-intensive and mechanical weeding. Since the introduction of herbicide-resistant GM crops, the United States has reduced fuel usage in tillage by 220 gallons, or a reduction of more than 2.2 million tons of carbon dioxide.[18] By reducing input costs, the balance sheet of a farm can become more stable.

Case Study O: An Apple a Day—Waste Reduction through Genetic Editing

Shockingly, between 30 percent and 40 percent of the annual food supply is wasted or lost in the United States annually, and a family of four spends $1,500 each year on food that ends up in the landfill.[19] New breeding techniques might further help reduce food waste, which is a significant challenge to our future. The perfect example of this is avocados. How many of us have chucked out an avocado that has gone brown once opened? Or discarded a pot of guacamole that sat on the table for too long? New gene editing technology is aiming to create the evergreen avocado with the goal of reducing food waste. Cutting an avocado exposes the cells to oxygen, which allows the polyphenol oxidase enzymes to rapidly oxidize the phenolic compounds in the fruit's tissues into ortho-quinones (o-quinones). Genetic modification might offer a solution to this by "switching off" or "silencing" the gene associated with oxidization. Such an intervention has been successful in other products. The Arctic Apple was approved by the FDA for sale in 2015.[20] The producers, Okanagan Specialty Fruits Inc., used genetic editing to ensure that their apples did not go brown after being sliced through a technique called RNA interference (RNAi). This approach enables silencing of PPO expression to less than 10 percent of its normal expression but does not change other aspects of the apple.[21] The producers claimed that it would encourage kids to eat more fruit since the apple slices in their lunchboxes would remain appealing and not toss it into the trash after mealtime.

GM Environmental Benefits

The use of GM/GE crops can also help improve soil health, which is closely linked to the changing climate. As discussed in Chapter 10, the world's soils are in crisis. Over the past century, soil fertility has declined dramatically due to intensive agricultural practices that deplete the ground of its nutrients. New varieties of GM crops could help to reverse this trend. Herbicides can be sprayed directly onto fields with glyphosate-resistant crops, killing the weeds but leaving the crops intact. This means that there is no need to plow up the ground to remove weeds or crop residue. This type of minimal tillage (known as min-till or no-till) benefits the soil by allowing a greater concentration of organic matter near the soil surface. This can help with soil structure, improve seedling success, and water retention, making them more resilient in the face of droughts or floods. Beyond mitigating the damage of industrial cultivation and climate change, no-till could allow agricultural soils to become a carbon sink. By not plowing, soils could help lock atmospheric carbon into the ground (known as carbon sequestration) by increasing soil organic matter over time. In this way, agriculture could move from being a climate problem to a climate solution. Governments around the world acknowledge this potential and support farmers who adopt no-till practices. It is worth noting, however, that it is possible, if harder, to practice no-till agriculture without glyphosate-resistant GMOs. Organic and regenerative farmers use alternative methods (roller crisper, weed suppression, hand-pulling) to control weeds without cultivation and without pesticides or GMOs. Research conducted by the Rodale Institute in their Farming Systems Trial showed that the organic manure systems that used standard or limited tillage had higher levels of soil organic matter (and, by extension, soil organic carbon).[22] Importantly, the organic matter was found to be higher at deeper levels of the soil, which was positive since the deeper the soil carbon, the more stable it is, supporting root systems and playing a critical role in long-term fertility and resilience. Organic fields with limited tillage also had higher crop yields than GMO herbicide no-till.[23]

Beyond the potential of GM to help sequester carbon within no-till cultivations, the next generation of GM crops might help to ensure the further reduction of emissions from the farming sector. New research is looking at whether the genetics of crops, in particular corn, could be modified so that they can make their own nitrogen.[24] For example, researchers have genetically modified bacteria from the roots of corn to maximize nitrogen fixation, even in the presence of excess nitrogen. Further down the line, these GM microbes could possibly replace synthetic fertilizers.[25] Currently, only legumes (peas and beans) can produce their own nitrogen. These plants have nodules in their roots that host bacteria that live symbiotically with the plant. The bacteria can readily convert atmospheric nitrogen into ammonia, which the plant can then use to grow. Farmers therefore currently have the choice of using legumes in rotation to add nitrogen to their soils or adding synthetic fertilizers. Many rely on the synthetic option. The synthetic fertilizer market was estimated to be worth $174.72 billion USD in 2023.[26] However, as discussed, the production of synthetic fertilizer is quite carbon-intensive, using natural gas as its major source of hydrogen. Consequently, being able to produce arable crops that would be able to produce their own nitrogen would be a game changer. Cutting reliance on synthetic fertilizer would slash carbon emissions and nitrogen pollution—while putting money back in farmers' pockets instead of into costly chemical inputs.

Biofortification

GM varieties could also help to reduce disease and improve health. These varieties have been developed with the goal of addressing the micronutrient deficiencies in poor communities in the Global South. Micronutrients are vitamins and minerals that the human body critically needs, but in very small amounts. They perform a range of functions, including enzyme and hormone production. Micronutrient deficiencies can cause visible and dangerous health conditions. It's worth noting that deficiencies can also lead to reduced energy levels, mental clarity, and overall capacity, which can lead to reduced educational outcomes, reduced work productivity, and damage economic opportunity. Deficiencies in iron, vitamin A, and iodine are common around the world, particularly in children and women, but low- and middle-income countries have the highest levels of micronutrient deficiencies. According to a study (the first study to provide global estimates of inadequate consumption of fifteen micronutrients critical to human health) by researchers, over half of the global population consumes inadequate levels of essential micronutrients, including calcium, iron, and vitamins C and E.[27]

Many of these micronutrient deficiencies are preventable through a combination of poverty reduction, education, and improved dietary practices. However, biofortification could play a role. This is when foods are modified to have higher rates of micronutrients. The most well-known of this category of

GM crop is golden rice, which aims to challenge vitamin A deficiency by raising beta-carotene consumption. The rice genome is modified by incorporating a gene from a daffodil (Narcissus pseudonarcissus) to produce almost twenty times higher levels of beta-carotene, a micronutrient that is converted into vitamin A in the liver. This crop could have a transformative effect. According to the WHO, an estimated 250,000–500,000 children who are vitamin A-deficient become blind every year, and half of them die within twelve months of losing their sight.[28] Deficiency of vitamin A is associated with significant morbidity and mortality from common childhood infections and is the world's leading preventable cause of childhood blindness. Vitamin A deficiency also contributes to maternal mortality and other poor outcomes of pregnancy. When considering how to use GM food crops to treat nutritional micro-deficiencies, rice was the preferred vector since it's a common dietary staple of many poorer communities in the Global South. Over 50 percent of the global population depends on rice for about 80 percent of its food requirements.[29] Approximately 95 percent of the global output of rice is produced and consumed in developing countries.[30] As a result, it is also a common food provided in emergency food aid packages from organizations, such as the World Food Programme. Thus far, golden rice has yet to have a transformative effect on global nutrition since its rollout has been halted by food safety regulators on the basis of the precautionary principle that we just discussed.

Another variety of modified rice aims to combat iron deficiency. It does this by introducing a number of different genes into the rice genome. Firstly, a ferritin gene from a bean (Phaseolus vulgaris) that produces a protein capable of binding iron, secondly, a gene from a fungus (Aspergillus fumigatus) that produces an enzyme capable of improving iron absorption, and, lastly, it over-expresses an existing rice gene that produces a protein that enhances iron absorption. The goal of this technology is to address the challenge of iron deficiency within Global South communities. Anemia, low iron levels, affects around 30 percent of the global population.[31] It is especially prevalent in lower-income regions, such as sub-Saharan Africa, South Asia, and the Caribbean, and also disproportionately impacts women, making it a gender equality issue. Wheat strains are also being adapted to produce flour with increased iron levels. The ongoing trial, which is being carried out at the John Innes Centre in the UK, has shown that the new grain contained double the amount of iron compared to normal grain.

In addition, biofortification might become of value in the Global North, in addition to the Global South. As discussed in Chapter 2, over the past sixty years, there has been an alarming decrease in a wide variety of nutritionally essential minerals and vitamins in fruits, vegetables, and other crops. The University of Texas compared USDA nutrient data from 1950 and 1999 for forty-three garden crops and found notable decreases in protein, calcium, phosphorus, iron, riboflavin (vitamin B2), and vitamin C.[32] The potential cause of the decline in the nutritional density of foods is the preference for prioritizing higher yield over maintaining high nutritional quality. The focus on breeding more productive genotypes that thrive in more industrialized farming environments has inadvertently resulted in a decline in nutritional value of many common crops. It's also worth noting that there is a suggestion that a further rise in atmospheric carbon dioxide levels could contribute to additional reductions in the nutritional quality of fruits, vegetables, and food crops.[33] Consequently, it might be that horticultural produce for the Global North could also benefit from biofortification in order to improve its nutritional value to consumers.

Solving the Wrong Problem?

On paper, biotechnology promises big benefits for the future of food. But while companies highlight investments in biofortified foods and climate-smart crops for the Global South, in practice most efforts are aimed at commercial gain. In fact, the majority of GM crops today aren't even grown for people to eat. The four main types of GMOs are soy (~50 percent), corn (~30 percent), cotton (~13 percent), and canola (~5 percent), and the vast majority of these crops are used for animal feed and vegetable oil products, including bioethanol.[34] The other direction of travel for GM crop development is aimed at Global North consumers. A prime example is the "pinkglow" pineapple. In 2020, Del Monte launched a GM pineapple that is pink in color by overexpressing a gene derived from tangerines and suppressing other genes.[35] Given the challenges our food system urgently faces and the need to tackle our food future, it's reasonable to question whether there is a societal benefit of creating novelty pink pineapples. Just because we can, doesn't necessarily mean we should. It is arguable that those resources could have been better served if they had been put toward nutrient-rich varieties for subsistence farmers, not novelty crops that appeal to Millennial Instagrammers.

If we genuinely want to address hunger in the Global South, we need to focus on supporting subsistence farmers, helping them to increase their yields in a sustainable (both economically and environmentally) way to produce healthy food for their families. This includes recognizing the need to support adaptation to the shifting climate that will inevitably impact food production, particularly in Global South countries. Smallholder farmers account for 84 percent of the farming globally and produce one third of the world's food.[36] One key reason that biotech crops have become a popular solution within Global South agriculture is a trend in the wider geopolitics. Following the 2008 financial crisis and the austerity programs that followed, many countries in the Global North cut back on their agricultural aid programs. For example, the UK folded the Department of International Development into the Foreign and Commonwealth Office and reduced aid spending from 0.7 percent to 0.5 percent of gross national income.[37] The World Food Programme that receives the majority of its funding from national governments in the Global North saw its funding reduce dramatically, leading them to clearly state, "Severe funding shortfalls are forcing WFP to scale back assistance and refocus efforts on the most severe needs," warning that, "unless resources are made available and unrestricted access granted, lost lives and the reversal of hard-earned development gains will be the price to pay." As WFP makes clear, the need was just as great as before, if not greater, and private philanthropy stepped in to fill the gap. Some of the largest of these private foundations accrued their wealth through the tech industry. The Bill and Melinda Gates Foundation, for example, gave $63 million USD to nutrition programs in the Global South, while the Bezos Earth Fund committed $194.8 million USD to their "Future of Food" program.[38] Given that these philanthropists are from the tech sector, it's understandable that they consider tech to be a positive force. They inherently believe that tech will provide the solution to the problems that the world faces. As a result, a lot of funding has been directed into biotech (and tech more broadly) as a solution to the dual challenge of hunger and climate. This has been dubbed "climate-smart farming." Many countries in the Global South are required to adopt new technologies in order to secure the aid that they need. While some of the conditionality of philanthropic funding on Global South leaders adopting biotech-friendly policy is explicit, some of it is insidious, where leaders are aware that they are more likely to secure funding from large funders by embracing the tech solutions that those billionaires espouse.

At the heart of the debate over whether biotechnology can deliver a positive future for food lies the tension between profit and public good. Biotech seed companies are, by their nature, commercial enterprises designed to generate returns for shareholders, not to solve world hunger or safeguard the environment. This profit orientation shapes which crops are developed, which markets are prioritized, and who gains access. As a result, the potential of biotechnology to act as a force for equitable change in the Global South is constrained, raising critical questions about whose needs the technology ultimately serves. GM seed is expensive in comparison to traditional seed.[39] Companies pour a huge resources into R&D to develop new crops. The cost to develop a genetically modified (GM) crop can range from $115 million USD to $136 million USD, and the process can take up to thirteen years.[40] This cost needs to be recouped. Companies justify the higher market price on the promise of higher yields and lower inputs.[41] However, these seeds cannot be saved (i.e., seeds kept from one year's harvest to be replanted next year) and new seeds are required to be purchased the year after. This is legally binding since the genetic makeup of the seed is considered to be the intellectual property of the seed company and therefore cannot be saved to be used the following year—despite seed saving being a long-standing agrarian tradition that we have discussed. These restrictions are enshrined in international laws—such as bilateral trade deals or in the International Union for the Protection of New Varieties of Plants 1991 (UPOV91), which contains provisions to grant exclusive rights of new plant varieties to the plant breeder.[42] The future of biotech could further this move away from seed saving if there is a widespread adoption of terminator genes being built into the genetic code of GMOs. Also known as Genetic Use Restriction Technology (GURTs), the genetics are coded to ensure that the seed is sterile. This technology, if widespread, would end the tradition of seed saving that has been practiced for millennia. Being required to purchase new seed every year can trap farmers into a cycle of debt that they struggle to escape—especially if the price of that seed increases down the line. Oftentimes, other products that are associated with the GM seed are required to be purchased alongside. GM seeds are rarely sold on their own; they are often bundled with other products that farmers are expected to buy. For instance, planting Roundup Ready crops typically means also purchasing the matching herbicide, Roundup. Increasingly, the package extends beyond chemicals to digital tools, as companies encourage farmers to adopt their agritech platforms, from data services to precision sensors that we will discuss in Chapter 10. This creates a system where access to one product locks farmers into a broader suite of technologies, deepening their dependence on a handful of powerful companies.

The impact that the high price of GM seed (and the associated inputs) can have on farming livelihoods came into sharp relief in India. As previously mentioned, GM cotton is the only biotech seed approved for sale in India, after gaining official authorization in 2002—although illegal planting was reported before then. The authorization was part of a wider economic liberalization move by the Indian government at the turn of the century, which also included a reduction in farming support that negatively impacted rural economies and subsistence farming communities. Environmentalists, including leading farming activist Vandana Shiva, claimed that the approval of Bt cotton had worsened the distress and debt that farmers were facing. This was because (1) Bt-resistant GM cotton seeds were considerably more expensive than ordinary cotton hybrid seeds, (2) because the promised reduction in pesticide application did not materialize for many farmers, and (3) farming the seeds often required expensive irrigation. Consequently, farmers were facing higher inputs with no reduction in costs to offset the additional expenditure of the seed. Worsening the situation further, the promise of higher yields also did not materialize. Instead of 3306 lbs per acre as promised by the seed company, the harvest was as low as 441 lbs per acre.[43] Many farmers were forced to turn to loan sharks to cover their costs. The move to GM cotton also led to a shift away from rotational farming that produced food to feed the local community toward monocultures of cotton for export, creating food insecurity. A mental health crisis exploded in the low-income farming communities, resulting in a rise in suicides. Between 1999 and 2013, 270,000 Indian farmers killed themselves.[44] Vandana Shiva and other activists attributed this rise in suicides to the introduction of GM into India, calculating that the farmers lost $11.5 billion USD due to crop failure in 2002.[45] While it is impossible to directly tie the rise in suicides to the introduction of GM cotton, especially when the global price of commodity crops dropped significantly during the period, the situation raises critical questions about whether biotech crops actually help farmers.

Case Study P: Gene Editing Teff in Ethiopia—Promise and Peril

Teff is the tiny, gluten-free grain that underpins Ethiopian cuisine and culture. It is both nutritionally rich and, as the foundation of "injera," Ethiopia's national staple bread, it sustains millions of households. Yet teff farming faces persistent challenges: yields are low, production is labor-intensive, and the crop is highly susceptible to lodging—when stems collapse under the weight of grain—causing yield losses of up to 25 percent.[46]

In response, researchers at the US-based Donald Danforth Plant Science Center, in collaboration with the Ethiopian Institute of Agricultural Research, have applied CRISPR gene editing to improve the crop. By modifying the *SD-1* gene, they created semidwarf varieties that are shorter, sturdier, and resistant to lodging.[47] Early trials suggest these plants maintain normal germination while offering the potential for higher yields. From a farming perspective, lodging-resistant teff could help improve incomes, enhance food availability, and reduce vulnerability to climate shocks.

Importantly, improved teff would address food security in a way that respects cultural traditions. Many food aid programs deliver imported staples such as rice or corn, which are not central to Ethiopian diets. By contrast, more productive teff varieties would allow households to continue preparing injera and other traditional foods. In this sense, gene-edited teff is not just about boosting yields—it also underscores the importance of aligning innovation with cultural practices, ensuring that food security efforts reinforce rather than displace traditional diets.

In 2023, the US Department of Agriculture determined that gene-edited teff did not count as a GMO under American rules, since the edits were small and could occur naturally.[48] Ethiopia, by contrast, has a stricter framework for biotechnology. Its 2009 Biosafety Proclamation was among the toughest in Africa, requiring government consent, public consultations, and liability measures.[49] A 2015 amendment loosened restrictions, allowing confined trials and limited commercialization, leading to approvals of Bt cotton and Bt corn.[50] Yet there is still no clear guidance on how gene-edited crops like CRISPR teff would be treated, leaving no pathway for domestic adoption.

The risks extend beyond regulation. Intellectual property claims by foreign institutions could restrict farmers' ability to save or share seeds, creating new dependencies. Advocates also warn that introducing gene-edited teff without robust community consultation could erode seed sovereignty, shifting control of Ethiopia's most important crop from smallholder farmers—who have stewarded teff diversity for centuries—to external actors. Such a shift risks not only economic dependency but also the erosion of cultural and agricultural traditions inseparable from Ethiopia's food system.

Environmental Impacts

As we have discussed throughout this book, our environment is in crisis. If we are to stop the environmental damage from getting worse, we need to act urgently to mitigate the damage from our farming practices. While biotech advocates might emphasize how the technology can help farmers to become "climate-smart" by adapting crops to more hostile environments, GM is closely associated with more industrial farming practices. For example, biotech crops tend (although not exclusively) to be grown in monoculture systems. Increasing the focus on biotech as the future of plant breeding would certainly further the expansion of monoculture plantations. As discussed in Chapter 2, monocultures tend to require the addition of synthetic fertilizer since the lack of rotation means that the soil gets depleted leading to erosion. Monocultures lead to a decline in natural resilience due to the lack of crop diversity, increasing the risk of disease spreading rapidly through the crops if there is an outbreak. This means higher spraying rates to reduce the risk of outbreaks. Beyond the higher pesticide rates associated with monocultures, biotech crops in themselves tend to be associated with high pesticide use. "Roundup Ready" crops can be sprayed with Roundup without damaging the crop. This type of production has been hugely popular with farmers—although insufficient PPE for agricultural workers can result in human health consequences, including being a "probable carcinogen" according to the International Agency for Research on Cancer.[51] As of 2019, over 99 percent of the GM crops grown in different parts of the world exhibit one or both of two traits—insect resistance and herbicide tolerance.[52] Globally, this has led to a 7 percent increase in pesticide use, with a correlator impact on local biodiversity.[53] Friends of the Earth has estimated that glyphosate usage would increase by 800 percent if GM glyphosate-resistant sugar beet, corn, and soybean were approved for cultivation in the EU.[54]

Glyphosate can harm the surrounding ecosystem. "Non-target" organisms may experience direct effects from the herbicide that are often fatal or indirect effects when species are affected by smaller changes to their ecosystems or food resources.[55] The chemical can be toxic to birds, fish, and insects if they come into direct contact with it.[56] Often buffer strips around the fields get inadvertently sprayed, destroying weeds that are important food sources for insect, bird, and animal species in agricultural areas, particularly in the winter. For example, the decline of the Monarch butterfly in North America has been linked (in part) to the use of glyphosate-containing herbicides on GM corn and soy crops.[57] Monarch caterpillars consume common milkweed as their primary food source, a key target species for Roundup. Common milkweed plants have been lost at very high rates from fields of glyphosate-resistant crops, and it is estimated that common milkweed has been largely eliminated from 100 million hectares of US cropland following the introduction of glyphosate-resistant crops.[58] While not directly toxic to adult butterflies, the use of glyphosate has a hugely detrimental impact when the Monarch is a caterpillar. This has contributed to the decline of the Monarch, which has declined by more than 80 percent since the 1990s in central Mexico, and by more than 95 percent since the 1980s in coastal California.[59]

Roundup may also contaminate soils in and around treated fields. We still have a limited understanding of the long-term impacts of glyphosate on soil health. In some instances, it binds to soil particles, particularly in clay soils.[60] Glyphosate can also form complexes with metal ions, potentially affecting the availability of nutrients in the soil.[61] For the microorganisms that live in the soil, upon which soil health rests, glyphosate can be a source of energy and nutrients, but it can be toxic to others.[62] For example, earthworms were found to avoid glyphosate-treated soils; growth rates of some earthworm species were reduced by applying glyphosate-containing herbicides and the hatching of cocoons was delayed.[63] Populations of microorganisms that suppress disease-causing fungi have been found to decrease in soils treated with glyphosate, while other fungal species increase in glyphosate-treated soils.[64] So the presence of glyphosate in the soil could change the balance of bacteria and fungi, altering soil ecosystems. Furthermore, the chemical is soluble in water, so it may quickly wash out of soils and into waterways, damaging aquatic habitats. Glyphosate residues have been detected in surface waters across the EU. The European Glyphosate Environmental Information Sources (EGEIS) summarized surface water monitoring data from 1993 to 2009 for thirteen European countries.[65] Over 50,000 samples were included, and glyphosate was found in 29 percent of these samples. Residues of glyphosate were found in 50 percent of samples. These findings are supported by other monitoring projects, some of which found glyphosate in virtually all samples tested. Amphibians are particularly vulnerable to glyphosate exposure because they can absorb water-borne chemicals through their skin. While research is ongoing, a number of studies have found worrying results about the effect of exposure to herbicides containing glyphosate on the growth and development of amphibians.[66] Freshwater mussels have been found to be acutely sensitive to glyphosate, and a study on the European eel concluded that "environmentally relevant concentrations of Roundup can pose a health risk for fish populations" and found that the herbicide damaged the DNA of the exposed fish.[67]

Admittedly, not all biotech crops are associated with glyphosate. Yet, the others still have the potential to cause environmental damage. One of these key challenges is that once you have planted these varieties in the wild, they can't easily be undone. It is almost impossible to put the genie back into the bottle. If you plant a genetically modified crop in an open field, it can easily cross-pollinate with wild varieties. A pollinator doesn't distinguish between a

GM brassica and a non-GM brassica. This means modified genetic material can rapidly become permanently and irrevocably part of the species genome. Exacerbating this problem is the fact that biotech varieties often outcompete wild varieties. Such is the instance with GM salmon. GM Atlantic salmon have a gene from the Pacific Chinook salmon that regulates growth and a promoter from an ocean pout. This modification gave the Atlantic salmon a year-round appetite, so they get bigger quicker. Because the males are bigger, they are more attractive to females, and therefore wild varieties tend to lose out to the GM varieties during breeding. While GM salmon has been deemed safe to eat by US food regulators, there are still questions about how modified genetic material might interact with the natural ecosystem in the long run. That said, it is worth noting that GM salmon are raised entirely in land-based tanks to avoid the risk of crossbreeding. This restriction on where they can be raised was put in place to mollify salmon producers who feared for their livelihoods. Wild salmon is a highly prized product and can generate significant income for fisherfolk due to its perceived health benefits and luxury status. If the genetic lines of wild varieties of salmon became crossbred with GM fish, the prestige of salmon would likely decrease, and that would devastate incomes for those communities. However, ultimately, the company that brought the GM salmon to market could not make it profitable and closed in 2024.[68]

Cross-pollination of GM wheat with conventional varieties can close key export markets due to strict biotech restrictions—this happened in 2013 when unapproved GM wheat was discovered in Oregon, prompting Japan and South Korea to suspend US wheat imports, while China, Thailand, the Philippines, and EU member states tightened inspections, severely damaging US wheat exports.[69] Organic producers are particularly concerned about this cross-pollination risk. Organic producers are able to sell their products at a higher price to consumers because some individuals want the option to purchase food that aligns with their ethics and are willing to pay a higher price to know that their food is produced in a certain way. The higher price, in part, recoups the higher production costs that organic farmers incur by opting for a low-input farming method. If GM technology were to be found in an organic producer's food, it could result in the immediate loss of their organic certification, which takes two years (and considerable expense) to get, and the recall of their products from retail and wholesale outlets. Such an outcome would be hugely costly and could spell the end for that business.

Secondly, the GM species are patented and are owned by the company that created them. Already our global food system demonstrates high levels of corporate control, as we discussed in Chapter 4. A further expansion and adoption of biotech seeds across the world will inevitably benefit these transnational companies. Since they own the intellectual property of the bioengineered seeds and financially benefit from their sale, their profit (and power) will increase as more farmers transition to their products and become yoked into a production system that requires purchasing new seed and associated agrochemicals each year. The dominance of the same few transnational companies would grow with the trend toward consolidation worsening the entrenchment of corporate control of our global food system. Recognizing the troublesome nature of genetic material becoming intellectual property and the danger of the further strengthening of the companies that own those patents, the EU's authorization of genetic editing banned patenting NGT plants (i.e., GE plants) to "avoid legal uncertainties, increased costs and new dependencies for farmers and breeders."[70] Whether other countries will follow this lead is yet to be seen. However, this might disincentivize the R&D into new varieties if companies don't see a way to recoup their investment.

A Redundant Future?

One of the most popular types of GM crops is glyphosate-resistant varieties, also known as Roundup Ready. However, farmer confidence that Roundup will help rid them of weeds is increasingly being eroded as glyphosate-resistant weeds appear with greater regularity. Nature has an incredible way of adapting. More and more weeds have found a way over the past few decades since the 1996 introduction of Roundup Ready crops to survive rates of herbicide treatment that would normally prove lethal. Biotech glyphosate-resistant crops cross-pollinate with nearby weeds, allowing the resistant gene to spread into the wild through vertical gene transfer.[71] These herbicide-resistant weeds, known as superweeds, contaminated fields, creating a feedback loop wherein farmers were forced to use ever-harsher herbicides to beat back increasingly stronger weeds. A 2013 survey found that almost half of US farms had glyphosate-resistant weeds that were spreading exponentially, and by 2021, there were forty-eight weed species that had evolved glyphosate resistance.[72] Monoculture farming has also exacerbated the emergence of super weeds since the lack of diversity gives weeds time to better adapt to certain crop species and become more threatening.[73] Ultimately, if this trend continues, this could make Roundup Ready products redundant.

Of course, glyphosate-resistant weeds do not only impact GM crops but also all crops that use glyphosate. Glyphosate is the most used herbicide in human history. Approximately 3.5 billion pounds have been sprayed across America with a further 15.4 billion pounds sprayed worldwide since 1974.[74] If glyphosate doesn't work, farmers may be forced to switch to stronger herbicides, such as Dicamba. Dicamba is a synthetic fertilizer that has been used since 1967 in both agricultural and non-agricultural settings to control weeds.[75] Its use has been increasing because of increasing resistance to glyphosate.[76] However, its use has been associated with an elevated incidence of various cancers.[77] In Australia, farmers are trying to identify solutions to glyphosate resistance. The Australian Herbicide Resistance Initiative (AHRI) is mixing and rotating different weaker herbicides with different combinations of the chemicals and switching them up annually.[78] AHRI modeled different strategies and found that herbicide rotation alone worked in the short term, although it had limited benefit in the long term.[79] Another solution Australian farmers have found is Harvest Seed Weed Control, which captures and destroys the seeds for super weeds by burning them off and, consequently, stopping their spread.[80] As the owners of Roundup, Syngenta have a vested interest in finding a solution to glyphosate-resistant weeds. They are hosting a trial site to consider blackgrass elimination, which is one of the super weeds that is causing trouble for farmers, particularly in the UK. Across five fields and sixteen hectares, they are assessing a wide range of integrated approaches to blackgrass control that range from the five-year cultivation matrix to novel approaches such as cricket covers, propane burners, hoovering, and interrow hoeing.

Biotech crops will only have a future within our food system if there is a market that is willing to buy them. For many farmers, the decision to adopt biotech crops will be solely based on whether they can sell those crops. Many farmers, particularly in the Global South, have a business model that is entirely export-based. This creates a specific challenge for the future of biotech crops since one country might approve them to be grown, but they can only be sold into a market that also approves their use. If another country bans, or even just limits, the sale of biotech crops, that can undermine the business model for cultivation. For example, the GM papayas that are currently grown in the United States cannot be sold in the EU since, as discussed, the European Food Safety Authority does not approve the fruit for sale. This might not be a significant consideration for US papaya growers since there is enough of a domestic market to generate demand, but it could hugely disincentivize horticultural exporters from the Global South from adopting GM papaya (or any other biotech products) if their primary export market would not accept them. Ghana, for example, exports 813 tons of papaya to the EU and UK each year.[81] This market would no longer be accessible if they switched to the GM variety, and therefore it is not pragmatic for Ghana to begin growing GM papaya even if the Ministry of Food and Agriculture approves it as safe. Some major retailers in Europe have committed to avoiding biotech products, even if they become authorized at a future date. In the UK, Sainsbury, Tesco, Marks and Spencer, Burger King, and McDonald's all announced their intention to avoid GMO ingredients, while in Spain, Pryca announced it would no longer use GMO ingredients in its branded products, and the French food retailer Carrefour adopted a similar policy. Such internal policies, even if not legally binding, limit the market for biotech products.

Yet, restrictions on the import of GM aren't static. Changes in regulation (for or against biotech) happen constantly due to shifts in government leadership, increases in evidence, and shifting public priorities. These shifts aren't always purely science-driven. For example, Mexico attempted to curtail the domestic cultivation of GM in recent years, enacting a controversial presidential decree that sought to ban genetically modified corn in 2024 and phase out glyphosate.[82] This decision was motivated by a desire to protect the domestic corn industry from US competition and also to ensure the survival of heritage corn since GM varieties outcompete landrace (wild and heritage) varieties. The United States successfully challenged the legality of the export ban under the terms of USMCA, yet Mexican President Claudia Sheinbaum has expressed a desire to push for the outright ban of GM corn cultivation inside Mexico.[83] That said, even if a product is approved by regulators for sale in a specific market, it doesn't necessarily mean that consumers will buy it. If there isn't demand from the consumer, it wouldn't make sense for farmers to grow it. Outside of the United States, consumer polling shows reticence to buy biotech products. Europe is notoriously apprehensive about GMOs. As discussed, Indian and Chinese consumers are also still hesitant about purchasing biotech products. That said, as time goes on, consumers have shifted opinion and are more accepting of biotech. According to a 2019 poll, while 27 percent of Europeans are concerned over GMOs, only 4 percent were concerned about genome editing, so perhaps there will be an increased acceptance (and therefore expanded market) for newer types of biotech in the future.[84] Whether approving the supply of biotech products will stimulate demand will now be tested since both the EU and the UK have approved the use of genetically edited products (although they refer to them as NGTs and NPBTs, respectively).

The Value of Landrace and Heritage Varieties

Biotech products aren't necessarily better than the traditional variety. No clearer is that then in the case of cotton in Burkina Faso. A GM variety of cotton that produced Bt toxins to kill bollworm pests was created by Monsanto and was approved for planting in Burkina Faso in 2008 with the promise that it would help with poverty alleviation. Burkina Faso is one of the poorest countries in Africa, and the desire to raise rural livelihoods was understandable. The new GM variety of Bt cotton had shown promising results with field studies conducted in India in which Bt cotton was compared with non-Bt cotton. The GM cotton demonstrated a 30–80 percent increase in yield from the non-GM crop. This increase was attributed to marked improvement in the GM plants' ability to overcome bollworm infestation, which was otherwise common. Studies of Bt cotton production in Arizona, United States, demonstrated only small gains in yield—about 5 percent—with an estimated cost reduction of $25–$65 USD per acre owing to decreased pesticide applications.[85] However, while infestations might have decreased, the cotton threads that were produced from the GM cotton were significantly inferior to the cotton varieties traditionally grown in Burkina, a product that has long been globally recognized for its high quality. The GM cotton had shorter-fiber lint, so ginning machines extracted proportionally less lint from harvested cotton bolls, which resulted in lower yields. Ultimately, it cost Burkinabè cotton companies $76 million USD in losses, and within only eight years the project had been abandoned.[86] In China, farmers gained access to Bt cotton in 1997, and the crop was an initial success. Bt cotton reduced pesticide use by 50–80 percent and increased farm earnings by as much as 36 percent. By 2004, however, farmers who had been growing Bt cotton for several years found that other pests, such as mirids, had increased. Farmers once again were forced to spray broad-spectrum pesticides throughout the growing season, such that the average revenue for Bt growers was 8 percent lower than that of farmers who grew conventional cotton.[87]

Admittedly, cotton isn't primarily a food, but similar situations have been seen in food crops. As we discussed earlier, there has been a significant emphasis placed on biofortified foods, in particular the potential that Golden Rice has to deliver high levels of vitamin A. However, it is possible to achieve levels of vitamin A in staple food crops without the need for genetic engineering technology. Orange maize is a conventionally bred crop (i.e., does not use biotechnology) that is gaining traction across sub-Saharan Africa. Maize, or corn, is a common staple for many poor families who cannot afford fresh fruits and vegetables and part of traditional African diets (unlike rice). Orange maize is a variety bred to have higher beta-carotene levels (giving it a distinctive orange color) and provides up to 50 percent of daily vitamin A requirements.[88] It also raises the Vitamin A content of breastmilk, providing important nutrients for newborns.[89] However, that's only useful if farmers choose to plant it and consumers choose to eat it. Helpfully, the seed has also proven to be more climate-resilient with better drought-tolerance and early maturity and is preferred by consumers for its taste, texture, and aroma.[90] The positive traits of orange maize have been achieved through hybrid breeding. A hybrid is created by crossing two different varieties of the same plant. Crossing involves taking the pollen from the male flower of one plant and transferring it to the female flower parts of a different plant. The offspring are often referred to as F1 and have uniform traits to give growers consistency. One notable drawback of F1 hybrid seeds is that they don't produce seeds that reliably "come true to type." So, if saving F1 hybrid seeds, the plants produced are likely to show a broad range of characteristics, often not resembling their parent. Recognizing its transformative potential of orange maize, the Consultative Group on International Agricultural Research has been promoting the variety through its HarvestPlus program. This program aspires to scale up production and consumption of biofortified staple crops that utilize conventional breeding methods. Harvest Plus uses a community-based approach to reach farmers across Africa, with the goal of engaging hard-to-reach farming households that are not readily serviced by commercial seed companies—for example, through women farmers' groups and school feeding programs. The program also provides technical assistance and peer-to-peer learning programs to support the adoption of the new seed. The project has been a success. Within the first four years of its introduction, 100,000 farmers had switched to orange maize.[91] This uptake has been achieved through collaborative partnerships between Harvest Plus, local NGOs, government agencies, and private companies within the food system. These partnerships help to build trust with communities and collaborate with local seed companies and breeding programs to ensure consistent access and ready supply of the grain. The success story of orange maize as a conventional hybrid seed demonstrates how biotech is not essential to deliver effective biofortification of crops to address micronutrient deficiencies in communities in the Global South.

Clearly, hybrid breeding has the potential to deliver positive outcomes, as demonstrated by orange maize, but new plant breeding techniques are starting to be developed that could further improve outcomes. While GE/GM are the most dominant types of novel plant breeding techniques, there are other options that do not involve manipulating the genome at a molecular level. In recent years, marker-assisted selection (MAS) has become a good option for improving plant breeding and increasing its efficiency without the need for genetic editing. Typically, breeders improve crops by crossing plants with desired traits, such as high yield or disease resistance, and selecting the best offspring over multiple generations of testing. A new variety can take eight to ten years to develop. However, MAS can speed this process up dramatically. MAS marks genes on the DNA that are associated with desirable traits, and therefore breeders can select a plant or animal for inclusion in a breeding program early in its lifetime. The breeders can make a decision based on the parent's marker pattern (genotype) rather than their observable traits (phenotype). This approach dramatically reduces the time required to identify which varieties or breeds have the genes that express the desired trait. For example, scuba rice was developed using MAS and helped solve the problem of rice harvests being destroyed due to flooding. Around 20 percent of calories consumed across the world come from rice, and it is the main food source for three billion people. Yet the places where rice is most often grown, including areas of India, Bangladesh, and China, are constantly at risk of flooding. Rising sea levels and increasingly intense tropical storms mean that this problem is worsening. Developed by the International Rice Research Institute (IRRI), scuba rice can withstand being soaked in floodwater for extended periods and remains harvestable. Although rice thrives in wet conditions, it can't survive if the whole plant is completely submerged due to the hypoxic conditions. Scuba rice has been engineered to react to complete submersion by becoming dormant and waiting until the floodwater recedes before continuing to grow again. The SUB1 gene (which tolerates the submersion) was identified in 1996, and scientists began transferring it into the farmer-preferred variety in the early 2000s ahead of its release in 2009. Scuba rice has enabled subsistence farmers in Asia and Africa to adapt to climate change and the associated increase in flooding.

As more common food crops have their genome mapped, MAS will become an increasingly straightforward method for breeding with greater efficiency. The first plant genome to be mapped was the Arabidopsis thaliana in December 2000, and since then, the number of sequenced and assembled plant genomes has been exponentially growing. Several major projects are currently in progress, aimed at deepening our knowledge of plant genomics and enhancing taxonomic representation. These include the Darwin Tree of Life Project, the 10K Plant Genomes Project, the Open Green Genomes Initiative (OGG), and the 3k Rice

Genome Project. As of January 2024, GenBank hosted 1781 sequenced plant species.[92] One of these mapped genomes is the avocado. Avocado breeding is difficult since avocados are highly heterozygous making conventional breeding a challenge since it's difficult to know what parent genes will be inherited and expressed in their progeny. However, in 2019, researchers mapped the avocado genome. The lead researcher Luis Herrera-Estrella said, "We needed to sequence the avocado genome to make the species accessible to modern genomic-assisted breeding efforts," adding "the long life cycle of avocado makes breeding programs difficult, so genomic tools will make it possible to create faster and more effective breeding programs for the improvement of this increasingly popular fruit."[93] Having mapped the genome, avocado breeders across the world can now better predict what traits will be bred into future avocado crops, allowing them to ensure that the popular fruit can better survive extreme weather events or disease outbreaks.

Conclusion

New plant breeding techniques have moved beyond the genetic modification of the 1990s to today's more precise gene editing. This new generation of biotechnology has already shown promise in boosting yields and could play an important role in the future of food. Yet these technologies are not without risks, and once deployed, they are effectively irreversible and difficult to contain. Their value will depend on whether they can genuinely support climate goals and help meet rising food demand. The history of GM crops offers a cautionary tale: they were largely developed for large-scale commodity production—animal feed and oils—while also bringing unintended ecological impacts. For gene editing to gain broad public trust and deliver real benefits, these lessons must be acknowledged and addressed. Cost is another barrier. Biotech seeds and inputs are expensive, requiring long-term investment and often pulling farmers away from local, community-based farming systems that prioritize resilience and self-sufficiency.

One alternative vision for more accessible biotech could be the creation of open-source food forests—shared landscapes where gene-edited trees provide free food on communal land. Imagine blight-resistant bananas planted alongside heat-tolerant avocados and papayas engineered to resist ringspot. But such initiatives can only succeed if they emerge from meaningful community consultation. Without that, they risk repeating a long history of top-down interventions from the Global North, where solutions are imposed on pastoral and farming communities. Until the role of biotech becomes clearer, sustenance farmers may be better served by affordable, low-input approaches like agroecology, which can raise yields while avoiding the negative ecological consequences seen with earlier generations of GM crops.

9

Feeding a City

Cities have long shaped how we consume food, and reflexively, food consumption has shaped cities. Historically, food was a limiting factor on the location and spread of cities. Food production determined where cities were built and how large they could grow. This all changed with the arrival of the railways and the invention of refrigeration, which allowed cities to expand without being constrained by the need to feed their population. Carolyn Steel, the leading expert in the history of food in cities, said in her first book, *The Hungry City*, "The ability to preserve food, as well as transport it long distances, has freed cities from the constraints of geography."[1] The industrialization of the nineteenth century allowed cities to expand beyond the constraints of the farmland that surrounded them. It created the conditions for the megacities that we know today. These cities (and their populations) need constant feeding, and, as Steel acknowledges, "feeding cities takes a gargantuan effort; one that arguably has a greater social and physical impact on our lives and planet than anything else we do."[2] Just as in the Victorian Era, the cities of the future will be intrinsically linked to the advancement of food technology and infrastructure. More and more people will live in urban areas in the coming century.

Currently, 55 percent of the world population live in urban areas—more than 4 billion—and this is set to grow in the coming years.[3] Based on current projections, 6.7 billion people will live in cities by 2050.[4] While the total number of people matters, what matters more is the density of populations. This will result in forty-seven megacities (populations over 10 million) across the world.[5] The majority of these people will be from rural areas, looking for greater economic opportunity as a result of resource-intensive farming reducing opportunities in rural areas. With the vast majority of the population living in urban centers, cities will determine how food is consumed. For example, the rise in the snack and convenience food sector is almost exclusively driven by urban consumers who look for food that they can eat on their commute. Yet, as the world increasingly urbanizes, new challenges will need to be overcome. Feeding a city is harder than feeding a rural community that has access to space to grow their own food. Feeding a city requires having strong supply chains that can bring food into urban centers. These supply chains need to be resilient so that shocks to the system (e.g., an extreme weather event, a terror incident, or a disease outbreak) do not result in the supply chain collapsing and people going hungry. We need to redesign food systems to integrate food production into our cities to ensure better access to local, healthy, and affordable food for all.

Eating in a City

Feeding a city requires careful planning and near-perfect logistics. Supermarkets today rely on a model of just-in-time readiness. This means maximizing efficiency by never keeping inventory unnecessarily. Deliveries are perfectly timed to be responsive to customer demand. This requires real-time data to be fed through the supply chain to ensure the smooth flow of food. These models endlessly iterate, and AI technology is further improving the delivery. There is no margin for error since there is very little hedging built into the model. However, this model also relies on physical infrastructure. The roads, rails, and ports all need to work seamlessly to ensure that food moves rapidly into the city and onto the shelves. The average American meal travels over 1,500 miles before it reaches the table, crossing international borders and passing through numerous stages on its way to the consumer.[6] However, the just-in-time system of urban food distribution that we have is starting to show cracks that suggest that it is not fit for purpose. The current urban food system might excel at delivering sufficient calories to urbanites, but it's not equipped to deliver a healthy diet for its citizens to thrive.

As discussed in Chapter 2 our diets are increasingly poor and damaging our health. This is even more true in cities. A 2025 review in the *Journal of Obesity and Metabolic Syndrome* found that urbanization is a driver of dietary change.[7] Urban food consumption is associated with on-the-go meals (e.g., Subway) for people working long hours and with ready meals (such as pre-prepared Mac and Cheese) for those with smaller kitchens, if any. As discussed in Chapter 2, urban residents in Global South nations tend to be wealthier than their rural counterparts and consequently experience higher rates of obesity and poor dietary health than rural areas.[8] Without adequate planning for city governments, a damaging food economy will likely arise. Food deserts are areas, predominately urban but not exclusively, where there is limited access to fresh produce. In the United States, this is defined as an urban area where at least 33 percent of the population live more than one mile from the nearest large grocery store.[9] In rural areas, it's more than ten miles.[10] This definition is open to debate. For example, in the UK, food deserts are considered neighborhoods housing between 5,000 and 15,000 people with access to two or fewer large supermarkets.[11] Food deserts are not accidental—they are the result of decades of policy choices, economic patterns, and structural inequality and reflect the long-term consequences of community neglect and unequal development. In the mid-twentieth century, discriminatory practices like redlining left minority communities without access to resources and investment, including grocery stores. Supermarkets popped up in the suburbs while inner-city neighborhoods, especially those with large minority populations, were increasingly neglected. Fast food chains and convenience stores flooded the void, entrenching poor dietary habits. Exacerbating this trend is the fact that urbanization across the globe is associated with higher UPF consumption. A study in Brazil found that the highest rate of UPF consumption across the country was among urban residents under thirty and accounted for approximately 46.3 percent of their calorie intake. Rural residents had a much lower consumption at 23.8 percent.[12] Cities have been instrumental in the rise of "snack foods" that arose in the Industrial Revolution when factory workers needed to access energy-filled food quickly. Long commutes, limited access to fresh produce, and increased time pressure have increased this trend over time. This trend is likely to continue with the snack food market projected to grow over 6 percent annually until the end of the decade.[13]

Addressing hunger in cities requires not just new grocery stores but deep investments in housing, food education, transportation, healthcare, and racial equity. Consequently, city planners are attempting to improve access to fresh produce and reduce food deserts through policy interventions. One popular approach is incentivizing grocery stores to open in low-income neighborhoods through tax breaks (such as in Philadelphia with the Fresh Food Financing Incentive).[14] This can have an immediate impact on health outcomes and offset the cost via savings to the city for public health and welfare programs that address dietary ill-health. A study published in *Health Affairs* examined the impact of a supermarket opening—this time in a Pittsburgh food desert.[15] Findings indicated a significant decline in food insecurity (−11.8 percent) and a reduction in new diagnoses of high cholesterol (−9.6 percent) and arthritis (−7.4 percent) among residents in the neighborhood with the new supermarket compared to a control neighborhood. This approach has been successful in tackling food deserts in rural areas across America. Rural food deserts are not unique to America but happen in higher frequency since residents are often more disparate—something discussed in Chapter 4. Tax breaks for rural grocery stores have been introduced in Kansas, where the smaller customer base makes retail operations less viable without subsidies.[16] Another successful program is offering free rides to grocery stores from key underserved neighborhoods. This has happened across the United States with success. For example, the Central Ohio Transit Authority introduced the Mid-Ohio Market Ride, in partnership with a local Columbus food bank. Another example is Austin, Texas, which introduced a dedicated free bus line to connect East Austin residents with local grocery stores.[17]

Lift-sharing companies are attempting to help solve the problem by offering subsidized rides for those residing in food deserts. For example, Lyft worked with city officials and food banks in Washington, DC, to offer $2.50 rides for families in wards that had been identified as facing limited access to fresh produce.[18] Uber has done something similar in Dallas. This intervention of increasing access to healthy food is also balanced with greater zoning restrictions on fast-food outlets in neighborhoods that already have an excess number. These interventions could include caps on the number within certain wards so as to deny additional approvals or minimum distances that these can be located from a school. For example, US cities could draw inspiration from London, where many boroughs have policies banning new fast-food outlets within 400 meters of schools.[19] Core to delivering this invention is the GIS spatial data that demonstrates where food deserts are and overlays demographic information. The primary example of these tools is the USDA Food Access Research Atlas.[20]

Case Study Q: From Backyard to Bioferia—Quito's Model for Urban Food Justice

In Quito, Ecuador, urban agriculture has become a key approach to improving food security and promoting social inclusion. Since 2002, the city's Participatory Urban Agriculture Program, known as AGRUPAR, has supported thousands of urban and peri-urban farmers in producing fresh, organic food within the city limits.[21] Many of these farmers are residents of low-income areas. They grow crops in family, school, and community gardens, as well as through vertical farming, transforming underutilized urban spaces into productive agricultural zones. A central feature of AGRUPAR is the creation of "bioferias," which are organic farmers' markets that operate in neighborhoods across Quito.[22] These markets are located in both wealthier areas and low-income communities, ensuring broad access to healthy, chemical-free produce. Once producers have achieved household food security, they can sell surplus produce through bioferias. This direct-to-consumer model shortens the food supply chain, reduces transportation needs, and keeps food fresher and more affordable for residents. However, forming microenterprises requires skill—AGRUPAR provides training in business planning, marketing, and accounting. When capital is lacking, support is provided "through grassroots investment societies." From 2009 to 2018, the volume of produce sold through the bioferias increased by more than 487 percent, and total revenues rose significantly, demonstrating the growing impact of the initiative on both production and community engagement.[23] AGRUPAR demonstrates how urban agriculture can reduce food insecurity, boost local economies, and build community resilience. Quito's model shows the importance of good policy and infrastructure in creating a more equitable and sustainable urban food system, offering valuable lessons for other cities.

Risk versus Resilience

Our perfectly orchestrated food delivery system is just one hiccup from collapse. Many urban consumers experienced what happens when just-in-time readiness cannot adjust quickly enough in the early days of COVID-19. Many supermarket shelves ran empty when citizens stockpiled goods out of fear and supply chains faltered. Lockdowns, labor shortages, and transport restrictions led to delays in production, processing, and distribution. The World Food Programme reported that the number of individuals at risk of acute food insecurity nearly doubled to 265 million due to the pandemic's effects on income and supply chains.[24] On a more regional level, another instance of food failure includes the blockage of the Suez Canal by the *Ever Given* cargo ship in 2021. The canal was blocked for six days and caused a backlog that cost the global economy $1.35 billion USD.[25] Several businesses issued warnings over potential food shortages. Costco faced a shortage of cheese, European coffee producers faced delays on Robusta beans from Vietnam, and Indian exporters feared the delay of grape exports to Europe and North America.[26] Fortunately, most of the panic proved hyperbolic, and the worst-case scenario never materialized. However, with the public still primed from the shortages caused by panic buying at the beginning of COVID-19, supermarkets soon found their shelves picked through as consumers rushed to stock up on their essentials. The food system's inability to adapt in these (and other) crises exposes the weaknesses of this system.

There is an inherent vulnerability within our current food supply system to guarantee cities with a consistent supply when crises hit. Many thought leaders, in particular Professor Tim Lang of City University, responded by emphasizing the need to actively build more resilient supply chains that can better withstand future disruptions. In response, governments are investing in infrastructure projects to ensure the smooth flow of food into our cities. President Biden passed the Infrastructure Investment and Jobs Act in 2021 to upgrade the US transport network and improve ports.[27] As Biden's Maritime Administrator, Ann Phillips, said, "Modernizing America's port infrastructure is essential to strengthening the multimodal network that supports our nation's supply chain. Approximately 2.3 billion short tons of goods move through U.S. waterways each year, and the benefits of developing port infrastructure extend far beyond the maritime sector. This funding enhances the flow and capacity of goods moved, bolstering supply chain resilience across all transportation modes."[28] Similar efforts to upgrade the port of Rotterdam, one of the largest ports in Europe, are ongoing. The Maasvlakte II project costs $1.17 billion USD and aims to expand and deepen the container terminal by reclaiming 2,000 hectares of land from the North Sea.[29] This will add to the capabilities of the Rotterdam Food Hub, a state-of-the-art logistics center, increasing ease of access and assuring sustainable, reliable, and quick refrigerated and frozen cargo handling. Ports across the Global South from Lamu in Kenya to Santos in Brazil are being expanded and upgraded to improve their capacity and make them more efficient. City planners (particularly those in the Global South that are seeing rapid urban expansion and sprawl) are considering how to future-proof food infrastructure to ensure that their population can be easily and consistently fed.

One key model for increasing resilience and improving dietary health is relocalizing food production and bringing food production into the heart of our cities. Our current food supply is not local. Food is shipped from around the world. The lamb in UK supermarkets is more likely to come from New Zealand than it is from Wales. Avocados in Oregon are more likely to come from Mexico than California. Increasingly, city planners are considering ways in which more local (or at least regional) food can feed cities. What is local food? Some propose a radius model of 100 miles.[30] This has some value since emissions are less, and short supply chains have fewer opportunities for disruptions on the way to the consumer. However, this approach fails to capture the ethical and social aspects inherent in the push to relocalize food. After all, an intensive poultry unit with 50,000 birds per shed could still be deemed "local" to a consumer within that arbitrary radius. The concept of "local food" rests on a much more intangible philosophy. It prioritizes fair pay for the farmer and values the work that they do as stewards of the land. It factors in the artisanal and environmental aspects of production and tends to see a duty toward the community. Considerable debate remains on how "local food" is defined, and the term is open to interpretation but is often associated with smallscale, family-run farms. The Local Food Growth Plan, developed by UK NGOs, defines it as "food that is produced, processed, sold, and eaten within the same region or local administrative area through transparent, SME-focused supply chains."[31] Regardless of the nuance, local food systems essentially eschew the current corporate nature of our food system, which is globalized and distributed through a limited number of transnational agrifood businesses. Local food advocates want to see more smallscale farmers and food producers—more markets, more butchers, more bakeries—ultimately creating more connections between the person eating the food and the person producing it. By shifting to a model with greater diversity, the hope is that the system will become more resilient. Local food has a high level of interconnection, so there are multiple routes for food to flow from farm to citizen. If one supply fails, there are other options to pick up the slack. This reflects the distributed network model that Paul Baran distinguished in his research into differing network types in 1962.[32] Unlike the decentralized and centralized networks, where one break has the potential to cripple the network, the distributed network can adapt to crises by finding another route for distribution. Baran's understanding of the resilience of distributed networks led to the creation of the internet. The same disrupting and innovative capability could lie in switching to a distributed model for the future of our food in cities. There are many benefits for cities in strengthening local food.

Economic

A thriving local food system is dependent on small and medium enterprises (SMEs) that create new employment opportunities. A study by the Federal Reserve Bank of St Louis found that farms engaged in the "local food" economy created nearly thirty-two jobs per $1 million in revenue, in comparison to only 10.5 jobs for farms selling through conventional supermarket supply chains.[33] It can also help with skills training for young people. There are many barriers facing high school graduates to getting into their first job and starting a career. Local food systems can create a pipeline of training opportunities from entry-level to career-level training that supports employment opportunities through training. Furthermore, it encourages local investment and ensures that money circulates within the regional economy. The New Economics Foundation estimates that every £10 ($13 USD) spent in a local food business generates almost £25 ($34 USD) in the local economy.[34]

Sustainability

Increasingly, people understand the huge environmental footprint that feeding a city has. In 2017, C40 found that food-related emissions amounted to 13 percent of all consumption-based emissions in cities.[35] One of the clearest sustainability advantages of local food is the significant reduction in transportation distance, which lowers greenhouse gas emissions from transport—an important factor, as food miles may account for nearly 20 percent of global food-related emissions.[36] Local food also requires less packaging, shipping, and external inputs, further reducing its environmental impact. Shorter supply chains mean fewer handling stages between farm and plate, lowering the risk of bacterial or viral contamination and allowing produce to ripen naturally.[37] However, the sustainability of local food goes beyond mere proximity. Local smallscale farms tend to adopt more sustainable practices—including crop rotation, reduced pesticide and water use, and biodiversity-friendly approaches.[38] For example, a study of vegetable production in Iowa found that locally grown vegetables used just 10 percent of the water and produced half the emissions of conventional systems.[39] By relying less on pesticides, herbicides, and other chemicals—which harm ecosystems and pose health risks such as cancer—local food systems help protect both environmental and public health.[40]

Community

Local food brings people together, building closer social bonds and a positive sense of place—whether that is through growing food together in community gardens, promoting shared spaces, such as farmers markets, or just through a chat at the local greengrocer.[41] This is especially valuable when urban residents are experiencing increased anxiety and depression due to isolation. Research has shown that those living specifically in cities experience a 21 percent increased risk of developing an anxiety disorder and a 39 percent increased risk of developing a mood disorder.[42] This is likely to worsen in the future as city populations age. In cities within the OECD, the percentage of residents aged over sixty-five is predicted to increase from 20.9 percent in 2020 to 27.9 percent in 2040.[43] This demographic experiences a higher rate of isolation and associated mental health impacts.[44] Investing in local food could help city governments reduce the increasing levels of support needed for these residents in the coming years. There is also a mental health benefit for those directly involved in food production. Engaging with soil, plants, and natural cycles can reduce stress, anxiety, and symptoms of depression. It provides a sense of purpose and routine—especially important for vulnerable individuals. The physicality also boosts mood-enhancing endorphins, while time spent outdoors improves vitamin D levels and sleep quality. A 2021 meta-analysis published in the journal *Preventive Medicine Reports* found that gardening is associated with a reduction in depression and anxiety.[45]

Food Connection

People are increasingly disconnected from food production since they never interact with the production side. A 2022 survey by American dairy producer Land O'Lakes found that nearly 40 percent of the US adult population has never met a farmer.[46] Additionally, a 2022 survey by Michigan State University found that "half of Americans say they never or rarely seek out information about where their food was grown or how it was produced."[47] Local food can help people reconnect with their food. Buying directly from farmers' markets, community-supported agriculture programs, or local cooperatives gives consumers a clearer sense of how their food is grown and who grows it. This not only builds trust and transparency but also fosters stronger community ties between producers and consumers. By shortening the distance between farm and plate, local food systems can help restore the connection many people have lost with the sources of their nourishment.

Urban Growing

What does relocalizing food production within the urban environment look like? We might more readily imagine farming happening outside of the city, but the separation of food production from urban areas is a fairly recent phenomenon. Historically, until the nineteenth century, farming happened inside the city. For example, Orchard Street in New York got its name from the productive fruit trees that grew there. Cherrybrook in Sydney, Australia, and Coco da Rocha in Recife, Brazil, can similarly trace their names to their agricultural past. That said, more commonly, farming happened in the areas immediately surrounding the city in the peri-urban environs. The food was then transported into the city for market day. These markets were the beating heart of the city and often the location of political and civil debate too. The farmers who came into the city to sell their produce weekly felt strong connections to the urban center—relying on it for security and commerce. The urban dwellers similarly felt connected to those who grew and produced their food, understanding their reliance on the hinterlands. However, this changed with industrialization and urban sprawl, increasing pressure on land. As cities expanded, farmland was pushed further away, and food production became increasingly hidden from view. Over time, this physical and social distance eroded the direct connection between eaters and growers, replacing it with a more transactional and less personal relationship to food.

In recent years, there has been a return of farming to the urban center—but in a more innovative, tech-focused way. Vertical farming systems allow food to be grown in the center of cities and without the significant footprint that is usually associated with agriculture. As Chloe Rutzerveld wrote in her book, *Food Future*, "Vertical farming technology enables us to grow perfect crops; it helps us make optimal use of resources and reduces agricultural land use, while enabling the production of local and fresh food,"[48] reducing transportation impacts and better connecting communities with their food. How does it work? Vertical growing systems use hydroponic, aquaponic, or aeroponic technology. Hydroponics is the most well known. In this system, plants are grown in a nutrient-rich water solution, often supported by an inert medium like perlite or clay pellets. The roots are submerged directly in water, allowing them to absorb nutrients efficiently. By controlling the environment—light, pH, and nutrient concentration—growers can produce crops year-round, using significantly less water than traditional farming. It's clean, precise, and scalable. These vertical units control everything to maximize growth. They adjust the light, atmosphere, and temperature to ensure that plants grow in the ideal conditions. For example, according to Chloe Rutzerveld, a little adjustment of the UV light levels can result in healthier fruits with brighter colors. High-intensity red light can increase the growth rate. Similarly, high levels of blue light can reduce photosynthesis, which might seem like an odd decision since stems, leaves, and fruits become smaller, but the amount of nutrients, antioxidants, and flavor increases.[49] Aquaponics, on the other hand, introduces a biological twist. It combines hydroponics with aquaculture, the practice of raising fish. Fish waste provides a natural source of nutrients for the plants, which in turn help purify the water before it cycles back to the fish tanks. This symbiotic system mimics nature, requiring careful balance but offering a sustainable, closed-loop model of food production. Plus, the fish can serve as a protein source. Superior Fresh in Wisconsin produces approximately ten tons of leafy greens and one ton of fish annually within a three-acre greenhouse.[50] Then there's aeroponics, the most high-tech of the three. In this method, plant roots are suspended in the air and periodically misted with a nutrient-rich solution. Without the need for any growing medium, aeroponics allows for maximum oxygen exposure to roots, often resulting in faster growth and greater efficiency. However, it demands precise control and constant monitoring, making it more technically complex.

Vertical farming offers a range of benefits that make it a promising solution for sustainable food production, particularly in urban areas. By growing crops in stacked layers within controlled environments, vertical farms drastically shorten supply chains, reducing food miles and the associated emissions from transport. This proximity to consumers also improves freshness and minimizes post-harvest losses. Among its biggest strengths, vertical farming directly confronts the crisis of water scarcity. Conventional agriculture consumes vast amounts of water, much of it wasted or embedded in exported crops. Controlled environment agriculture (CEA) systems used in vertical farming can reduce water use by up to 95 percent, recirculating and reusing water far more efficiently than open-field methods.[51] This makes it ideal for regions facing water stress. Although energy usage in vertical farms—especially for artificial lighting—remains a challenge, improvements in LED technology and renewable energy integration are making systems more efficient. Importantly, the energy is used to maintain optimal growing conditions, allowing crops to be produced year-round regardless of weather. CEA also enhances food safety and biosecurity by isolating crops from pests, diseases, and contamination. As a result, vertical farms can minimize or even eliminate the use of pesticides, herbicides, and synthetic fertilizers. Studies have shown that vertical farms can be over twenty times more productive per square meter compared to traditional farms.[52]

What would vertical farming mean in practice? The reality is that they yield very little when compared to the huge number of people that need to be fed in a city. Singapore has placed huge investment into vertical farming. As we discussed in Chapter 6, the island nation placed priority on improving food security through innovation and consequently has provided grants and seed funding for vertical farming projects. Multiple vertical farming companies have been funded by Singapore government investments, including Sky Greens, ComCrop, and Genesis One Tech Farm.[53] Yet, despite the government investment, Singapore only produces 10 percent of leafy vegetables through controlled indoor farming methods.[54] To gain a meaningful share of the nation's vegetable market, vertical farms will need significant further investment. The question is whether that investment of public money is justifiable, given the limited food security that vertical farming can deliver. Another land-pressured city that is dabbling in vertical farming is New York. The Vertical Farm Project at the city's Columbia University envisioned a vertical farm with a footprint of one New York City block, rising thirty stories. It would equate to approximately 3 million square feet and could feed 50,000 people.[55] On this basis, the city would need almost 1,000 of these units to be spread across the city by 2050 to meet the demand for food. By comparison, Mumbai (predicted to be the world's biggest city by 2050 with 42.4 million people) would require over 4,000 units to feed the population.[56] Given how expensive property is in New York (and Mumbai), is that a realistic proposal? The average price per square foot in Manhattan was approximately $1,370 as of January 2025.[57] The cost of the property would need to be covered by the sale of the food, making vertically grown produce an expensive endeavor and only affordable to an elite clientele.

Many high-income urban customers prefer organic produce. In the United States, households earning over $100,000 annually are nearly twice as likely to purchase organic products regularly compared to households earning under $50,000.[58] However, despite this preference among their key clientele, there is strong debate whether vertical farming systems can be certified organic. The debate over whether hydroponic growing systems can be certified as organic is a contentious issue that strikes at the core of how "organic" is defined. Proponents of hydroponic certification argue that these systems can align with organic principles by avoiding synthetic pesticides, herbicides, and genetically modified organisms. They also highlight benefits such as efficient resource use, reduced water consumption, minimal land impact, and the ability to grow food in urban or climate-challenged regions. From this perspective, hydroponics represents a modern, sustainable solution that supports food security and innovation. Critics, however, argue that organic agriculture is fundamentally rooted in soil-based ecosystems. They emphasize that soil health, biodiversity, and the natural cycling of nutrients are core tenets of organic farming. Hydroponic systems bypass soil and associated ecological interactions and, critics contend, fail to support the environmental and ethical foundations of organic agriculture. This disagreement has led to divergences: the United States permits hydroponically grown produce to be certified organic under USDA rules, while the EU and countries like Canada and Mexico will not certify hydroponic production as organic.[59] The divide reflects deeper questions about whether organic standards should prioritize production methods or broader ecological values. However, it may act as a limiting factor on the growth of the vertical farming market if their key demographic wants to see the organic stamp of approval before purchase.

Productive Green Urban Spaces

Other attempts to return horticulture to the city look more traditional. City planners have historically recognized the need for access to green space for urban residents. While there were public green spaces in the ancient cities of Athens and Rome, the first publicly funded civic park was in 1847 in England. This led to city parks arising across the world as city leaders acknowledged the public benefit. However, few of these public parks include productive growing space. As Steel said, in Hungry City, "Living in cities, we have learnt to behave as if we did not belong to the natural worlds; as if we were somehow distinct from the 'environment,'" but there have been attempts to reverse this trend and try to connect urban dwellers back to the natural world.[60] Community gardens and allotments differ from public parks since they are directly linked to food production. A community garden is a neighborhood space where locals grow and harvest vegetables together. There is a sense of collaborative working and shared ownership. In the UK, this model has deep roots in the Enclosures Acts and the resulting urbanization, going back to Nottingham's Hunger Hill Gardens in 1765. Community green spaces increased as a method of delivering food security for the working poor after the Swing Riots and formalized in the 1908 Small Holdings and Allotments Act. In the immediate postwar years, many were paved over for development (a threat faced again today), but they gained traction again in the counterculture of the 1960s.

Allotments are different. Plots are assigned to individuals, and there is little shared group activity. The focus is almost exclusively on productive horticulture, while community gardens tend to have a greater focus on aesthetics with flowers integrated into the design. Allotments have a long history in Europe, linked to land reform. They were often granted to the poor via private landowners for subsistence farming. In Germany, they are known as kleingarten. In France, they are jardins ouvriers. In Russia, they take the form of dacha. They have seen a resurgence in recent years as urban populations look to reconnect with the land and with their food. Managed by local city councils, the UK had over 157,000 applications pending in 2023, and some urban areas report wait times extending up to fifteen years.[61] This heightened demand is driven by rising food prices, increased awareness of food sustainability, and the mental health benefits associated with gardening. Housing and land pressure have also increased the popularity of allotments and community gardens. They provide a semiprivate green space for people to garden, which is appealing to younger people living in densely built areas where housing does not commonly feature gardens. As a result, there has been a greater focus on integrating growing spaces into urban developments.

Recognizing the benefits of productive green space in cities, many state and local governments are trying to encourage more of them within urban areas. For example, California established the Urban Agriculture Incentive Zones Act in 2014 to promote urban farming.[62] It allows landowners to receive significant property tax reductions if they commit their vacant or underused land to agricultural use for a minimum of five years. By making urban agriculture financially attractive, the act supports sustainable land use, community engagement, and food access in cities, particularly in areas lacking fresh produce and public green space. Some cities are going further to ensure future space for growing. The Nairobi City County Urban Agriculture Promotion and Regulation Act was passed in Kenya in 2015 to increase food security.[63] This law acknowledged the critical role that urban agriculture can have in delivering consistent food for Nairobi's nearly 5 million residents.

Yet, one of the major challenges that urban and peri-urban growers face is access to finance. The majority of agricultural support and finance programs are geared toward rural areas. Oftentimes, these urban farms are deemed too small to qualify for funding. Such is the case in the EU, where the majority of member states have a threshold of between 0.3 and 4 hectares to be eligible for support under CAP.[64] Often financial infrastructure (bank offices, relevant expertise) is not in place to support the urban grower to apply for a loan, even if they would be eligible. There are, of course, many seed funding programs for SMEs, but these are not often tailored to urban agricultural businesses. Consequently, many urban growers are unable to expand since they have to self-finance. Some US states have established programs to better support urban growers, such as Pennsylvania's Urban Agriculture Infrastructure Grant Program. Since 2019, the program has supplied more than $3.2 million in grants, funding 160 projects including the purchase of greenhouses, cold storage, and off-grid energy and water systems.[65] Outside the United States, Kampala offers an example of how a city government can support local food production with their Urban Agriculture Program. The city's Urban Agriculture Program was enacted in 2006, and prior to this urban farming was illegal in Kampala.[66] The program requires individuals who farm within the city boundaries to obtain a permit to farm and a license to sell produce by paying a small fee and putting their name on the Kampala Capital City Authority's Urban Agriculture List.[67] Legalized urban agriculture has provided jobs to women and the urban poor and has increased access to quality food for those living in Kampala.[68] The Kampala Capital City Authority is now also providing training centers to teach Kampala residents about vertical farming practices.[69]

Case Study R: Cape Town's Urban Agriculture—Food Justice and Economic Opportunity

With the laudable goal of embedding agriculture into an urban fabric shaped by inequality, food insecurity, and contested land use, Cape Town established the Urban Agriculture Unit (UAU) in 2007.[70] The UAU aimed to make food production part of city planning, seeing urban farming both as a safety net for struggling households and as a way to create income and jobs. Its mandate included providing training, facilitating access to land, supporting infrastructure, and coordinating across stakeholders. For marginalized households in informal settlements and townships, this support showed that urban farming was officially recognized as a valid and legitimate way to produce food. However, UAU faced limited resources, and shifting administrative structures meant it declined as a political priority.[71] It was overtaken by the Food Gardens Policy that narrowed the ambition to focus exclusively on nutrition and domestic food security, as opposed to UAU's goal of economic growth and livelihood creation.[72] Despite its short tenure, the UAU created a policy framework that allowed NGOs and community groups to anchor their projects within a sanctioned municipal agenda.

One of the most successful grassroots projects to emerge from the UAU approach is Oranjezicht City Farm (OZCF). Founded in 2012 on the site of a disused bowling green, OZCF transformed a neglected space into a productive organic farm and educational hub. The farm quickly became a vibrant gathering point, offering workshops, school programs, and volunteer opportunities. In 2013, the associated OZCF Market was launched and has since grown into Cape Town's largest farmers' market, supporting dozens of small farmers and artisanal traders while drawing thousands of visitors each weekend. Beyond food sales, OZCF pioneered composting initiatives, turning household and business food waste into soil-enriching compost, and has supported the creation of twenty more community gardens across Cape Town. At its heart, OZCF has been a social justice project, creating opportunities for local farmers, connecting consumers with marginalized producers, and reshaping access to nutritious food in a highly unequal city. It demonstrates how urban farming can bridge divides of class and race while building more inclusive food systems.

Food justice gardens differ from community gardens or allotments. They are community-driven spaces that address systemic inequalities in the food system and tend to be located in historically Black neighborhoods. Rooted in the principles of equity and empowerment, these gardens provide access to fresh, healthy, and culturally appropriate (not just nutritionally sufficient) food in underserved areas. Unlike traditional community gardens, food justice gardens focus explicitly on social justice, racial equity, and reclaiming land and agency for historically marginalized communities. They often become hubs for community organizing, youth engagement, and education. Trailblazers, such as Soul Fire Farm in New York and Green Grounds in Los Angeles, have inspired many more food justice gardens to pop up across the United States. The concept is gaining traction in other countries—one particularly interesting instance is Abalimi Bezekhaya in Cape Town, which is using food growing as a way to address post-Apartheid contexts. Within these food justice gardens, education and youth programs tend to be central. Core to addressing the public health crisis is tackling children's food. As we discussed in Chapter 2, child obesity is an increasing challenge in both the Global North and the Global South and disproportionally affects low-income communities and communities of color. Food justice gardens are attempting to reverse this trend by educating kids on how to grow food and reconnect them with the earth. Roy Finley works in Los Angeles to increase awareness of the importance of growing food. The self-titled "gansta gardener" teaches children how to cultivate vegetables in their neighborhoods. In relaying the necessity of education for food justice, Finley claimed, "We need food education in our community. Black people need to be more self-sufficient, especially when it comes to what we put in our bodies. We need to change our value system. Illnesses like diabetes and heart disease can be fought with the growth and consumption of healthy, naturally grown food. It's my mission to disseminate that message."[73] Finley's approach to food education teaches children and communities that they can prioritize health in what they eat, and it encourages communities to take power over what they consume.

Urban Food Halls and Markets

In cities worldwide, markets and food halls are experiencing a powerful resurgence, transforming how urban people shop and eat. Historically, these were key locations where the whole city came together to purchase and consume food together. Food markets were locations where farmers from the peri-urban areas formed connections with city dwellers and sold their produce. They also serve an important social and political function. In the Western tradition, the Ancient Athenian Agora was simultaneously a public market and the birthplace of democracy and civic engagement. While food markets tend to focus on the sale of produce and food for home preparation, food halls tend to serve freshly prepared, cheap meals for workers. For example, the South-East Asian tradition of the hawker market. As supermarkets grew in popularity and spread, the local market and food halls declined. The supermarket arrived with the Piggly Wiggly in Memphis, Tennessee, in 1916 that allowed consumers (mainly women) to walk through the aisles of the store and select their own items, where previously they would have requested the items from the clerk behind the counter at the general store.[74] This model iterated with the founding of King Kullen in Queens, New York, in 1930. With a bigger footprint, it offered a huge range and variety of products and would be more recognizable as the supermarket we know today. The suburbanization of the 1950s turbocharged the supermarket expansion, and soon it became the default location to shop for the middle-class families who had moved out to the suburbs. Shortly after, the shopping mall, in all its glory, arrived and was the final nail in the coffin of many food markets. As city centers saw population decline, urban renewal projects saw many markets be torn down. Famously, the iconic iron-and-glass pavilions of Les Halles in Paris (known as "the belly of Paris") were demolished in 1969.

While many saw the closure of these markets as modernity sweeping away the unsanitary past, some saw the cultural and social value of these spaces and acted to protect and revitalize them. Boston's Faneuil Hall Marketplace was founded by a local merchant in 1742 and expanded in 1826.[75] It sold fresh produce from local farmers but was also a gathering place for the whole city and often featured Revolutionary speakers, such as Samuel Adams. However, by the 1960s, it had fallen into disrepair. The City Government partnered with James Rouse to revitalize the market. He envisioned a "festival marketplace" that would blend local food retailers with restaurants and entertainment.[76] There was a focus on Boston's culinary history (cream pies, lobster, and baked beans) but also incorporated cuisines from immigrant communities across the city. This was highly innovative and an entirely new approach to urban food markets. Its popularity saw the model copied across America and then the world. What these food halls and revitalized markets offer is not food, but experience. For example, no one in their right mind goes to Borough Market in London to do their weekly shop. Instead, it's about the sensory enjoyment of the event. Urban renewal and development that focuses on creating public spaces that are vibrant and inclusive is known as placemaking, and placemaking through food is one of the most successful modes. These food halls are highly curated, indoor spaces. While they might have concessions for fresh produce, the focus is on prepared meals, often in quite a trendy and luxury manner. Think $20 smash burgers and fancy chicken wings, not sacks of potatoes and cabbages. The placemaking through food has been so successful that Borough Market has revitalized an entire region of the city and made it a thriving and popular place to live.

The rise of these food halls echoes the revival of farmers' markets. Farmers' markets began to reappear in the 1970s, closely tied to the growing environmental movement and a renewed interest in seasonal, locally grown food. They became gathering places where urban dwellers could connect directly with producers, support sustainable agriculture, and push back against the anonymity of industrial food systems. Over time, these markets evolved into cultural institutions, celebrated not just for their food but for the sense of community, transparency, and authenticity they brought back into city life. In some instances, this revival was supported by local and national governments, such as in the United States, where Congress passed the Farmer-to-Consumer Direct Marketing Act of 1976.[77] However, the rise in public interest in organic and "local" food saw many new markets open in the 1990s and early 2000s with the Farmers Market Promotion Program passed in 2002.[78] The USDA published its first national directory of farmers markets in 1994, and the number of US farmers markets grew from nearly 2,000 in 1994 to nearly 3,000 by the year 2000.[79] They have now become a common sight in many towns and cities, and an event for many millennials on a Saturday morning before brunch. Today,

US farmers markets number nearly 9,000.[80] USDA programs have also aimed to make farmers' markets more accessible to low-income communities. The USDA has been working to expand the network of farmers markets that accept SNAP benefits via an EBT credit. Individuals swipe their EBT card, and the credit is exchanged for wooden tokens that vendors across the market will accept as payment. Some states (including Michigan, Indiana, Kansas, and Missouri) have programs that double the value of farmers market tokens received for EBT credit in an attempt to advance healthy eating.[81] After seeing the successful implementation, there is hope that "Double Up Food Bucks" will become a federal policy.

That said, both the revival of farmers markets and food halls are deeply middle-class and geared toward affluent consumers. They tend to be located in middle-class areas or in development regions that are looking to attract middle-class residents. Traditional working-class neighborhoods and urban centers experienced the same closure of food spaces from the 1950s onward that we discussed but did not yet experience the revitalization. This has led to the aforementioned food deserts arising. There has been growing attention in recent years to redress this problem. Cities have set up initiatives to increase access to fresh fruit and vegetables in city regions that have otherwise struggled. For example, Portland's Produce for People and California's Farm to Family that allows community gardeners to donate surplus food to food banks. One of the key challenges is affordability since fresh produce is often more expensive than UPF, as discussed in Chapter 2. To tackle this, there have been schemes introduced to subsidize the cost for consumers. For example, Rose Vouchers, an initiative by the Alexandra Rose Charity in the UK, helps families on low incomes access fresh fruit and vegetables. The program provides weekly vouchers—£6 ($8.16 USD) per child under one year old and £4 ($5.44 USD) for older children—that can be redeemed at local markets and greengrocers.[82] Since its inception in 2014, the project has distributed over three million vouchers, totaling £3 million (approximately $4 million USD) spent on healthy food by more than 10,500 families.[83] These vouchers not only support families in providing nutritious meals but also bolster local markets by encouraging regular trade. Results after eight months showed that 90 percent of participants experienced improved physical health, doctor visits were nearly halved, and adherence to nutritional guidelines increased from below 30 percent to nearly 80 percent.[84] Additionally, 75 percent of participants lost or maintained their weight, and over half reported improved mental health.

Conclusion

When considering the future of farming, we often think of rural areas, but the future of food will be defined by cities. Cities consume the majority of the world's food and define food consumption trends. As a result of urbanization, UPFs and snack foods have become more and more popular, contributing to a rise in obesity and associated impacts on public health. As the world increasingly urbanizes, we need to redesign our food systems to improve access to healthy and sustainable food in our cities. As cities expand, planners and local governments are increasingly aware that food insecurity, obesity, and their knock-on effects on economic growth cannot be ignored. Urban design must prioritize the smooth flow of food into city centers, with investments in roads, rail, ports, and new technologies to improve distribution. This is especially urgent in the Global South, where megacities are growing at unprecedented rates. Without careful planning, these cities risk becoming even more vulnerable to food shocks. Building resilient urban food systems is essential to ensure reliable access to food, particularly in an era of frequent crises and geopolitical instability. Part of building up the resilience of urban food systems will be greater emphasis on how to relocalize. When considering the future of urban food, we need to design a system that improves access to local, healthy, and affordable food for all. Vertical farms are an innovative solution that is gaining traction, but there are more traditional methods of urban food production and distribution that are coming back into fashion. These, connected with a facilitating policy environment that supports SMEs in the food sector to set up and thrive, have the potential to transform urban food culture and restore a connection between food and citizens that is at risk of being severed.

10

Agri-Technology

We have just explored the innovations needed to feed the cities of the future, where technology and urban design are often seen as natural partners. But the story does not end in the city. Rural landscapes are also at the frontier of technological change, with farms becoming testing grounds for some of the most advanced tools of the digital age. From precision agriculture and AI-driven monitoring to drones, robotics, and gene-edited crops, technology companies are increasingly recognizing farming as a sector ripe for disruption. The future of food, therefore, is not divided between urban and rural innovation but shaped by the interplay of both.

Technology has always disrupted farming practices. The invention of the steam-powered thresher transformed arable farming—and provoked a backlash in the UK with the Swing Riots in 1830. The invention of the assembly line at the dawn of the twentieth century allowed for the processing of livestock in huge numbers across the American Midwest. The discovery of the Haber-Bosch method of producing nitrogen fertilizer after the Second World War allowed for crop yields to skyrocket. The Technological Revolution of the twenty-first century has been no different. The arrival of the internet and computers has changed every aspect of our global society, creating new avenues for innovation. AI holds huge potential to transform farming systems. Today's farmers are at the forefront of these advancements, and technological innovation will revolutionize agriculture in the coming decades.

Tech is already playing a significant role in farming with big data and precision agriculture becoming widely adopted in the Global North, particularly for large-scale commercial producers. Yet, there are more innovative practices on the horizon that rely on data-driven mapping and predictive modeling that hold the potential to change the quotidian realities of farming. These technologies can transform the productivity of enterprises by allowing farmers to gain greater insight to preemptively adjust to maximize yield and improve efficiency.

Smaller-scale and regenerative farmers are also embracing new technologies, which can help them reduce their environmental footprint. Data can help them limit their use of pesticides and fertilizers, helping the environment and lowering production costs. Robotics can help reduce the reliance on migrant labor, which has become inconsistent and expensive in recent years, as discussed in Chapter 4. In the Global South, tech can also help subsistence farmers adapt to climate change by advising new crops or changing weather patterns, helping protect harvests and ensure livelihoods. However, new technologies are expensive, and many farmers operate on tight budgets and aren't able to invest in smart equipment. Additionally, the age and experience of most Global North farmers create significant barriers to adoption. For some, they have been managing their land in a certain way for years and don't necessarily see the need to change. Those belonging to an older generation can often be hesitant or confused by new technologies.

Precision Farming Is Making Huge Differences Already

"Precision agriculture" is a major trend in agriculture today. This term refers to a more targeted approach to agrochemical application and irrigation. Commercial farmers around the world have embraced the technique in the last decade. As discussed in Chapter 1, the diffuse spraying of agrochemicals, popularized in the 1950s, causes environmental damage. Spraying herbicides and pesticides has a wide impact on local insect and wild plant populations. Extensive applications of fertilizer can see much being washed away when it rains, running off into waterways and causing eutrophication—also discussed in Chapter 1. This is not only environmentally damaging but also cost-ineffective, since expensive chemicals were merely washed away. The adoption of precision agriculture aims to minimize this damage by applying these chemicals in a more targeted manner. After assessing the particular conditions and topography of each part of their fields, farmers can more effectively utilize their agrochemicals and maximize their yield while minimizing cost and environmental damage. Similarly, the approach of irrigating fields through drip irrigation, as opposed to huge sprayers, can help plants receive the water they need without a lot being lost to evaporation or run-off. While drip irrigation had been used in traditional farming practices, its popularity has been increasing since the 1950s when Dr. Daniel Hillel and Simcha Blass adopted the method to bring water to the Negev Desert in Israel. By the 2000s, the approach had gained huge recognition for its environmental benefits, and Dr. Daniel Hillel was awarded the 2012 World Food Prize for his contribution to agriculture. Unlike other forms of irrigation, such as rotary sprinklers that are only 65–75 percent efficient, drip irrigation is 90 percent efficient at allowing plants to use the water applied.[1] Some farmers adopted precision agricultural techniques because they saw their yields declining. Some did so because of increased government regulations that aimed to limit the diffuse pollution from agriculture. Others because they felt that adopting a more environmentally sensitive approach was their moral duty in light of the climate crisis. The climate motivation is all the more apparent in the Global South, where farmers are on the frontlines of the climate crisis. For those farmers who have the means to adopt precision agriculture techniques, the impact can be dramatic. Iraqi farmer, Mohammed Sami, spoke of the impact of installing drip irrigation systems, saying, "I now irrigate 10 donums (approximately 10 acres) with the same amount of water that I used for one donum before," adding that his wheat harvest had shot up from seven to twelve tons per year.[2]

Technology has helped precision agriculture to evolve to become more data-driven with the creation of mapping tools that use different data points to build up a nuanced picture of the entire farm. This is known as smart farming. By collecting data (e.g., soil pH) from the fields, farmers can gain greater insight into their land and know exactly where to apply pesticides, or when a crop needs a bit more fertilizer, or apply a precise amount of water. Most farmers know their land, inside and out. They can tell you about each tree and each hedge. They know how their soil responds because they have held it in their hands. However, as we all know, it is easy to overlook how your immediate environment changes over time when you see it every single day. Similarly, it is easy to fall into the habit of doing something routinely without stopping to consider whether it is the best course of action. Data can allow farmers to use technology to consider their land and see what is required without bias. This data is compiled through GPS and advanced mapping tools. Farmers can add their own data sets to these farm maps to make the algorithms even more accurate. These highly detailed farm maps can allow them to visualize the layout and specifics of their farms and support data-driven analysis of their land. Numerous variables can be monitored including crop growth rate, biological activity in the soil, moisture content, nutrient availability, disease and pest damage, and livestock location. Machine learning and modeling can then incorporate this data with weather patterns, market trends, and the performance of other farms to supply farmers with key insights. For example, FieldBee, a Dutch company founded in 2014, created a mapping device for smart tractors that gathers data as it goes and feeds it into the system. Similarly, livestock farmers can use GPS to remotely monitor their herds in real-time, allowing for a more rapid response if one starts to be separated from the rest—although sometimes a thick hedge might flummox the GPS signal!

In-field sensors are part of the picture for precision farming, collecting precise data to feed into the algorithm. These sensors allow for the real-time collection of a huge amount of data from across the entire enterprise and faster interventions. Agricultural sensors are being developed for every aspect of farming.

Soil Sensors

Soil sensors allow farmers to collect data on root zone moisture, salt accumulation, light quality over time, and temperature variations in soil throughout the area. The information is generated every hour, which allows for the detection of subtle variations in all these different areas. This technology can be used to measure trends in temperature, light, salt, moisture, among others, over seasons and years. It would be physically impossible to accomplish this level of data collection without the technology.

One company that is doing this is Spiio, which sells sensors that monitor "hyperlocal soil information" and collect readings every hour.[3] They can also be used to "create seasonal comparison studies for drainage or shade projects."[4] Building these trend charts and doing comparative studies can be useful for tracking which plants to grow, expected yields, and so on. This information would allow a farmer to better predict what varieties might thrive on their land, improving the success of the harvest. Hyper-specific information can allow farmers to understand what crops will succeed in which areas, minimizing waste and maximizing yield. Similarly, CropX partnered with Dacom Farm Intelligence in 2021 to build a platform specializing in soil health monitoring by leveraging real-time data to help users review and compare vital parameters alongside crop performance. Farmers gain insights into soil type and vegetation indices like normalized difference vegetation index (NDVI), soil adjusted vegetation index (SAVI), and soil moisture index to optimize crop management strategies. CropX reports that its solutions have led to a 57 percent reduction in water usage, a 15 percent reduction in fertilizer usage, and up to 70 percent yield increase.[5]

Water Sensors

Farmers can also use water sensors to track the water they're using and minimize waste. This allows them to irrigate plants if they are starting to get thirsty, long before they have started to wilt. Eureka Water Probes are sensors that measure the temperature, pH, conductivity, dissolved oxygen, turbidity, and depth of the water on farms.[6] This is helpful for farmers who need to detect algae blooms and pollutants in water that they use for irrigation. Livestock farmers in particular require a lot of water—one cow requires forty–fifty gallons per day.[7] With rising droughts and water shortages, farmers need to minimize the amount of wasted water. Fertilizers that seep into the water supply encourage the growth of algae blooms, which can negatively affect soil health if irrigation water becomes contaminated with algae.[8]

Pest Sensors

Fighting pests and disease in their crops and livestock is a main occupation for a farmer. Often when the signs are visible to the human eye, it's too late and the crop is lost. Sensors can help farmers to detect diseases and pests at the very early stages so that they can act preventatively to solve the issue. In 2020, a lab at MIT developed "tiny carbon nanotubes wrapped in polymers" that act as sensors. These nanotubes detect hydrogen peroxide, which plants emit when they are under attack from insects and bacterial infection, and salicylic acid, which is "involved in regulating many aspects of plant growth, development, and response to stress."[9] When the polymers detect hydrogen peroxide or salicylic acid, they emit fluorescent light, which can be seen using infrared cameras. This allows farmers to track real-time changes to their plants. It is useful in detecting stress caused by disease and infection in plants early on. Another model was created by Trapview in 2019, which has a device that traps pests and identifies them using the camera in the device. By leveraging Trapview's database, AI identifies over sixty pest species, such as the codling moth, which afflicts apples, and the cotton bollworm, which can damage lettuce and tomatoes. Once identified, the system uses location and weather data to map out the likely impact of the insects and pushes the findings as an app notification to farmers. As AI technology improves, so will Trapview's transformative potential. Horticultural producers are not the only ones to benefit from disease sensors. Livestock sheds can be fitted with sensors that detect early signs of disease such as changes in animal movement and other behaviors. AGXIO Limited, an AI company located in the

UK, makes collars that collect data (temperature, pulse, etc.) on animals that can predict any issues that arise in health and safety for the livestock.[10]

Case Study S: Digital Innovation—E-Commerce Revives Rural China

In China, rural residents are using technology to revitalize agricultural markets. Live streaming has transformed farming into a viable, profitable, and modern career. There is a growing movement of rural entrepreneurs who use digital platforms more commonly associated with consumer goods (such as cosmetics) to reach wider markets and secure direct sales. Xing Jianli, a 38-year-old flower grower from Shandong province, doesn't rely solely on local markets, but instead, embraces live streaming to sell her peonies. Broadcasting directly from her fields to viewers on their smartphones, Xing can earn up to 3,000 yuan (around $400 USD) a day—more than a month's wages at her former factory job—through e-commerce.[11] Beyond selling goods, some streamers market a romanticized rural lifestyle. Influencers like the Huanong Brothers and Li Ziqi have built massive followings by sharing scenes of traditional farming, food preparation, and scenic village life.[12] Their success brings attention—and income through e-commerce—to remote areas once overlooked by China's rapid urbanization.

Beyond the farmers' use of the technology, the rise of rural live streaming has also sparked new business models, with streamers buying directly from farmers and reselling the food via live broadcasts. When the season ends in one region, the streamers move to another. In 2023, Chinese consumers placed approximately 1.4 billion orders for agricultural products on Kuaishou, a popular rural streaming app, up 56 percent from the previous year.[13] Alibaba reported 441 billion yuan ($60.5 billion USD) in sales from villages on its platforms between 2021 and 2024.[14] Recognizing the economic potential, local governments are encouraging this digital transition. Local officials in the Heze region offer training programs, teaching farmers online sales techniques. The technology has the power to transform rural livelihoods in China, but as more join the digital fray, competition intensifies.

Drones in Agriculture

Sensors work at the micro level, gathering data on an individual m2 of a field. However, drones allow for the macro view, flying across the entire farm while capturing data and entering it into the farm map. This is a huge time-saver for farmers, since some US farms are over 2,000 acres. Imagine needing to drive across that each day—you wouldn't have much time for anything else—and you can see why drones would provide an appealing alternative. Instead of cruising down a hot, dusty road for hours, you could sit in an air-conditioned office with a coffee in front of a computer screen. Drones allow farmers to efficiently and effectively monitor livestock, assess plant health, scout pests, and manage irrigation systems. Beyond imaging, drones can be used to perform basic agricultural tasks, like spraying crops with pesticides and fertilizer.[15]

While drones are the vehicle, it is the camera on them that delivers the real benefit, allowing farmers to stop diseases from spreading by catching them early on in remote areas that they might accidentally overlook otherwise. Because plants reflect sunlight across different wavelengths as their health changes, using cameras to detect the changes in wavelengths being emitted can determine how healthy the crops are. Multispectral imagery detects subtle variation in plant health before visible symptoms appear. The data from these cameras is fed into the map to help advise on irrigation, growth rates, and disease intervention. Because it can be done remotely, people don't need to go out into the field. Companies are pushing this technology further and have developed hyperspectral cameras, which cover more bands and have higher resolution. One such company is IMEC.[16] Their website proudly states, "Hyperspectral snapshot cameras enable real-time inspection of crops, paving the way to a new level of precision agriculture."[17] The technology produces images that reveal spectral information about every pixel of the image, allowing farmers to inspect their crops remotely and in more depth than the human eye can see. IMEC's camera allows the user to look at the image as a set of image layers, where each layer corresponds to a different wavelength of the electromagnetic spectrum.[18] While many hyperspectral cameras are bulky and can only take photos, IMEC's can take videos using a striped spectral filter pattern that allows for more high-resolution hyperspectral imaging. Similarly, Cubert, a German company, uses a technique they call "pansharpening" in their agricultural drones and fuses images together to enhance them. The company claims that "this enables analysis at the leaf level of vegetation even from a UAV, helping to detect diseases and other issues that require action at an early stage."[19]

Though the plethora of benefits that drones offer in the agricultural sector is clear, not everyone is convinced. There are still regulatory barriers that farmers face when attempting to use drones. In the United States, this is particularly onerous. Since 2016, there have been regulations in place to limit drone usage.

In order to operate drones commercially, farmers must obtain a FAA Part 107 certification, which requires a training course that can take anywhere from twenty to eighty hours, and generally costs $150 to complete.[20] Farmers also need to register their drones with the FAA using a remote ID, which "broadcasts their identification number" so that the FAA can easily identify violators, which farmers must comply with under the penalty of heavy fines which can go up to tens of thousands.[21] However, recognizing the benefit of the technology, California is trying to enable more widespread agricultural drone usage. The California Assembly passed legislation (Bill 1016) in 2023 to modernize the certification process for farmers to use drones to spray pesticides and beneficial biological treatments on their farms, directly in response to the bureaucratic red tape that farmers experienced under the FAA policy.[22]

However, the enormous regulatory hurdles faced by drone users in the United States are nothing in comparison to the challenges facing farmers in Hungary who want to use them. To operate the most basic drone that can spray fields, Hungarian farmers are required to obtain an unmanned pilot certificate that is highly technical, requires almost expert knowledge of aircraft systems, and demands four separate licenses.[23] After these qualifications, a farmer can start training as a drone pilot for a specific field within agriculture, such as crop spraying or video surveillance. If a farmer wants a drone to spray crops (one of the most basic drone functions), they must get several further licenses that would allow them to transport hazardous material and apply pesticides aerially. Further, before a flight by a drone begins, the airspace above a field must be reserved under a specific category. All of this becomes more complicated if the farm is near a residential area, as this also requires more permissions from local authorities. For insurance purposes and in case the government needs to investigate misconduct, logbooks and other documents must be kept to show all data from each flight.

On the other side of the coin, Japan has embraced agricultural drones. The Japanese government has recognized that its aging and decreasing population of farmers needs the help of technological innovation. Since 2019, Chinese technology company DJI (Da-Jiang Innovations), which specializes in drones, has been working with the Japanese government to update the Japanese regulations.[24] Previously, Japanese law made drone registration extremely onerous and prohibited the use of cameras on agricultural drones. Updates in legislation allowed Japanese farmers to use drones to spray pesticides and monitor crops with video cameras. Moreover, it allows them to use drones at night, and by using special cameras, farmers can see damage caused by nocturnal animals such as boars and deer. This has not only made farming activities easier, but it has also attracted young people to farming since it appears more modern and exciting. According to 29-year-old farmer, Shota Chiba, "People still have a strong stereotypical image of farming as a dirty and hard-labor job, but it's no longer true thanks to gradual mechanization. New technology like diagnostic drones could help change this old image and attract more young people to farming."[25]

Increasingly, farm maps overlay publicly accessible maps (such as a map of the local water authority's hotspots) to their own data to create a more well-rounded image of the land. The algorithms pull data in real time to adjust constantly and advise on best practices. For example, they can use meteorological data from NOAA to improve predictions of rainfall and warn of extreme weather events, allowing farmers to gain greater insights. In the future, as more and more data sets become available, smart farming systems could also allow farmers to monitor the microclimate of their farm and immediate area. This could allow for personalized crop insurance options, allowing greater control over premiums, since they would have lower risk. This could allow farmers who might otherwise be exempt from coverage to gain access to financial protection. Such is the case with crop insurance programs in India, where farmers are particularly vulnerable to severe income shocks caused by crop failure. While there have been crop insurance programs across India for many years to help support the livelihoods of the over 100 million Indian farmers, technology can create more tailored offerings. One example is the Kshema App, which launched in 2018. A fully digital insurance company, Kshema uses machine learning to assess risk and offer tailored protection programs to Indian farmers. Farmers can choose between protection plans that cover specific issues such as droughts, earthquakes, or flooding based on the issues their region faces, offering a shield against income shocks. The precise data collection also helps farmers make insurance claims, as they can gather evidence of exactly what parts of their fields were damaged.

The data from the drone cameras and the in-field sensors are fed into the farm map. Once sufficient data points have been compiled to create a detailed map, the algorithms can guide the farmer's decision-making. This technology can make precision agriculture sharper by telling farmers exactly where and when to use agrochemicals for maximum effect with minimal waste. It can also guide bigger decisions, such as selecting more suitable crop varieties or identifying unproductive areas that might be better restored through reforestation. And it is easy to see how it might further progress with AI. The AI agriculture market is projected to grow from $1.7 billion USD in 2023 to $4.7 billion USD by 2028, highlighting the pivotal role of advanced technologies in this sector.[26] As AI advocate Ganes Kesari wrote in Forbes Magazine, "AI could become a master gardener, perpetually monitoring and fine-tuning every growth stage in the farm, from seed selection to harvest and beyond. It can help adjust farming practices in real time to climatic shifts, ensuring optimal crop health and yield."[27] An example of the potential for AI in farming includes the startup Carbon Robotics. It leverages deep learning algorithms to identify weeds using data from high-resolution cameras that scan the fields in real time and then uses robotics and lasers to deliver high-precision weed control. It can eliminate up to 5,000 weeds per minute at 99 percent accuracy and reports reducing weed control costs by up to 80 percent with a potential return on investment in one to three years—and all without the farmer lifting a finger![28]

Case Study T: Under Glass—Greenhouses Gone Smart

Many farmers in Global North countries are just starting to deal with climate change, but in Kenya and other parts of sub-Saharan Africa, its impact is already severe. Over the past three decades, agriculture has been central to Kenya's economy, contributing 28 percent of GDP and 65 percent of total export earnings.[29] Poverty is more prevalent in rural than in urban areas, and more than 80 percent of Kenya's rural workers are employed in the agricultural sector, accounting for around 18 percent of total formal employment.[30] In response to shifting weather patterns and high rural poverty levels, what mechanisms can Kenya and other sub-Saharan countries implement to protect farmer livelihoods and ensure food security?

One solution is the adoption of technology. Synnefa, a Kenyan startup operating for over a decade, uses hardware and software to help smallholder farmers grow more while using fewer resources and reducing climate-related risks. Its flagship product is smart greenhouses—protective structures that shield crops from pests and heavy rain while enabling internal climate control, including optimal humidity. Unlike traditional models, Synnefa leverages technology to create ideal growing conditions, minimizing losses and boosting yields. For instance, smart sensors paired with automatic drip irrigation systems water crops only when needed. These sensors connect to an all-in-one app, allowing farmers to remotely monitor irrigation and manage other farm inputs and outputs.

Synnefa's mission is to provide modern farming solutions to one million smallholder farmers across Africa. By mid-July 2025, Synnefa says that over 21,000 farmers are using its products; however, upfront costs remain high. The smallest metallic greenhouse (20 sq.m.) with a 500L water tank and drip kit that they sell costs 109,000 KSh. In comparison, in 2022, the average monthly national expenditure per adult in Kenya was approximately KSh 8,030, with food accounting for over 59 percent of total consumption.[31] In rural areas, the proportion of total consumption spent on food was even higher than the national average, exceeding 65 percent.[32] To make greenhouses more accessible, Synnefa announced flexible payment options. While Synnefa aims to lower barriers for vulnerable farmers, the long-term survival of this impact-driven business model depends on its financial sustainability. Questions remain as to whether Synnefa can make smart greenhouses truly affordable for African farmers.

New Formulations in Agrochemicals

Just as the application of agrochemicals is becoming more precise, so are the chemicals themselves. New technology, developed by a lab at UT Austin in 2023, could help to sequester nitrate from runoff and turn it into ammonia, allowing it to be recycled back into fertilizer. The UT technology uses a copper-based hydrogel to capture excess nitrate waste from fertilizer runoff and transform it into ammonia that can be reused. In tests, the system equals or increases crop yields over traditional methods while also minimizing environmental impacts. This would solve a major problem in agriculture by turning harmful waste into helpful fertilizer, reducing costs, and increasing sustainability.[33] Similarly, AgZen, an MIT lab focused on droplet behavior, has developed a new formulation to minimize pesticide loss by getting droplets to stick on leaves, minimizing runoff, and maximizing the effectiveness of the pesticides. Cofounder of AgZen, Jayaprakash says, "Across the board, we were able to save between 30 and 50% on chemical costs and increase crop yields by enabling better pest control," adding, "By focusing on the droplet-leaf interface, our product can help any foliage spray throughout the year, whereas most technological advancements in this space recently have been focused on reducing herbicide use alone."[34]

We discussed in Chapter 9 the push to make a gene-edited corn that can "fix" nitrogen to reduce the emissions associated with producing synthetic nitrogen. Another innovation that could help to reduce the footprint of fertilizer is green hydrogen. Green hydrogen production splits water through electrolysis (using electricity from renewable energy sources) and results in only two outputs: hydrogen and oxygen. The hydrogen can be stored safely under pressure in fuel cells, and the oxygen can be vented to the atmosphere with no negative impact. The green hydrogen can then be fed into the traditional Haber-Bosch process to make ammonia. Since this green ammonia is 100 percent renewable and largely carbon-free, switching to green ammonia would greatly reduce the carbon footprint of nitrogen fertilizer production. That said, as discussed in Chapter 1, ammonia from fertilizers is a major cause of air pollution.[35] Even if it is possible to produce a "greener" form of synthetic fertilizer at scale, green ammonia-based synthetic fertilizers will still cause environmental problems associated with current nitrogen fertilizer use, including fertilizer run-off into streams, which creates dead-zone in rivers, lakes, and even some oceans through eutrophication.[36] So, perhaps the gene-edited solution is preferable?

Robotics

Data-driven decision-making and integrated mapping tools aren't the only ways in which technological innovation is transforming farming. Robots are increasingly key to our farming future. As discussed in Chapter 10, there is an ongoing labor crisis for farming with access to migrant labor becoming more and more challenging, particularly against a backdrop of more restrictive immigration policies and with the challenges that COVID-19 presented. With farmers struggling to access the workers that they need, it is logical to turn to technology to solve the problem. Robotics presents a huge opportunity for farms to reduce their dependence on human labor. Robotics has already had an impact on arable production through autonomous sowing and harrowing for cereal crops, such as robots that automate the sowing of seeds, ensuring optimal depth and spacing. In livestock, robots are also changing traditional practices, such as milking robots that allow cows to choose when they need to be milked.[37] Giving cows the freedom to decide when to go for milking improves animal welfare levels. It reduces the cow's stress level since she isn't being poked and prodded when herded into the parlor with all the others. Lower stress levels result in high milk production.[38] They also reduce the chance of mastitis (and associated antibiotic usage) since she can decide when her udders are full and isn't left feeling uncomfortable waiting for the next milking.

However, the biggest opportunity for robots is in horticulture, which is hugely reliant on migrant labor. As we discussed previously, many horticultural enterprises have been struggling to find the people that they need to manage their operations and have also been weighed down by rising wages with manual labor costs making up 50 percent of horticultural enterprise's expenses.[39] Consequently, there is significant scope for robotics to transform the horticultural industry in the coming years. Potentially, the ability of robots to remove the need for the more arduous and unpleasant side of farm work might actually make the agricultural sector more attractive to skilled workers and graduates. Already, robots are performing mundane heavy labor tasks—grafting seedlings and moving plants from one spot to another. The most promising innovations are for fruit picking, where robots could be designed to identify and pick ripe fruit, such as strawberries or apples, reducing the need for seasonal labor. For example, Advanced Farm, an American robotics company, is working to develop robots capable of recognizing and picking apples. Over four meters tall, their robots are equipped with cameras that identify ripe apples and six mechanical arms that pluck apples from the branch through suction cups. Washingtonian orchardist Kent Karstetter voiced his optimism for the future of such technology to resolve current labor shortages, hoping that he can manage to "figure out how we can keep enough employees to get our crops picked until this technology evolves."[40] Thus far, programmers have not been able to give robots the softness required to pick stone or soft fruits without bruising or damaging them but have achieved success with harder tree crops, such as olives. Nor have they achieved the speed required. "There's no robot in the world right now that can harvest or thin peaches as well as people can, said Ai-Ping Hu, an engineer at Georgia Tech Research Institute working on horticultural robots. "The technology's not quite there yet."[41] With time, they will surely develop the softness required and pick up the pace.

Another opportunity for horticultural robotics is weed management. On conventional farms, robots can be deployed to apply herbicides with pinpoint accuracy, targeting only the weeds rather than blanket-spraying entire fields, which reduces chemical use, lowers costs for farmers, and minimizes environmental impact. For agroecological farmers robots can help with mechanical weeding. Weeding is one of the most labor-intensive aspects of organic horticulture and is backbreaking work with laborers going through the fields from row to row to remove weeds. New technologies can help ease the burden by using robots to do the hard work. Launched in 2022, the LaserWeeder from Carbon Robotics identifies weeds in the fields using sophisticated machine vision systems to distinguish weeds from crops and then eliminates them with lasers, burning them off.[42] Mechanical weeders, such as Odd Bot, work differently and identify weeds before plucking the unwanted plants from the soil.[43]

While much of agricultural robotics has been designed with large-scale, commercial producers in mind, these innovations also hold potential for agroecological and organic farmers, helping them boost productivity while staying true to their philosophy of working in harmony with nature. Being organic does not mean rejecting technology; it simply means avoiding reliance on synthetic agrochemicals to drive yields. Robots that can identify and mechanically remove weeds, monitor soil health in real time, or harvest delicate crops with minimal damage offer exciting possibilities for organic systems. In the Global North especially, such tools could help close the persistent productivity gap between organic and conventional farming, enabling organic farmers to remain competitive while still protecting biodiversity, soil health, and ecosystem resilience. By adapting cutting-edge technologies to serve ecological goals, robotics can become an ally in reshaping organic farming for a more sustainable future. As the Soil Association, the UK's leading organic certifier, said in their report on technology in agroecological production,

> Small robots have the potential to operate in more heterogeneous systems, suiting them to agroecological practices such as strip agriculture and agroforestry. The mechanization and automation of more complex systems may make them more viable at scale and more competitive with conventional monocultures. Robotics could unlock highly diversified mid to large scale farms that have previously been unviable due to their incompatibility with conventional technology.[44]

Similarly, robotics and the other technologies discussed in this chapter have the potential to transform farming in the Global South. To date, the majority of agritech innovation has originated in the Global North, with technologies largely designed for commercial producers in those same regions. This concentration reflects not only the geographic location of most investors and developers but also the greater profitability of targeting Global North farmers, who generally have higher purchasing power and more capital to invest in new tools. As with many other technological sectors, Silicon Valley and similar innovation hubs play an outsized role in shaping the direction of agricultural technology, reinforcing a Global North-centric model of development. However, just as with other technologies, agritech will eventually go global as the cost slowly gets cheaper. That is an exciting premise since the power of these technologies to transform farming will be even more beneficial to the farmers in the Global

South. Those farmers have the tightest margins and face the greatest climate challenges. Technologies that can help them increase their resilience and adapt to face an uncertain future will allow them to survive and hopefully thrive. Adapting these new innovations for smaller plots and a wider geography while lowering the price will make them accessible to more farmers and, therefore, transform the future of smallscale growing—whether in Northern California or South Africa. The Soil Association's report sums this up neatly by saying, "Access to sustainability enhancing technology, and particularly data gathering solutions, should be maximized across all scales of agriculture. Regulation should be restructured, and incentives put in place to encourage companies to develop innovative modular, flexible, interoperable technologies accessible through rental and service-based schemes. This will increase access to cutting edge agricultural technologies."[45]

Nanotechnology

Nanotechnology is another field with promising applications within the agricultural process. Nanotechnology utilizes particles of nanometer size (one billionth of a meter), and their small size and high surface area give them desirable properties for various uses. For example, nanofertilizers have proven to be quite effective. Nanoparticles of iron oxide (Fe_3O_4) can be applied as a fertilizer to improve plant growth, increase overall yield, and enhance the quality of the plants.[46] In goldenrod, which is used for medicinal purposes, application of iron oxide nanofertilizers to the leaves was found to increase the amount of flavonoids and essential oils of the plant, which are the compounds that confer health benefits, including as antioxidants, anti-virals, and anti-inflammatory agents.[47] Iron in its natural form as an iron ion (Fe^{3+}) is insoluble and plants cannot absorb it, but the nano form enhances uptake of the fertilizer into the plants.[48] This is, in part, because their incredibly small size allows for better penetration into the plant. In addition, their small size gives them a higher surface area, which increases the rate of reaction since they are more exposed.[49] Nanotechnology can also be utilized in the form of nanopesticides. Similar to nanofertilizers, nanopesticides work more efficiently to kill insects because of their high surface area. This means that much less pesticide would need to be used to achieve the same effect. In addition, nanopesticides can be encapsulated in nanocarriers that allow for a controlled release of the pesticide under certain environmental stimuli. For instance, some nanocapsules can respond to heat, humidity, or light. In addition, nanopesticides are more stable than conventional forms, reducing the need for constant reapplication.[50] Another use for nano is as biosensors since nanoparticles can change their properties based on atmospheric conditions, such as in gas composition.[51] However, there is a significant environmental health risk associated with these agrichemicals that use types of nanotechnology since they can make their way into the natural environment. Once in the natural ecosystem, they enter the organisms living there. Just as in humans, when nanoparticles enter wildlife they cause cell damage and interfere with the gut microbiome and immune system. These nanoparticles can also leach into waterways, not only impacting aquatic wildlife but also humans who may use and drink that water, making it a challenge for water companies to tackle.[52] A new regulatory framework will need to be created to assess these risks before the use of nanotech in agrochemicals can be approved.

Technology outside the Field

But it is not just in the field that technology is transforming farming. Previously farmers were locked into contracts with supermarkets and had little opportunity to directly communicate with their customers. However, technology is changing this. Social media has allowed farmers to speak directly to the public. Joshua Westerfeld broadcasts snippets from his daily life on his Texas ranch to an audience of over 28 million people on TikTok. Many consumers think that farming is red barns, hay bales, and pitchforks—social media allows farmers to display what it really looks like and help the public gain a more honest understanding. It can also create a sense of community for an individual who often finds themself 8,000 acres away from their nearest neighbor. "It's a support system … I've seen times where farmers will post a video with a problem, like some piece of equipment that they're not sure how to fix, asking for help. And you'll see the comment section just filled."[53] This allows for peer-to-peer learning and informal knowledge exchange. Recognizing the benefit of this type of informal knowledge exchange between farmers across huge distances, online platforms have sprung up to try to commercialize that interaction—such as the French Terre-Net. This online platform connects farmers and allows them to communicate directly and share advice and guidance.

And once the harvest is done, technology can continue to connect the farmer directly with his customers. New websites have created an opportunity for direct sales and act as digital food hubs. Websites like Crowd Farming or Wylde allow farmers to sell directly to consumers who live thousands of miles away—in essence, a digital farmers' market. As these marketplaces grow, there is more focus on the need to build integrity into the system. Blockchain is increasingly playing a role in underpinning the provenance of products and ensuring that claims made by producers can be trusted. Given such long supply chains and reliance on humans who are prone to error, it is difficult to have confidence in the origin of a product from so far away. Can you feel confident that your pineapple is organic? Are you sure your sourdough is genuine? With such distances and lack of oversight, there is a risk of corruption and false claims. Blockchain, an unalterable, transparent digital ledger, is increasingly being used as a way to ensure that the consumers can trust in what they are buying, particularly claims over sustainability. The retailer, and even perhaps the consumer, would be able to know exactly how the fruit had been produced and its journey. It is therefore logical to see the argument that blockchain could help to build a stronger sense of provenance since the data would be unimpeachable. Palm oil certification is a prime example of this. Today, palm oil is in nearly everything—almost 50 percent of the packaged products we find in supermarkets contain the oil due to its beneficial properties.[54] However, our dependence on palm oil has led to significant environmental damage, particularly in Southeast Asia, and has resulted in huge swaths of rainforest in the region being cut down. In Southeast Asia, 45 percent of palm oil plantations are in areas that were forests in 1989, and every year it is estimated that between 1,000 and 5,000 orangutans are killed in palm oil concessions.[55] A certification scheme for sustainable palm oil was created in 2004. In the past, there were questions over the legitimacy of the certification scheme, but blockchain can help to ensure its validity by guaranteeing that the data cannot be corrupted once entered.[56]

What's the Price?

While few can doubt the transformative benefit of these technologies, many are out of reach to the farmers who might most benefit from them. The price of these technologies is incredibly high—especially given that farmers are surviving on tight margins. For example, a single Digitanimal GPS collar for a heifer is $769 USD, while a milking robot can cost $200,000 USD and can be used by fifty cows.[57] That's a pretty huge investment for a small dairy farmer with 200 cows. On the arable side, the leading autonomous electric tractor from Monarch costs $75,000–100,000 USD—a significant price since a second-hand tractor can cost as little as $1,000 USD. The company has been lobbying for the US government to create a trade-in scheme to guarantee an 80 percent discount if they turn in an old diesel tractor.[58] This is an issue for farmers and producers in the Global North, but the issue is even more acute for subsistence farmers in the Global South. Many might be eager to use technologies to improve productivity but are priced out of progress. "That's the issue," said Abdulai Jalloh, Director of Research and Innovation at the West and Central African Council of Agricultural Research and Development, "Because farmers don't have resources, they cannot buy inputs, so they can't capitalize on the innovations that are available to them."[59]

Furthermore, a lot of these technologies are proprietary and owned by large transnational agrobusinesses, such as John Deere and Monsanto. When a farmer buys these new innovative systems, the farmer is increasing their reliance on that individual company and creating more dependency of their production on the products and systems that the agribusiness sells. When this happens across the world, that means that more and more of our food production is reliant on the products sold by a few specific companies, which increases the corporate control of our food system. It also makes it challenging to convince agroecological growers who have a philosophical objection to the power of transnational agribusinesses to adopt these technologies, since they have a distrust of expanding the power of these businesses further—even if they would improve their sustainability. Perhaps they are right to fear the further expansion? Recognizing that technology will be the future of farming, these companies are competing to buy up the smaller tech companies that have brought these products to market in order to maintain their market dominance. For example, in 2023, John Deere purchased Smart Apply, Inc., which had developed an intelligent spray control system that can improve the precision and performance of virtually any air-blast sprayer.[60] In 2021, the company bought Bear Flag Robotics, a startup based out of Silicon Valley that had developed autonomous driving technology compatible with existing machines.[61] And, just as there is a growing conversation around privacy concerns online, so is there a concern about the privacy of farmers' data. Farm mapping technologies gather and store huge quantities of data about a farm, its crops, its assets, and so on. Some farmers are concerned about the information being sold or used against them. For example, if you buy your pesticide from the same company that owns the system that holds all your farm data and knows that you're currently facing an infestation of cotton bollworm, you might worry that you would get gouged for the treatment. As Robert Saik, a leading advocate of technology in agriculture, said, "Agriculture has traditionally been a siloed sector where a farmer is reluctant to share information with a contractor because there might be a loss of bargaining position."[62] This lack of willingness to share their data openly will be a limiting factor on the power of smart farming technology since the more data sets are shared and publicly available, the better the algorithms will become.

Connectivity and the Digital Gap

Another challenge to the adoption of these new technologies is the age of farmers in the Global North. In 2022, the average age of US farmers was 58.1 years old, while in the UK, 35 percent of farmers were over 65.[63] Some might have embraced technology, but most remain hesitant (or at least timorous) with tech and are happy to continue doing things their own way. It can be hard to convince them to start using mapping tools, for example, let alone to maximize their potential. To truly achieve a digital revolution in farming, it might be that these technologies will need to wait to be adopted and implemented by a younger generation. However, there are fewer young people coming into the industry, as discussed in Chapter 4. In the UK, under thirty-fives make up less than 3 percent of all farmers, and in the United States it is only slightly better with 9 percent.[64]

Moreover, all the benefits of these smart technologies are dependent on internet access. While the benefits of these innovations are often said to have the power to transform the fortunes of the farmers in the Global South, many of them still do not have access to the internet, let alone broadband. However, despite the advances made by some countries, only 35 percent of the populations of developing nations have access to broadband as opposed to 80 percent in developed countries.[65] As one might imagine, while cities are better connected, the rural areas where the majority of farmers work have lower rates of connectivity. The World Bank has shown that a 10 percent increase in broadband access causes a 1.4 percent increase in GDP in developing countries.[66] That said, there is a significant amount of investment into this issue, since broadband connectivity can be a pathway out of poverty. India, which has between 90 and 150 million farmers, increased its public subsidies for broadband access across the country starting in 2013, which has caused users to increase from 200 million to 900 million. However, in much of the Global South, China dominates foreign investment into broadband access.[67] While the investment is very beneficial to the developing countries, the loans can be very difficult to repay, causing a power imbalance. Further, when China finances and builds broadband infrastructure in developing nations, it raises risks to data privacy and sovereignty since companies like Huawei are legally bound to cooperate with Chinese intelligence, creating long-term dependence, potential surveillance, and geopolitical leverage over information flows. South Africa, on the other hand, has relied entirely on the private sector to build their broadband networks. They have a competitive open access model, which allows broadband companies to sell affordable subscriptions (at $20 USD per month) without government subsidies.[68]

While one might think that the lack of internet access is a problem faced solely by the Global South, Global North nations are dramatically behind in delivering high-speed internet access to rural areas. Although broadband greatly improved throughout the 2010s and 2020s in America, 19 million Americans still do not have access to fast, reliable internet.[69]This issue is concentrated in rural America; one in four rural Americans still report that reliable internet access is a major problem in their community.[70] Because houses are so spaced out in rural areas, the cost to build the infrastructure is too great, and major companies often do not see much return on their investment in those areas. The $42.45 billion Broadband Equity, Access, and Deployment Program, created in 2021 under President Biden's Infrastructure Investment and Jobs Act, aimed at expanding broadband so that more rural communities can get reliable access.[71] Poverty is rampant throughout rural areas, which, combined with the lack of people, causes fewer people to subscribe to broadband services in these underserved areas. When faced with costs like medical bills, car payments, or rising food costs, a household may choose to drop their subscription to broadband. Due to a disproportionate number of Black Americans living in affordable rural rental housing, which is the least likely to receive broadband services, Black rural Americans are the least likely population to have broadband access.[72] Efforts to promote access to broadband and other farming innovations must take into account both underserved countries and underserved groups.

Conclusion

Due to innovation and the digital revolution, there is incredible tech that is now available to farmers and food producers. This technology allows for more sustainable production that is informed by data and will also make it easier for farmers. However, this technology is only as good as its implementation. Currently, this tech is cost-prohibitive for the majority of farmers, especially those in the Global South who might best benefit from it. Secondly, it is dependent on rural internet access that is spotty at best, and uptake from farmers who are predominantly elderly and less tech-savvy than younger generations. Until these three problems are overcome, it will be difficult for the digital revolution to revolutionize farming, and moreover, the impact of failure could be devastating for a farmer who operates on tight margins. As Robert Saik sensibly put it, "Farmers don't get to test and iterate like the tech entrepreneurs releasing new products every eight months or eight days. Farmers have to wait months for a crop"; so the tech needs to work effectively from Day One, and if it's going to be adopted, there needs to be strong security.[73] In an increasingly unstable world, there is a risk that food supply could be weaponized, and a cybersecurity attack could take aim at the technology that underpins and automates production. A successful strike on these systems could paralyze supply chains, disrupt harvests, or even contaminate data that farmers rely on to make daily decisions. As we move deeper into an era of agrotechnology, building resilience and safeguards into our food systems is not optional—it is essential.

11

Combating Food Waste with Technology

When considering how to meet the future demands for food against growing populations and climate change, reducing food waste is a necessary first step. There's enough food to feed every person on the planet, but currently one-third of all the food produced for humans never actually gets eaten.[1] When thinking about food waste, the first image that comes to mind is probably scraping the remnants of dinner off a dirty plate, but there is loss at every point along the food chain from the field to the fork. Recognizing the need to reduce wastage and loss at every single stage of the food system is critical if we want to achieve a sustainable and equitable future. New innovations can help tackle the issue of food waste. These innovations include revolutionizing the supply chains to maximize and improve storage and transportation and advancing in-home tech to help extend shelf life.

What Is the Problem?

Shockingly, between 30 percent and 40 percent of the annual food supply is wasted or lost in the United States, and a family of four spends $1,500 USD each year on food that ends up in the landfill.[2] And it's getting worse—Americans waste 50 percent more food today than in the 1970s. But America is not the only culprit. Globally, roughly 1.3 billion tons of food are lost and wasted per year, and it's worth around $1 trillion USD.[3] Beyond the food wasted that could be used to feed communities, there is an environmental impact of food waste. According to the NRDC, the food wasted in the United States accounts for 21 percent of the agricultural water usage and 18 percent of US fertilizers.[4] There is a carbon impact too. Piled high in a landfill, food doesn't break down as you would normally imagine. Given the mountains of waste, oxygen is limited, and the environment becomes anaerobic. As food rots, broken down by bacteria, it emits methane. As we discussed in Chapter 1, methane is a potent greenhouse gas. While we attribute most of the methane produced by the food system as the by-product of livestock production, food waste is a significant emitter too. According to the WWF, about 6–8 percent of all human-caused greenhouse gas emissions could be reduced if we stop wasting food.[5] In the United States alone, producing food that is ultimately lost or wasted food generates the equivalent of 32.6 million cars' worth of greenhouse gas emissions.[6]

Yet, addressing food waste is more complicated than merely saving scraps from the bin. When thinking about food waste, there is a key distinction that needs acknowledging. As defined by the FAO, "food waste" refers to the discarding of edible foods at the retail and consumer levels, mostly in developed countries, while "food loss" refers to the decrease in edible food at the production, post-harvest, and processing stages of the food chain, mostly in developing countries.[7] Diving deeper, the World Resources Institute (WRI) defines food loss as food that has become unfit for consumption before it reaches the consumer and food waste as the discarding of food that was fit for consumption. WRI starts measuring food loss and waste at the moment that crops are ripe, animals are slaughtered, milk has been taken from the udder, or wild fish have been caught. According to the World Food Programme, it's estimated that close to a third of food in high-income countries is "wasted," while roughly 40 percent of food in low-income countries is "lost."[8] The two different concepts have different causes, and therefore different solutions.

Food Loss on Farms in the Global South

In the Global South, food is lost on farms because of the challenges that farmers face with growing food. Unlike in the Global North, much of this wastage is unintentional. Shockingly, many of the hungriest people in the world are actually smallholder subsistence farmers, who should have the capability to grow their own food. However, living in poor, rural areas of the Global South, they often lack the right infrastructure and resources to succeed. Much of their harvest is lost, thereby perpetuating their hunger. Farmers face challenges from climate change that can easily wipe out an entire harvest or damage some of the fields. Oftentimes, these weather events are accompanied by disease and pests, exacerbating the problem as food can be destroyed by infestations or outbreaks. Supporting efforts to reduce on-farm loss will help farmers improve yield rates and meet new climate realities. Farmers in the Global South also struggle to obtain capital investment and therefore are not able to afford the best equipment. Without access to new technologies, harvesting and storage are often inefficient, failing to maximize the potential.

Improving Global South access to agritech (such as those mentioned in Chapter 10) will help to reduce this yield gap by improving efficiencies. Even if farmers are able to harvest the produce, there is a significant loss due to inadequate storage. This means that even good and healthy crops can be lost to mold, vermin, bug infestations, and the like. According to the World Food Programme, sub-Saharan Africa loses the same amount of food post-harvest as it receives in food assistance.[9] Improving storage and transport could therefore have a significant impact on reducing dependence on aid programs. There are some incredible low-cost innovations in packaging that could transform farmer livelihoods in the Global South and reduce hunger by reducing spoilage. One of the best examples of this is the cowpea intervention in the 1980s by the USAID. At the time, on-farm storage of cowpeas (a common staple) was so poor that six months after harvest about 70 percent of the seeds were damaged and virtually unfit for consumption.[10] The project worked to educate farmers on the value of storing cowpeas in airtight bags after harvest to stop insect infestation. This type of simple, low-cost, and environmentally friendly technology for managing post-harvest pests needs to be replicated. A similar intervention happened in Bangladesh with great impact. Farmers traditionally transported their produce to market in plastic mesh bags. However, there was a significant level of spoilage since it was easy for the fruit and vegetables to bruise and get damaged as they bashed together. Switching to more structured plastic crates reduced waste by an average of 50 percent for each trip, meaning that the farmer was able to sell more produce and better support their family.[11] Supporting the rollout and adoption of more modern packaging can help reduce food loss. The World Food Programme is committed to low-cost interventions, like donating airtight storage containers (at the cost of $3.27 USD per family) that dramatically reduce the risk of infestations or mold, cutting food loss to less than 2 percent.[12] While food loss is a major concern currently affecting many farmers in the Global South, the adoption of a variety of new technologies can and is addressing this challenge.

Case Study U: Saving the Harvest—the Fight for Equality in Senegal

Women around the world play crucial roles in agricultural value chains, managing both sales and household consumption. Yet, they often face significant barriers to accessing storage facilities and formal financial services like loans. Senegal is actively working to address these challenges. While both male and female agro-producers in Senegal share some obstacles—such as lacking formal business registration needed to qualify for loans that require proof of income or collateral—the gender gap remains strikingly wide. A total of 14 percent of men have access to formal loans, while only 4 percent of women do.[13] Additionally, women's land ownership is significantly restricted, with only 28 percent of women owning land compared to 65 percent of men.[14]

These disparities are compounded by broader structural issues. Due to limited financial access, agro-producers often rely on immediate sales after harvest to secure income for daily necessities. However, during harvest time, oversupply typically drives prices to their lowest. This issue is further exacerbated by the lack of reliable storage facilities, which prevents producers from storing their crops until prices rise. As a result, potential financial gains are lost, and unsold crops spoil, contributing to food loss and waste.

To address the dual challenges of inadequate storage infrastructure and limited financing options for vulnerable populations, the AgResults Senegal Crop Storage Finance Challenge Project was created. This initiative was a partnership among the Australian Government, the Bill & Melinda Gates Foundation, the Government of Canada, UK Aid, USAID, and the World Bank Group.[15] Launched a few years ago, AgResults was designed to encourage warehouse owners and smallholder producer groups to upgrade their existing warehouses and establish collateral-based storage finance schemes.

Furthermore, recognizing that women bear the brunt of these systemic gaps, the initiative, with a USD 6.8 million prize fund incentivized women's producer organizations, among others, to adopt Warehouse Receipt Systems (WRS), allowing them to store crops in certified warehouses.[16] This system allowed women to use their stored crops as collateral for loans, giving them the flexibility to delay sales until market prices became more favorable. The initiative aimed to enhance economic stability, reduce post-harvest losses, and support women's financial independence.

"The advantage of WRS is that we have access to bank financing under the production guarantee," said Bineta Gueye, a member of a thirty-woman producer group in Senegal.[17] "It means we can wait for the market to evolve positively, increasing our sales … This helps us take better care of our children's food and education."[18]

By linking storage to finance, the project helped women farmers gain greater economic control and reduce post-harvest losses. Despite positive feedback from women, AgResults was discontinued in 2024. However, the model offers a compelling blueprint: when women are equipped with tools to overcome structural barriers, entire food systems become more equitable and resilient.

Food Loss of Farms in the Global North

In the Global North, food loss is caused by different factors. Crops go unharvested because the price might not cover the costs of bringing the product to market. There are many instances where farmers in the Global North might choose to plow crops back into the soil because it doesn't make financial sense to harvest them. And sometimes even if they want to harvest them, they aren't able to. As discussed in Chapter 4, labor shortages have become a common issue for farmers in the Global North. When they don't have the people to harvest the crop, they often have no option but to plow it back into the ground—potentially at the loss of thousands of dollars. With crops, farmers at least know that the lost vegetables are helping to restore soil fertility for the next season. When the dairy sector faced labor shortages and supply chain disruptions due to COVID-19, they had to dump gallons and gallons of milk down the drain since they didn't have the manpower to process it.[19]

In the Global North, most on-farm food loss comes from horticultural crops. The World Resources Institute has calculated that 24 percent of food calories produced for human consumption are lost or wasted. That is a huge amount of food that could be used but isn't. For fruit and vegetables, it is even higher. Fresh produce makes up 44 percent of food waste by weight.[20] In order to avoid produce being damaged, farmers are increasingly using the technologies discussed in Chapter 10 and carefully selecting seed varieties (from Chapter 8) that have the best chance of success. Beyond the future of plant breeding that we have already considered, some plant breeders are considering how they might be able to use their techniques to help reduce food waste. If you think of the shape of many fruits and vegetables, they aren't exactly well designed for transportation. In a shipping container, you want items to fit neatly together, like Tetris, to maximize the number that you can fit into the box, but also to better protect them from bumping against each other. The odd spherical shape of tomatoes, for instance, doesn't make life easy. They take up a lot of empty space and can easily be damaged in transport. Tomatoes and watermelons have been successfully grown in molds to make the square and easier to pack into boxes without a load of empty air around them, ensuring that they would be better protected. Interestingly, the ability to protect itself from bruising and other damage is why the Hass became the most popular variety of avocado. Unlike the softer green-skinned varieties that were popular in the past, the thick exocarp of the Hass reduces the risk of damage. The hard shell provided by its knobby dark skin provides a hard shell, protecting it in transit from damage and providing a barrier to bugs. This dominance of the Hass has helped reduce the level of avocado spoilage.

Challenging superficial standards and redirecting food to consumers who are willing to tolerate a less perfect apple is a quick way to reduce food loss. Almost a third of what's grown in the United States never gets harvested because it doesn't meet the aesthetic standards of the grocery store.[21] Products are often discounted because of issues with their ripeness, size, color, weight, and blemishes. A recent study in Minnesota found that up to 20 percent of fruit and vegetable produce is typically too large, too small, or otherwise too cosmetically compromised to meet prevailing commercial procurement standards.[22] A California peach grower once explained: "In the middle of the season, I have about 200,000 pounds each week of stone fruit I can't sell. For 8 of 10 of them, if you looked at them, you couldn't tell me what's wrong with them."[23] To help tackle this problem, certain states in the United States offer tax credits for donations of farm surplus, which helps reduce hunger in the community and alleviates pressure on landfills. Colombia does something similar with national tax incentives for individuals and corporate taxpayers who donate safe surplus food to food banks or registered nonprofit organizations, making donation a more affordable option.

New companies are entering the space and launching schemes that will deliver a box of wonky fruit or veg to your door. In the typical way that a CSA box scheme works, each week an assortment of produce is delivered to your home. The produce has been deemed not to have met the aesthetic standards to sell in conventional retail, or the supplier just has too much of it. These schemes have grown in success around the globe. However, there has been some pushback, especially in Northern California. While customers thought that they were choosing a more sustainable option by selecting a veg box of rescued food that would otherwise be wasted, the reality was that the produce came from large-scale industrial producers in the region whose production model is commonly associated with environmental and social degradation, as discussed in Chapter 1. Those producers had the highest levels of waste because (1) they produced the most and (2) because they grew food for supermarkets that had to meet highly stringent aesthetic standards. While it was good to stop their waste from being plowed back in or trashed, it caused consternation to the smaller community and organic growers. They saw subscriptions to their box schemes drop as people switched over to the wonky veg box options because of their enticing marketing campaign, while larger producers with lower environmental standards saw a new income stream. Misfits Market recognized this concern, and the Pennsylvania startup decided to only partner with organic certified producers. Starting in Philadelphia, they have now expanded nationwide. The website identifies inconsistencies in the product to inform the consumer. For example, "slightly scarred." They charge a lower price than what the perfect produce would fetch in the grocery store and also offer delivery to help provide affordable access to healthy food for all Americans. As they have grown, they have moved out of just fresh produce and work with other suppliers in the food space that have damaged or misshapen foodstuffs that are still perfectly fine to eat. Some other CSA box schemes have also gotten in on the act by offering their customers the option to buy surplus produce, in addition to the regular offering.

Reducing Food Waste in the Supply Chain

Processing in the Global South has not seen adequate investment. Many farmers are required to send primary goods across hundreds of miles before they can be processed. This raises the risk of damage. Development banks have recognized this and have started to support the development of processing plants and packing infrastructure closer to farmers in the Global South. This not only reduces food loss during transport, but it also supports economic growth in the region since exporting goods that have already been processed can bring more money into the community. An example of this is the investment in the processing of avocados in Ethiopia. Avocados have long been grown in Ethiopia for the local market but have not been a significant export since the dominant domestic variety has thin skin, bruises easily, and does not transport well. To address this, the US-owned Nutiva worked to support the development of processing plants in-country. These plants are located close to the avocado orchards to press the fruit into avocado oil before exporting it in liquid form from Ethiopia to the United States. By processing it in-country before shipping, there was a higher success rate of the product reaching the export market without damage. Furthermore, the avocado growers were able to charge a higher price for the product, improving local livelihoods. Even in the Global North, where there has long been investment and innovation in the supply chain, there is still significant food loss within the processing phase. Food manufacturers in the Global North lose about 4 percent of their product during processing.[24] Some sectors are better than others in mitigating waste. In general, the meat industry is adept at using the entire animal. As we discussed in Chapter 7, some parts of the animal are not commonly eaten in the Global North, such as chicken feet, but are eaten in our cuisines. This means that parts of the animal can be exported to other countries that consider them delicacies, while other parts are processed for pet food or become non-food products, such as cosmetics.

Once food has left the farm or processing plant, there is still a huge journey that it needs to make to reach consumers. This journey can be all the more challenging in the Global South. Trucks regularly break down because they are older and in need of repair—or their refrigeration can fail. This can leave produce sitting in the baking sun for hours, at high risk of spoilage. Even if the truck is in good condition, many of the roads in the Global South are poor, increasing the risk of accidents or blockages that can delay journeys. Imagine trying to drive milk across Nigeria, for example. It is easy to envision all the problems that could arise and cause the milk to spoil. Investment in the transport systems can help. For example, to improve the transport of fruit and vegetables from Mexico over the US border, Mexican President Calderón established the National Infrastructure Program in 2007. This created a network of high-speed, tolled roads that crisscross the nation. A major highway now runs from Michoacán to Ciudad Juarez and across the border to El Paso, Texas. Another route, known as the NAFTA Highway, connects Mexico City with Nuevo León and the US ports of Corpus Christi and Galveston. Prior to this investment in the country's infrastructure, it was three times more costly to transport food from Sinaloa to Mexico City via rail than it was to ship from Sinaloa to New Orleans on container ships. Today the total value of food crossing the US border from Mexico accounts for over $17.6 billion USD, in comparison to $3.5 billion in 2000.[25] Similar investments are being made in rails and roads across Africa. As of 2022, the African Development Bank had financed twenty-five transport corridors and had constructed over 18,000 kilometers of roads, twenty-seven border posts, and sixteen bridges for a total amount of $13.5 billion USD. These include the Kazungula Bridge connecting Botswana and Zambia; the Maputo

Corridor linking Mozambique to South Africa; and the Abidjan-Lagos Corridor connecting five West African countries. There is also a new plan to connect Angola, Zambia, and the Democratic Republic of the Congo, making it easier for food to move between the countries.

Beyond road and rail, much of our globalized society is reliant on cargo shipping across vast oceans. In total, 95 percent of all manufactured goods in the world are transported via container ships, equivalent to $4 trillion USD.[26] According to the USDA, container-based shipments account for just around 24 percent of US agricultural export volume but command over 50 percent of the export value.[27] Given these vast distances, there are significant risks to the food. Shipping containers traveling across the ocean run the risk of being exposed to extreme temperatures, and there is a constant threat that the saltwater spray might infiltrate the container, damaging the produce inside. Innovation is trying to improve the success rate. A new system called TRAXENS has launched to upgrade standard containers into smart containers, using IoT. Until now, the crew has had to inspect and provide updates to their counterparts on shore, and 80 percent of ports continue to rely on manual solutions (such as whiteboards or spreadsheets) to manage critical services.[28] If widely adopted, this new technology will automatically provide updates on the ship's location, temperature, humidity level, vibrations, impacts, access points, customs clearance status, and more. For containers holding perishable goods, the system will be able to remotely adjust the temperature of refrigerated containers as required to ensure lower spoilage and reduce food loss. The hope is that it will make transportation more effective and streamline processes.

Once docked, international borders present a particular issue for food loss. Long waits at border crossings and delays in inspection, particularly at the choke points discussed in Chapter 3, can result in spoilage. Border delays and the resulting impact on the food system have become a key part of the negotiations between the European Union and the United Kingdom after the country's exit from the bloc. Without the infrastructure, personnel, or technology in place to ensure a streamlined process, ports have routinely seen delays and blockages. As a result, the UK government has repeatedly had to waive inspections (potentially endangering food safety) in order to ensure that food doesn't spoil on the docks. However, this issue is not isolated to the UK. Shipments are also rejected on delivery, oftentimes due to delays in delivery deadlines, at ports around the world. This unwanted food is commonly dumped at a local landfill if another buyer cannot be found quickly or if the customs query cannot be quickly resolved. Recognizing this waste, Food Cowboy was launched to help truck drivers in the United States identify nearby food banks so they can donate rejected loads. The app helped find a home for almost two million pounds of food between 2013 and 2015.[29]

How Innovative Packaging Can Help

As discussed, the majority of food waste occurs within the supply chain, as it is moving from the farmer to the consumer. Food companies are considering how new packaging innovations might help protect food in transit to reduce the amount that becomes spoiled. The Californian-based company, Apeel Sciences, has developed a plant-based protection for fruit and vegetables. Apeel is an invisible coating made up of purified monoglycerides and diglycerides, edible fatty acids commonly found as natural compounds in the peels, seeds, and pulp of fruits, vegetables, and other plants. The company uses these food products, such as grape peels from a winery, and presses out an oil that is rich in the fat lipids, which it then turns into a colorless, odorless, and tasteless powder. Their coating ensures produce lasts at least twice as long as untreated produce by locking moisture and oxygen inside and creating a protective barrier. According to the company's research, application creates a 45 percent reduction in "shrivel and decay" after twenty-eight days for apples and 94 percent less shriveling of limes treated after the same period.[30] They claim to have saved 44 million pieces of fruit from the trash pile since the company launched.[31] A major German grocery chain, the Edeka Group, successfully trialed the tech in 2021, and Tesco has similarly used it effectively on citrus fruits sold in UK supermarkets. However, while the product is being used by organic certified producers in the United States, the production of the powder has been deemed incompatible with UK and EU organic standards, meaning that no organic producer in those regions can use Apeel and keep their certification. A spokesperson for the leading UK organic certifier, the Soil Association, said: "We understand Apeel contains E471 Mono- and diglycerides of fatty acids; this is not permitted under GB or EU organic regulation. So, this would not be permitted for use on organic produce in Britain."

In the coming years, new technologies might provide even more innovative packaging solutions, helping to reduce wastage. Nanotechnology provides huge opportunities to better protect produce and improve food safety. Nanoparticles are either utilized in the food product itself or in the food packaging in order to achieve this objective.[32] Leading research into food technology David Julian McClements considered the potential of nanotechnology in his book, *Future Foods: How Modern Science Is Transforming the Way We Eat*. He highlighted how "the chemical degradation of food ingredients can be slowed down by encapsulating them inside tiny particles (nano) that isolate and protect them from their environments—like a good rain coat protects us from the rain." This explains how nanotechnology could be used to create food packaging materials by utilizing their properties, including antimicrobial activity, oxygen absorption, and UV protection.[33] He continues,

> Rather than creating a package from a single material, it can be assembled by sandwiching together multiple layers of ultra-thin materials. This nano-laminated packaging combines the beneficial features of different materials—one layer provides mechanical strength, another gas control, and another protection from light. Alternatively, nanoparticles can be incorporated into packaging materials to increase their strength or modulate their permeability.[34]

Some metal compounds such as silver nanoparticles can act as antimicrobials. They function by generating reactive oxygen species, which are highly reactive compounds that can damage bacterial cell walls and organelles.[35] As a result, silver nanoparticles can not only reduce foodborne illness but also increase shelf life by inhibiting bacterial spoilage. This can decrease food loss by allowing for more time to finish consuming a food product before it goes bad. Nanotechnology can also be implemented in food packaging to create more ideal barriers that can create optimal flow for environmental elements such as gasses (like CO_2 and O_2), water moisture, and light. For example, fruit and vegetables require things like water, CO_2 and O_2, and light because they are still undergoing biological processes like respiration.[36] Thus, these more comprehensive barriers could also contribute to the food's preservation and so reduce food waste. Furthermore, nanotechnology could create "smart packaging" of foods as well. Nano sensors in food packaging can monitor a myriad of conditions such as humidity, temperature, levels of different gasses, and the presence of toxins and microbes in the food inside.[37] Changes in these parameters alter properties of the particular nanoparticle detector, which can then be relayed to the consumer.[38] For example, the food packaging could change color as a result of changes in a particular parameter inside. This can give an idea of when a food is no longer safe to eat, which can help avoid throwing out food prematurely.

Case Study V: Wrapped in Silk—the Surprisingly Simple Way to Fight Food Waste

The food on a supermarket shelf has already gone through a labyrinthine supply chain, resulting in most fresh foods, such as grapes, mandarins, and chicken, having very limited time left before they need to be removed from the shelves and become food waste. However, a nature-based technology could extend the longevity of food, allowing supermarkets to offer products for a longer period before they spoil and enabling consumers to store them longer at home.

Mori utilizes the power of silk: not with silk threads used in the textile industry, but with an invisible, water-based solution made from silk molecules. Mori uses a technology that employs only water, salt, and heat to extract a naturally derived silk protein from cocoons, which is then used to create a protective layer around food. The solution can be applied to foods of any size, shape, or texture, from produce to proteins, at stages where food is washed, sprayed, glazed, formulated, or dunked. The layer is ultrathin and only as thick as two red blood cells.[39] It is edible, tasteless, and odorless.[40] Aiming to double the shelf life of food, it slows down key spoilage mechanisms such as dehydration, oxidation, and microbial growth by limiting the gas exchanges that cause decay.

Mori's patented technology can be integrated at different points in the supply chain, from farm to shelf, offering food producers, processors, and retailers various advantages, including financial opportunities. On the one hand, it extends sales periods to help sell food that would otherwise go to waste and to reach new markets by allowing the transport of these foods over longer distances. On the other hand, its technology can help decrease the use of single-use plastics by providing a natural protective layer as a substitute. In addition to tackling both plastic pollution and food waste simultaneously, the technology can also support reducing reliance on chemicals, wax, and fungicides typically used in food production. However, since silk protein is derived from silkworm cocoons, making it an animal-based material, it is disqualified from being considered vegan. Additional ethical concerns may arise if Mori cannot guarantee that no silkworms are killed during the production process, which could make it a questionable product even for some vegetarians. As Mori scales its innovation, its success will depend not only on its ability to reduce waste and plastic use, but also on how transparently and ethically it navigates the complexities of consumer trust.

Food Waste in Retail

Once the food has made the journey, it reaches the supermarket. Despite being a finely tuned machine for food sales, supermarkets generate huge amounts of waste. The concept of "just in time readiness" is core to our grocery stores today, with food being brought to the shop floor just as it ripens. It doesn't take long for it to move past its best. To try and shift food that is on the turn or close to its sell-by date, RTC labels, denoting slashed prices, aim to encourage customers to select food that might not be at its best. The problem with this is that it is labor-intensive—it requires shop workers to go item by item and check the date and condition of the food before labeling the discounted products. It also requires consumers to go to the specially designated area to find the RTC option. Smart technology could help streamline this process by automatically adjusting prices to stimulate demand for surplus produce and reducing the price of food nearing its sell-by date. Such dynamic pricing not only helps reduce food waste and generate profits for food that would otherwise not have been sold but is also less labor-intensive due to automation. Startups like Wasteless, founded in 2017, are developing automatic and dynamic pricing systems, with some implementations already completed for retailers in countries such as Spain and Italy.

It is also worth considering the sell-by date itself. This is a particular issue in the United States. Date labels on food are not regulated federally in the United States—infant formula is a lone exception. The public might think that they denote food safety, but there is little to support that presumption. Given confusion over inconsistent language (use by, sell by, best before, etc.), consumers often err on the side of caution and discard food as soon as it reaches the labeled date. For example, the term "sell by" indicates when a product should be sold, allowing time for that product to have top-quality shelf life once in a consumer's home. However, many think to discard food on its sell-by date. In the United States, more than 80 percent of consumers report that they discard food prematurely due to confusion over expiration dates, exacerbated by the fact that many states have different requirements.[41] To remedy this, in December 2016, the USDA FSIS released new guidance. This guidance recommends that manufacturers and retailers use "Best if used by" as the universal terminology to indicate quality, which was adopted by the two largest food industry associations. However, this scheme is only voluntary. The Food Date Labeling Act of 2016 would have seen federal legislation to harmonize date labels, but it did not gain enough traction in Congress.

But the reality is that, regardless of RTC schemes and voluntary labeling, supermarkets in the Global North will always generate waste. Some argue that switching to digital-only supermarkets, like the UK's Ocado, can help reduce waste since there is greater efficiency and streamlining. However, consumers like to have the option to browse the aisles, so supermarkets have tried to find a way to redirect food waste from the trash. One option is to share it with local communities by encouraging partnerships with local food banks and soup kitchens. However, food that has passed its use-by date cannot legally be donated to food banks, and there are other regulatory barriers that hinder donations. Firstly, there may be ethical considerations. For example, giving underprivileged members of the community the least desirable, perhaps least nutritious, foodstuffs that no one else would buy. Secondly, liability issues are one example since companies fear that they might remain legally responsible for food that they donate—especially if it is close to its end date. In 2018, to address concerns over potential legal exposure, Argentina expanded its Food Donation Law to provide the necessary liability protections for food companies and facilitate donations.[42]

Startups are also stepping into this space to see whether technology can help solve the problem, such as Too Good To Go, which was founded in 2016 in Copenhagen to fight food waste. Supermarkets gather up their unwanted food at the end of the day and divide it into bags that app users then purchase on the app before collecting from the store. Users aren't able to pick and choose what they get—their bag is a surprise, allowing supermarkets the freedom to offload a range of products without the headache of sorting and listing items individually. The app has been a hit. It turns out that many people are willing to pay $5 USD for a bag of pastries from their favorite bakery, even if they don't get to pick exactly how many croissants versus pains au chocolat they get. In eight years, the company has grown to over 95 million registered users and 160,000 active business partners across eighteen countries in Europe and North America. The organization claims to have leveraged technology to help save over 330 million meals from being wasted, the equivalent of 891,000 tons of CO_2 equivalent avoided.[43]

One new trend is the creation of new products that utilize food that would otherwise be wasted. This trend is termed upcycled food and is defined as utilizing ingredients that otherwise would not have gone to human consumption. In 2021, the upcycled food products market size was valued at $53.7 billion USD, and it is estimated to reach $97billion USD by 2031, according to Allied Market Research.[44] A prime example of upcycled food is Toast Ale in the UK, which created a craft beer with surplus bread. The company worked with sandwich manufacturers to retrieve and divert crusts and the heels of loaves. Other brands that have redirected lost food to make entirely new products include Pulp Chips that create their product using the leftover pulp from cold-pressed juice, Dash Water that flavors their sparkling water with fruit that would otherwise have been discarded for aesthetic reasons, and Rubies in the Rubble that makes chutneys and jams. The Upcycled Food Association was established in 2019 to advocate for this micro-sector of the food industry and launched an Upcycled certification program in 2021 to increase consumer confidence.[45] Today, there are ninety-three companies with more than 480 products and ingredients certified, and UFA estimates that these products have diverted an average of 390,000 tons of food waste annually.[46] This trend of creating new products from waste is not limited to the Global North. When faced with a banana crop that was ripening too early and risked becoming a worthless harvest, women in the farming community of Karonga, Malawi, turned the fruit into wine and created a new product. The drink, produced by the Twitule Cooperative, has created a new revenue stream for farmers and created additional economic resilience to climate insecurity for a vulnerable region.[47]

Restaurants and Waste

Restaurants globally lose an estimated $2.5 trillion USD in food waste—45 percent wasted in preparation, 34 percent from customer leftovers, and 21 percent to spoilage.[48] A key culprit of wasted food in restaurants is the potato - lots of uneaten chips and mash.[49] The much-maligned breadbasket only accounts for 16 percent.[50] However, the all-you-can-eat buffet that generates the highest levels.[51] Some restaurants are working on reducing their waste themselves through their cooking. In 2015, Chef Dan Barber launched a pop-up at his Manhattan Blue Hill Restaurant to cook with ingredients typically deemed unusable. The restaurant served a rack of black cod (using bones and fish left over after filleting), deep-fried skate cartilage, and broken razor clams. While Barber's showcase was short-term, Silo is the world's first permanent zero-waste restaurant. Their vision started with the concept of nose-to-tail dining (as discussed in Chapter 7) but goes far beyond it. They have termed it "root to leaf" and focus on maximizing the potential of every ingredient, whether animal, vegetable, or mineral. To this end, they own their own flour mill and churn their own butter. They require that all their suppliers deliver products to the restaurant in reusable, returnable containers, and only work with those committed to regenerative and agroecological farming. They even have a fermentation unit underneath the restaurant to convert food waste into an umami food additive. However, while the ambition to challenge our culture of food waste and demonstrate the ways in which the hospitality industry can make changes is laudable, there is cost tension. A meal there costs $95 USD per person, and that's before you add wine. While the price reflects the incredible talent of the chefs, it also reflects the high overheads and additional costs associated with their zero-waste approach. It's an expensive business model. Not many restaurants are going to be able to charge that much for a meal. The salad chain, Sweetgreen, tried to do something similar but at a more reasonable price point. They offer a salad of broccoli stalks, bread ends, and carrot peels.

Other restauranteurs are installing technology that can help them monitor their waste and adjust their purchasing to reduce excess. Winnow Vision has a weighing scale installed under the kitchen waste bin and a camera to identify the types of food that are being discarded. The AI-powered system logs the weight, and a camera mounted over the bin automatically identifies the food type. Once a baseline has been established, the system then provides a daily or weekly report on what has been tossed into the trash. Management can then adjust their ordering for the following week. For the first few weeks of the analysis period, clients are requested to collect data without modifying behavior. This establishes a baseline for future comparison. Winnow uses information on food sales during the same time period to analyze the return on investment and profits. The analysis also provides other useful estimates, including the volume of avoided carbon emissions, which is becoming an increasingly important data point as businesses are forced to achieve net zero. Winnow claims that a typical client can expect to reduce waste by 40—70 percent, with an average reduction of about 50 percent. Less than a decade after its launch, Winnow has over 1,400 users across the world and is only growing.[52] That corresponds to a saving of 36 million meals each year (or $43 million USD in food cost savings) and, from an environmental perspective, 61,000 tons of CO_2e prevented per year and 1.2 billion gallons of water that would have otherwise been wasted.

Food Waste at Home

Many citizens are doing what they can to reduce their food waste at home. Technology is increasingly playing a role in this. Smart fridges and food tracking apps are ways to nudge consumers into wasting less. Just as companies are perfecting refrigeration in the shipping of food, new domestic smart fridges can help ensure that food is kept at the optimal temperature and humidity at home to preserve freshness by removing excess moisture and reducing condensation. They let you view what is inside without opening them via smart glass and internal cameras, ensuring consistent temperatures. They connect to Amazon's Alexa and other home devices to connect to your shopping list. But, as smart as our current fridges are, AI will soon make them even smarter. New innovations are conceptualizing fridges fitted with sensors: gas sensors to detect when food starts to spoil and optical sensors to monitor visual changes to spot when you misplace food that might shorten its shelf life. Machine learning can then encourage you to use up your tomatoes or spinach sooner rather than later with reminders pinging through to your phone. Or it could sound the alarm when you leave your celery outside of the crisper to encourage you into storing your food in the best possible way. It could even use the data from the sensors, combined with your purchasing history to send recipe ideas that maximize efficiency and minimize waste.

Many of these technologies are currently too expensive for the average consumer—$3000 USD is a lot to pay for a fridge—but companies are selling basic devices that cost much less. For example, FridgeCam can be easily stuck into any fridge to convert it into a smart fridge. This small camera can be installed to allow you to view the inside of your fridge via your phone. In theory, that means you'd avoid duplicating purchases when going around the supermarket since you'd be able to check whether you already had strawberries. Another novel technology that is helping to reduce food waste is Shelfy. Shelfy is a purifier that slows down the ripening process of fruits and vegetables by breaking down ethylene (which induces ripening) and reduces the microbial load. It reduces bacteria by 99 percent and odors by 80 percent, making food last up to twelve days longer and delaying the appearance of mold or odor.[53] With a rechargeable battery and a washable filter, it can last forever. Similar to Shelfy is the Bluapple that is designed to sit in your fruit bowl at home and absorb ethylene gas for three months.

Out of the Trash Bin

Inevitably some food will be wasted. For the food that is going to be scraped there are better options than just throwing it into the garbage. The US Government's

Environmental Protection Agency has created a Wasted Food Scale that ranks options from green to red, starting with "prevent" and then "donate."[54] The third best is for animal feed. Traditionally, our food waste would have been recycled by pigs. Pigs (and chickens to a certain extent) often lived in the backyards of homes to consume the food waste that was discarded. After a few months, those pigs would have become dinner. Beyond the fact that not many people in the twenty-first century want to have a pig living in their backyard, the rise in zootrophic disease and increased concern for food safety have resulted in more restrictions on feeding food scraps to animals. In the UK, for example, the rise of BSE in the 1990s is commonly thought to have resulted from cattle being fed mammalian meat and bone meal and resulted in restrictions on what could be fed to livestock.[55] Similarly, the outbreak of African Swine Fever in China in 2018 saw swill (food waste redirected for animal feed) being highly restricted to stop the disease's spread.

With proper safeguards in place, using food waste as animal feed has significant benefits. Recognizing this, the Japanese government supports pig farmers who want to use food waste as feed. Japan's 2007 Food Waste Recycling Law made recycling food waste into animal feed a priority with the goal of making livestock and poultry industries less reliant on imported feed grains.[56] The resulting pork was sold at a premium as eco-pork on Japanese supermarket shelves. Taiwan and South Korea have similar programs, and other countries may soon follow suit. In the United States, about 10 percent of surplus food is already being sent to feed animals, but there is room for improvement since another 14.7 million tons of food waste in the United States could be safely used for animal feed.[57] Individual programs are working in partnership with farms. For example, Rutgers University in New Jersey is partnering with a local farm, Pinter Farms. Every day, Pinter Farms collects about one ton of food scraps from Rutgers' four main dining halls and feeds it to the farm's hogs and cattle. Diverting food scraps to Pinter Farms costs Rutgers half the price of sending the scraps to the landfill. However, for other nations, particularly in Europe, there needs to be adjustments in legislation to allow for more food to be diverted from the landfill into livestock feed. The EU is slowly relaxing its regulations on food waste as animal feed, and the UK is also considering whether to adjust the strict legislation that was put in place after BSE to allow for more food waste to be fed to pigs. As our food culture shifts, the option of using food waste as feed might become more viable. As mentioned in Chapter 5, advocates of insects as an alternative protein source for the future highlight that food waste could be used to feed larvae. Ohio's Enviroflight feeds six tons of food waste (primarily byproducts and leftovers) per day to huge numbers of black soldier fly larvae. In just two weeks, the larvae can be processed as feed for commercially raised fish.

After animal feed, the EPA points to compost as the fourth-best option. As we have discussed in Chapter 1, our soils are increasingly degraded. Food waste could be composted and returned to the soil to help restore fertility. Rather than letting food decompose anaerobically in a landfill that results in methane emissions, compost piles are aerated and decompose in a sustainable way. Composting has the potential to increase soil fertility by increasing the amount of organic matter in our soils. When soil is healthy, crop yields are higher, which means that people are able to produce enough food to feed themselves and their families.[58] Using compost allows fertility levels to be raised without the need for expensive agrochemicals, which will help ensure that hunger levels reduce globally and raise farmer livelihoods. Higher levels of soil organic matter also make soil better able to retain water, which makes it more resilient to drought conditions.[59] Because water is better absorbed by healthy soils, improving soil health also reduces water runoff and therefore reduces agrochemical pollution of waterways.[60]

Many people already have compost piles in their gardens. However, it is harder to manage composting in urban environments. New startups are hoping to provide composting solutions for city dwellers. Lomi is a countertop compost maker from Canada. Lomi heats and grinds food waste while sensors measure and control the moisture content, temperature, and airflow. Designed to be odor-free, it turns a 3lb bin of food scraps into carbon-rich pre-compost overnight, which can then be put straight into the garden or plant pot. Another option is to convert the food scraps into biochar. "Biochar" is a catch-all term describing any organic material that has been carbonized under high temperatures (572–1832°F), in the presence of little or no oxygen, in a process called pyrolysis. Once everything has been burned off, it leaves a solid residue of at least 80 percent carbon, which is termed biochar. The concept arose from observing that natural wildfires would bring fertility into the soils—something that indigenous communities have known for thousands of years. Applying biochar to the soil can also help sequester carbon and lock it in the ground—at a much better rate than compost.[61] This potential as a climate mitigation method was noted by the Intergovernmental Panel on Climate Change in 2018 when they considered mitigation pathways.[62]

Fifth on the EPA's list is anaerobic digestion. Anaerobic digestion is a process through which bacteria break down organic matter—such as animal manure, wastewater biosolids, and food wastes—in the absence of oxygen. Municipal food waste is collected and fed into reactors that contain complex microbial communities that, over the course of a month, break down the waste and produce two products: biogas and digestate. Biogas is composed of methane, which is the primary component of natural gas, at a relatively high percentage (50 to 75 percent), carbon dioxide, hydrogen sulfide, water vapor, and trace amounts of other gases. The energy in biogas can be used like natural gas to provide heat, generate electricity, and power cooling systems. The digestate can be used as a fertilizer to grow next year's crops. Many cities and towns use anaerobic digestion to manage their waste while reducing their energy bills.

Conclusion

There is a critical difference between food loss and food waste. Food loss refers to edible food mass that is discarded or lost along the journey from production, postharvest, storage, transport, and processing stages and is primarily a problem in the Global South. Food waste refers to discarded or wasted food at the retail and consumer level and is primarily a problem in the Global North. Perhaps misrepresented in popular culture, food is lost or wasted at every stage of the food cycle, and not just by the consumer, and there needs to be strategies to reduce food loss or waste at every stage. Smallscale and site-specific interventions can have a big impact. Food loss in Global South countries could be dramatically improved with better storage, for example. New innovations are trying to divert food from the landfill in order to address hunger and reduce the carbon footprint of our food system. Sadly, it is inevitable that some food will be wasted, but there are still preferential options for managing that food waste by ensuring it is used to feed the soil and by using it to generate sustainable energy via anaerobic digestion.

Conclusion—Next Steps on the Path

Our food system has never been under such intense, overlapping pressures. Climate change is altering growing conditions, disrupting supply chains, and amplifying the risks of crop failure. Agriculture itself is a key driver of this crisis, responsible for nearly 30 percent of global greenhouse gas emissions, with industrial farming, livestock methane, and land-use change among the largest contributors.[1] Yet the impacts go beyond the atmosphere. Livestock production, particularly when concentrated in industrial systems, not only generates methane but also drives deforestation and habitat destruction. The expansion of croplands and pastures, fueled by demand for animal feed and fertilizer-intensive monocultures, is one of the leading causes of biodiversity loss worldwide. Nutrient runoff from synthetic fertilizers creates dead zones in rivers, lakes, and oceans, further degrading ecosystems. These interlocking pressures mean that food systems are both vulnerable to and responsible for accelerating climate change and ecological collapse.

Furthermore, the public health crisis of diet-related disease exposes vulnerabilities in how we produce, distribute, and consume food. While hunger has declined in many regions, obesity and diet-related noncommunicable diseases, such as diabetes and heart disease, have surged since 2000.[2] These trends are closely tied to the rise of a globalized commodity food system, which prioritizes scale, uniformity, and profit over nutrition. Subsidized staples like corn, soy, and wheat are processed into cheap sweeteners, refined oils, and filler ingredients that form the backbone of ultra-processed foods. These products dominate global markets, displacing traditional diets and crowding out healthier options. Their affordability makes them the default choice for low-income households, disproportionately harming vulnerable populations and deepening cycles of poverty and ill health. Children, in particular, are growing up in environments where unhealthy food is the most accessible food, setting the stage for lifelong health challenges. Our next generation is facing a bleak and unhealthy future—one that reflects not individual failure, but the structural consequences of a food system designed around commodities rather than communities.

Layered on top of this are shifting economics of smallscale farming and its potential collapse. In the United States, the number of farms has plummeted from 6.8 million in the 1930s to just 1.9 million in 2022, with Europe experiencing similar declines.[3] As small family farms disappear, rural communities hollow out, and younger generations turn away from agriculture as an inviable livelihood. Farmers' share of the food dollar keeps shrinking, and a single climate disaster or policy shift can wipe out a year's income.[4] At the same time, market consolidation concentrates power in a handful of agribusiness giants, squeezing out diversity, stifling innovation, and making the system less resilient. The result is a widening gap between what society demands from food and what the current agricultural model can deliver—a gap that must be addressed if we are to build a food system capable of enduring the crises ahead.

Strained by converging crises, the global food system stands at a breaking point. These vulnerabilities overlap and amplify one another since food is shaped by a vast web of politics, economics, culture, environment, technology, and biology. Geopolitical instability further undermines food security. The globalized supply chain means that wars, trade disputes, and political upheaval—from Ukraine to US trade tensions—reverberate worldwide, disrupting production and distribution, and threatening the steady flow of food across borders. This heightens the risk of simultaneous shocks with devastating consequences for hunger.

The most urgent challenge in shaping the future of food is the future of protein. As the centerpiece of the food debate, protein sits at the intersection of environmental, social, and ethical trade-offs. It is essential for human health, yet today's livestock-heavy production drives climate change, biodiversity loss, pollution, and animal welfare concerns. Livestock production is responsible for 14.5 percent of global greenhouse gas emissions, largely from methane, manure, and deforestation.[5] Industrial systems (such as CAFOs) confine animals in cramped conditions, rely on routine antibiotics—fueling antimicrobial resistance—and heighten the risk of zoonotic diseases such as avian flu and African Swine Fever. Growing public awareness, along with links between red meat and cancer or heart disease, is shifting some diets toward poultry, plant-based options, and meat-reduction campaigns, while policies such as Denmark's groundbreaking methane tax aim to address environmental impacts.

The search for sustainable protein is transforming the sector, with solutions emerging across traditional, alternative, and high-tech fronts. Alternative proteins—ranging from farmed fish and shellfish to algae and insects—offer nutrient-dense, lower-impact options. Aquaculture now produces more fish than wild catch, but environmental challenges like sea lice, pollution, and unsustainable feed use undermine its green credentials, prompting innovations such as land-based recirculating systems and algae-based feeds. Bivalve farming and seaweed cultivation stand out for their restorative benefits, improving water quality and biodiversity without external feed inputs. Insects, a staple in many traditional diets, deliver high-protein, low-emission nutrition for humans and livestock, though cultural aversions and welfare concerns remain barriers in the Global North.

Technological plant-based meats increasingly mimic animal products using textured vegetable protein and engineered additives like heme, while cultivated meat grows animal cells in bioreactors, potentially avoiding slaughter and reducing land, water, and antibiotic use. Microbe-based proteins, such as mycoprotein and novel strains like Fy or Solein, promise major efficiency gains in land and water usage. Yet questions remain over consumer acceptance, cost, regulation, and environmental performance at scale, as well as ethical and religious classifications. Countries like Singapore and Israel are leading regulatory approvals, positioning lab-grown meat as part of food-security strategies.

Alongside these innovations, sustainably raised meat aims to turn livestock from a climate problem into a climate solution. Regenerative and agroecological systems use rotational grazing to restore soil carbon, enhance biodiversity, and reduce inputs, with holistic grazing revitalizing degraded landscapes worldwide. These approaches improve animal welfare, produce nutritionally richer meat, and sustain rural livelihoods rooted in livestock culture. Advocates promote a "less but better" approach, sourcing from grass-fed, pasture-raised animals and investing in local processing infrastructure to maintain welfare and reduce emissions.

Given the significant emissions associated with protein, it makes sense that it's the primary focus for many looking to create a more sustainable future for our agricultural systems. Yet, there are a host of other factors that will define the future of food, shaped by a convergence of innovation, resource efficiency, and cultural change. These include how we rethink land use, reduce waste, and reimagine diets. Ultimately, the choices society makes will determine whether we achieve transformation or remain locked in current, damaging cycles.

Agri-technology is already transforming farming, from precision agriculture and AI-driven soil, water, and pest sensors to drones and smart mapping that optimize yields while reducing environmental harm. In the Global North, these tools help large-scale producers cut chemical use and water waste, while in the Global South they can bridge the yield gap, helping smallholders adapt to climate change. Emerging innovations hold the potential to democratize access, lowering costs and making advanced tools available even to the smallest subsistence farmers. With the right investment and collaboration, agri-tech can accelerate a global shift toward farming systems that are both productive and sustainable.

Advances in plant breeding are unlocking major gains for farmers, consumers, and the environment alike. By developing pest- and disease-resistant varieties, breeders help reduce reliance on chemical inputs while improving crop resilience to climate stress. Traits such as slower ripening and improved storage life extend shelf stability, cutting losses during transport and reducing food waste. Even subtle innovations—like breeding thicker-skinned avocados that travel better and bruise less—translate into higher returns for farmers, greater efficiency in supply chains, and more reliable access to fresh, nutritious food for consumers. Ethical and environmental questions remain around gene editing and biotechnology, but the potential to feed more people with fewer resources is driving rapid development.

All these advances are meaningless unless we stop food from being wasted at each step along the supply chain. Food waste remains one of the most solvable challenges: one-third of all food produced never gets eaten, wasting water, land, and labor while generating methane in landfills.[6] In the Global South, loss often occurs on farms due to inadequate storage, processing, and transport; in the Global North, stringent cosmetic standards and market economics leave crops unharvested or discarded. Solutions range from low-cost innovations—airtight storage bags, structured crates—to infrastructure investments in roads, processing plants, and "smart" shipping containers. Redesigning supply chains, changing procurement standards, and incentivizing surplus donation can redirect edible food to people instead of waste streams.

A key consideration in preparing our food systems for the future is adapting them to the realities of our rapidly urbanizing world. By 2050, nearly 70 percent of humanity will live in cities, making urban areas the defining force in the future of food.[7] Cities determine not only what we eat but also how food is produced, transported, stored, sold, and ultimately valued. Yet the modern urban food system—built on global, just-in-time supply chains—delivers calories more efficiently than it delivers nutrition, fueling rising obesity, heavy reliance on ultra-processed foods, and deepening health inequities. Food deserts, often rooted in historic discrimination, leave low-income and minority communities with limited access to fresh produce, while wealthier districts enjoy farmers' markets, gourmet food halls, and vertical farms. Equity demands re-embedding food production and distribution into the urban fabric through community gardens, food justice initiatives, and local supply chains that keep value circulating locally. But production is only part of the equation. The structure of supply chains and retail environments determines whether nutritious food is affordable, accessible, and appealing. Smarter distribution networks, shorter supply chains, and technology-driven traceability can reduce waste and improve access, while retail choices—from supermarkets to online platforms and street markets—shape diets and environmental impact. Policies like tax incentives for urban agriculture, restrictions on fast-food clustering, subsidized access to fresh food, and investment in equitable infrastructure—ports, markets, storage facilities, and distribution networks—can transform cities from engines of dietary decline into hubs of sustainable, inclusive, and resilient food systems. In the future, we can deliver urban food systems that ensure every resident has access to affordable, nutritious, and culturally appropriate food, supported by supply chains and retail landscapes designed for health and sustainability.

As acknowledged at the beginning of this book, there are many futures for our food system that radiate out from our present. Envisioning a guiding principle for the food system of the future requires wrestling with big topics. It is entangled with questions of equity, ecology, economics, and culture. It is up to us to choose which future we want. Some futures can bring us closer to equity. Others might reduce the carbon footprint. We could pursue a path of greater corporate consolidation that aims at efficiency and productivity, or one that can relocalize our agriculture and develop smallholder equity. The book has aimed at exploring the full landscape of possibilities and pitfalls in the various futures of food. We've examined the interlocking challenges and emerging opportunities. With the full picture now in view, the choice of what type of food future we want is before us.

What is also clear is that the window to transition toward a more sustainable future for food is closing rapidly. Every season of delay deepens our vulnerabilities: farmland degraded beyond recovery, species driven to extinction, and communities pushed to the brink by failing crops and fisheries. Acting now is not simply a matter of environmental stewardship; it is an investment in food security, public health, and economic stability. The choices we make today will determine whether we adapt in time or face cascading crises that overwhelm our capacity to respond. Solutions exist—regenerative farming, diversified protein sources, restored ecosystems, smarter land use—but their benefits depend on wide-scale adoption and coordinated action. Waiting until impacts are unavoidable will be too late for many communities, particularly in the Global South, where resources for adaptation are most limited. Urgency must guide policy, innovation, and consumer behavior alike, because the longer we hesitate, the narrower—and costlier—the path to a sustainable, equitable food future becomes.

That said, for all the talk of food future, these new food technologies are no longer on the horizon—they are here. This is not a theoretical debate about a distant future. Cell-cultivated products are already on supermarket shelves in some countries, with more approvals and product launches happening every month. If they're not in your local grocery store yet, they soon will be. The pace of change is accelerating, and the need to act is immediate. We have only a narrow window to decide what kind of future we want for our food system, and that decision is ours to make. Not engaging with the choices and playing dumb is in itself a choice. Maintaining the status quo is picking a future. Once we have decided what food future we want, it is required to put in place the policies, regulations, and ethical frameworks that will shape it. Waiting passively will not preserve the status quo; it will simply allow the market and a handful of powerful actors to decide for us. Inaction is not neutral. Failing to engage now means accepting, by default, whatever trajectory industry and market forces set. That will define not just what is on our plates, but how food is produced, who controls it, and who benefits. The choices we make—or fail to make—in the coming years will lock in patterns that could endure for generations. The time to shape that future is now, before the decisions are made for us.

Facing the immediacy of the food system crisis, it is easy to opt for the quick and simple option. The danger lies in reaching for the "silver bullet" fix before fully understanding the scope of the problem. Such responses risk treating symptoms while leaving root causes untouched, locking us into pathways that may prove neither effective nor equitable in the long run. In a system as dynamic as food, tackling one challenge without recognizing its connections to others can backfire—triggering future crises or sparking new ones. The lure of a quick fix is strong, yet oversimplified solutions often deepen the very problems they aim to solve. Without a clear-eyed diagnosis of the system's weaknesses, responses risk becoming little more than temporary patches that allow deeper issues to fester.

Technology will undoubtedly play a major role in shaping the future of food—through innovations such as cultivated meat, vertical farms, gene-edited seeds, and precision farming—but food is far more than a technical challenge.

It is bound up with culture, economics, ecology, and power, and no single invention can repair a broken system. The crucial question is not whether to use technology, but how. This means deploying new technological tools with care and purpose, ensuring they strengthen rather than undermine the vision of food that we want for the future. This needs to be combined with structural interventions such as policy reform and shifts in subsidies to align the economics and ethics of food production. Once we have decided the food future that we want, we can then create a framework through which technology can help us achieve that goal. The task is to choose wisely. With thoughtful design, strong oversight, and clear intention, technology can help build food systems that are not only high-tech but also healthy, equitable, and genuinely sustainable—surely the optimal outcome.

One of the big challenges working to dictate the future of food is that cheap ultra-processed food is politically appealing. In the face of a cost-of-living crisis and rising inflation, the public is understandably resistant to politicians advocating for higher food prices. Even when policymakers recognize the public health impact of inexpensive food, it's difficult to win elections on a platform that raises grocery store prices or limits consumer choice. Despite acknowledgment of the need for reform, meaningful change often seems unattainable. Cheap food remains politically popular, and the public is rarely encouraged to reconsider the true cost of what ends up in their shopping baskets. Proposals such as taxing sugary drinks or regulating fast food portion sizes are often met with backlash. In this environment, the path of least resistance and popular support tends to prevail. In the current economic turmoil, it would be impossible for political leaders to introduce policy to increase the price of unhealthy food. Therefore, the focus needs to be on finding a way to make healthy food more affordable.

In an age of overlapping crises—climate disruption, volatile markets, pandemics, and political instability—the most important quality our food systems can have is resilience. A food future that aims at adaptability, rather than perfection, might be our best option. A resilient food system is one that can absorb shocks, adapt to change, and continue to provide affordable, nutritious food for all. It is a principle that, if embedded from farm to fork, can address both the urgent threats of today and the inequities that have been entrenched for decades. Resilience means diversifying what we grow and how we grow it. It means shorter, more local supply chains that are less dependent on volatile global markets and vulnerable transport links. It calls for infrastructure—cold storage, processing facilities, local markets—that gives communities the capacity to store surplus, manage scarcity, and maintain access to food when external systems falter. In cities, this translates into urban agriculture, food hubs, and neighborhood-level distribution networks that can operate independently if wider supply chains fail. Moreover, real resilience must be grounded in equity, ensuring farmers can make a fair livelihood, consumers can access affordable and nutritious food, and no community is left behind. It also requires respecting environmental limits—protecting soils, water, and biodiversity so that productivity today does not undermine fertility tomorrow. Finally, it rests on systemic stability, diversifying supply chains, spreading market power, and embedding safeguards against geopolitical and economic shocks. Without these foundations, even the most brilliant innovations will eventually falter, but with them, farming can become not just more resilient but more just, sustainable, and secure.

If managed properly, food is a powerful lever for climate, health, and justice—but only if we make the right changes. The future of food is ours to shape, and that is a source of both responsibility and hope. Every decision—whether made in our kitchens, communities, boardrooms, or parliaments—has the power to move us toward a system that nourishes everyone while restoring the planet. We have the tools, the knowledge, and the creativity to create food systems that are fairer, healthier, and more resilient than anything we have known before. What's needed now is the shared will to act: to champion policies that reward sustainability, to support producers who care for land and people, and to celebrate the diversity of foods and cultures that enrich our lives. By choosing to be active stewards rather than passive consumers, we can leave a legacy of abundance, equity, and ecological balance. The opportunity is before us; together, we can ensure that the future of food is one we are proud to hand to the next generation.

Notes

Introduction

1 UNEP, "Emission Gap Report 2023," UN, 2023, https://www.unep.org/resources/emissions-gap-report-2023.

2 Rezaei, E. E., H. Webber, S. Asseng, K. Boote, J. L. Durand, F. Ewert, P. Martre, and D. S. MacCarthy, "Climate Change Impacts on Crop Yields," *Nature Reviews Earth & Environment* 4 (2023): 831–46. https://doi.org/10.1038/s43017-023-00491-0

3 Ritchie, Hannah, "Food Production Is Responsible for One-Quarter of the World's Greenhouse Gas Emissions," OurWorldinData.org, 2019, https://ourworldindata.org/food-ghg-emissions.

4 NCD Risk Factor Collaboration, "Worldwide Trends in Hypertension Prevalence and Progress in Treatment and Control from 1990 to 2019: A Pooled Analysis of 1201 Population-Representative Studies with 104 million Participants," *The Lancet* 398 (10304): 957–80. https://www.thelancet.com/article/S0140-6736(21)01330-1/fulltext; WHO, "Global Report on Diabetes," WHO, 2026, https://www.who.int/publications/i/item/9789241565257.

5 UNICEF, "The State of Food Security and Nutrition 2024," UN, 2024, https://data.unicef.org/resources/sofi-2024/.

6 Umair, Irfan, "The World Is Producing More Food than Ever—but Not for Long, Wired, 2025, https://www.wired.com/story/the-world-is-producing-more-food-than-ever-before-but-not-for-long/.

7 Willett, Walter, Johan Rockström, Brent Loken, Marco Springmann, Prof Tim Lang, Sonja Vermeulen, Tara Garnett, David Tilman, Fabrice DeClerck, Amanda Wood, Malin Jonell, Michael Clark, Line J. Gordon, Jessica Fanzo, Corinna Hawkes, Rami Zurayk, Juan A. Rivera, Wim De Vries, Lindiwe Majele Sibanda, Ashkan Afshin, Abhishek Chaudharyw, Mario Herrero, Rina Agustina, Francesco Branca, Anna Lartey, Shenggen Fan, Beatrice Crona, Elizabeth Fox, Victoria Bignet, Max Troell, Therese Lindahl, Sudhvir Singh, Sarah E. Cornell, K. Srinath Reddy, Sunita Narain, Sania Nishtar, and Christopher J. L. Murray, "Food in the Anthropocene: The EAT–*Lancet* Commission on Healthy Diets from Sustainable Food Systems," *The Lancet* 393 (10170): 447–92, https://www.thelancet.com/journals/lancet/article/PIIS0140-6736(18)31788-4/abstract.

8 Willett, Walter, et al., "Food in the Anthropocene: The EAT–*Lancet* Commission on Healthy Diets from Sustainable Food Systems," *The Lancet* 393 (10170): 447–92,

https://www.thelancet.com/journals/lancet/article/PIIS0140-6736(18)31788-4/abstract.

9 Willett, Walter, et al., "Food in the Anthropocene: The EAT–*Lancet* Commission on Healthy Diets from Sustainable Food Systems," *The Lancet* 393 (10170): 447–92, https://www.thelancet.com/journals/lancet/article/PIIS0140-6736(18)31788-4/abstract.

10 Dator, Jim, *Advancing Futures: Futures Studies in Higher Education*, Praeger, 1998.

Part 1

1 "Emissions Gap Report 2023," United Nations Environment Programme, November 2023, https://www.unep.org/resources/emissions-gap-report-2023; "Climate Change," World Health Organization, October 2023, https://www.who.int/news-room/fact-sheets/detail/climate-change-and-health.

2 "Extreme Weather," World Meteorological Organization, accessed: July 12, 2025, https://wmo.int/topics/extreme-weather#:~:text=The%20number%20of%20disasters%20has,of%20deaths%20almost%20three%2Dfold.

3 Ritchie, Hannah, "Food Production Is Responsible for One-Quarter of the World's Greenhouse Gas Emissions," Our World in Data, November 2019, https://ourworldindata.org/food-ghg-emissions; How Climate Change Is Stressing the Global Food Supply and Public Health," Columbia Mailman School of Public Health, September 2024, https://www.publichealth.columbia.edu/news/how-climate-change-stressing-global-food-supply-public-health.

4 "Importance of Methane," Environmental Protection Agency, accessed: July 12, 2025, https://www.epa.gov/gmi/importance-methane.

5 "Living Planet Report 2024," World Wildlife Fund, 2024, https://livingplanet.panda.org/en-US/.

6 "Living Planet Report 2024," World Wildlife Fund, 2024, https://livingplanet.panda.org/en-US/.

7 "World Hunger Facts: Seven Things You Should Know," International Rescue Committee, August 2024, https://www.rescue.org/uk/article/world-hunger-facts-seven-things-you-should-know.

8 "Health Depends on More than Health Systems," The Lancet, 2019, https://www.thelancet.com/pb-assets/Lancet/infographics/gbd-2019/gbd2019-1648208502440.pdf.

9 Vandevijvere, Stefanie, Lindsay Jaacks, Carols Monteiro, Jean-Claude Moubarac, Martin Girling-Butcher, Arier Lee, An Pan, James Bentham, and Boyd Swinburn, "Global Trends in Ultraprocessed Food and Drink Product Sales and Their

Association with Adult Body Mass Index Trajectories," National Library of Medicine, May 2019, https://pubmed.ncbi.nlm.nih.gov/31099480/.

10 CDC Nutrition, "Get the Facts: Sugar-Sweetened Beverages and Consumption," CDC, https://www.cdc.gov/nutrition/php/data-research/sugar-sweetened-beverages.html?CDC_AAref_Val=https://www.cdc.gov/nutrition/data-statistics/sugar-sweetened-beverages-intake.html.

11 Millar, Helen, "What to Know about Obesity and Poverty," Medical News Today, April 2023, https://www.medicalnewstoday.com/articles/obesity-and-poverty#:~:text=In%20developed%20countries,%20obesity%20rates,an%20increase%20in%20health%20problem; Ravi, Ethan, and Frank Herman, "The Wealthier, the Healthier; How Socioeconomic Status Affects Eating Habits," *The Evanstonian*, January 2023, https://www.evanstonian.net/uncategorized/2023/01/21/the-wealthier-the-healthier-how-socioeconomic-status-affects-eating-habits/#:~:text=Overall%2C%20wealthier%20people%20have%20the,reflects%20in%20their%20overall%20health.

12 "Overweight & Obesity Statistics," National Institute of Diabetes and Digestive and Kidney Diseases, accessed: July 12, 2025, https://www.niddk.nih.gov/health-information/health-statistics/overweight-obesity.

13 Norris, Pippa, "It's Not Just Trump. Authoritarian Populism Is Rising across the West. Here's Why," *Washington Post*, March 2016, https://www.washingtonpost.com/news/monkey-cage/wp/2016/03/11/its-not-just-trump-authoritarian-populism-is-rising-across-the-west-heres-why/.

14 "When Will China's GDP Overtake America's," *The Economist*, June 2023, https://www.economist.com/graphic-detail/2023/06/07/when-will-chinas-gdp-overtake-americas.

15 "Civil War in Sudan," Center for Preventive Action, April 2025, https://www.cfr.org/global-conflict-tracker/conflict/power-struggle-sudan.

16 Walker, Nigel, "Conflict in Ukraine: A Timeline (Current Conflict, 2022–present)," House of Commons Library, April 2025, https://researchbriefings.files.parliament.uk/documents/CBP-9847/CBP-9847.pdf.

17 Abbott, Chuck, "Smallest Number of U.S. Farms since 1850, Says Ag Census," Successful Farming, February 2024, https://www.agriculture.com/smallest-number-of-u-s-farms-since-1850-says-ag-census-8580393.

18 Madeline, Beatrice, "Comment le monde Agricole s'est peu a peu fragmente," le monde, 2024, https://www.lemonde.fr/economie/article/2024/05/16/comment-le-monde-agricole-s-est-peu-a-peu-fragmente_6233554_3234.html.

19 Goldin, Nicole, "A Tale of Twin Demographics: Youth in Cities," World Bank Blogs, October 2016, https://blogs.worldbank.org/en/jobs/tale-twin-demographics-youth-cities#:~:text=In%20today's%20global%20landscape%2C%20two,for%2090%25%20of%20urban%20growth.

20 "Farm Sector Income & Finances—Farm Sector Income Forecast," United States Department of Agriculture, February 2025, https://www.ers.usda.gov/topics/farm-economy/farm-sector-income-finances/farm-sector-income-forecast?utm_source=chatgpt.com.
21 "The State of Food and Agriculture," Food and Agriculture Organization of the United Nations, 2020, https://openknowledge.fao.org/server/api/core/bitstreams/6e2d2772-5976-4671-9e2a-0b2ad87cb646/content.
22 Horton, Helen, "Farm Income Falls in England after Extreme Weather and Subsidy Cuts," *The Guardian*, November 2024, https://www.theguardian.com/environment/2024/nov/14/farm-income-england-2023-extreme-weather-subsidy-cuts.
23 "Climate Change Impacts on Agriculture and Food Supply," Environmental Protection Agency, accessed: July 12, 2025, https://www.epa.gov/climateimpacts/climate-change-impacts-agriculture-and-food-supply.
24 O'Hare, Ryan, "More than One Billion People Now Living with Obesity, Global Analysis Suggests," Imperial News, February 2024, https://www.imperial.ac.uk/news/251798/more-than-billion-people-living-with/#:~:text=From%201990%20to%202022%2C%20global,from%2016.7%25%20to%2010.8%25.
25 Wallace, Tim, "Thousands of Farms Shutting as Industry Braces for Inheritance Tax Raid," *The Telegraph*, February 2025, https://www.telegraph.co.uk/business/2025/02/21/thousands-farms-shutting-industry-braces-inheritance-tax/.

1 Environmental Challenges

1 "Emissions Gap Report 2023," United Nations Environment Programme, November 2023, https://www.unep.org/resources/emissions-gap-report-2023; "Climate Change," World Health Organization, October 2023, https://www.who.int/news-room/fact-sheets/detail/climate-change-and-health.
2 Siegfried, Kristy, "Climate Change and Displacement: The Myths and the Facts," United Nations High Commissioner for Refugees, November 2023, https://www.unhcr.org/us/contact-us/privacy-policy/unhcr-verify-plus-privacy-notice.
3 Guterres, António, "Secretary-General's Special Address on Climate Action 'A Moment of Truth' [as delivered]," United Nations, June 2024, https://www.un.org/sg/en/content/sg/statement/2024-06-05/secretary-generals-special-address-climate-action-moment-of-truth-delivered?_gl=1%2A17omfbb%2A_ga%2ANTIzOTM5ODQuMTczNjI2MzY4Ng..%2A_ga_S5EKZKSB78%2AMTczNjQzODM5OC4zLjEuMTczNjQzODQ0MC4xOC4wLjA.%2A_ga_TK9BQL5X7Z%2AMTczNjQzODIyNy40LjEuMTczNjQzODM5OC4wLjAuMA.
4 "Sustainable Development Goals," United Nations Development Programme, accessed: July 23, 2024, https://www.undp.org/sustainable-developm

ent-goals#:~:text=The%20Sustainable%20Development%20Goals%20(SDGs)%2C%20also%20known%20as%20the,people%20enjoy%20peace%20and%20prosperity.

5 "How Climate Change Is Stressing the Global Food Supply and Public Health," Columbia Mailman School of Public Health, September 2024, https://www.publichealth.columbia.edu/news/how-climate-change-stressing-global-food-supply-public-health.

6 Ritchie, Hannah, "Food Production Is Responsible for One-Quarter of the World's Greenhouse Gas Emissions," Our World in Data, November 2019, https://ourworldindata.org/food-ghg-emissions.

7 Gerber, Pierre, Anne Mottet, Carolyn Opio, Alessandra Falcucci, and Felix Teillard, "Environmental Impacts of Beef Production: Review of Challenges and Perspectives for Durability," Meat Science, November 2015, https://www.sciencedirect.com/science/article/abs/pii/S0309174015300139?via%3Dihub#bb0115.

8 "Importance of Methane," Environmental Protection Agency, accessed: July 23, 2025, https://www.epa.gov/gmi/importance-methane.

9 Bhagat, Sudheer, A. K. Santra, Sharad Mishra, V. N. Khune, M. D. Bobade, Ashutosh Dubey, Aayush Yadav, Anupam Soni, Sweta Banjare, and Gajendra Yadav, "The Water Footprint of Livestock Production System and Livestock Products: A Dark Area: A Review," *International Journal of Fauna and Biological Studies*, 2020, https://www.faunajournal.com/archives/2020/vol7issue1/PartB/7-1-18-591.pdf.

10 "Oecd-Fao Agricultural Outlook 2021–2030," Organisation for Economic Co-operation and Development-Food and Agricultural Organization of the United Nations, 2021, https://openknowledge.fao.org/server/api/core/bitstreams/09e88a46-b005-4d65-8753-9714506afc38/content#:~:text=Growth%20in%20global%20consumption%20of,and%20sheep%20meat%20(5%25.

11 "Where Will We Find Enough Food for 9 Billion?," National Geographic, accessed: July 23, 2025, https://www.nationalgeographic.com/foodfeatures/feeding-9-billion/.

12 Ritchie, Hannah, "Deforestation and Forest Loss," Our World in Data, February 2021, https://ourworldindata.org/deforestation.

13 Eyshi Rezaei, Eshan, "Climate Change Impacts on Crop Yields," Nature Reviews, November 2023, https://www.nature.com/articles/s43017-023-00491-0.

14 Heggie, Jon, "Why Soil Matters (and What We Can Do to Save It)," National Geographic, 2023, https://www.nationalgeographic.com/environment/article/paid-content-why-soil-matters-and-what-we-can-do-to-save-it.

15 Heggie, Jon, "Why Soil Matters (and What We Can Do to Save It)," National Geographic, 2023, https://www.nationalgeographic.com/environment/article/paid-content-why-soil-matters-and-what-we-can-do-to-save-it.

16 Mount, Ian, “Farmers Use $60 Billion of Pesticides Each Year. 2 MIT Scientists Have Developed a New Technology That Could Cut That Number in Half,” MIT MechE, December 2022, https://meche.mit.edu/news-media/farmers-use-60-billion-pesticides-each-year-2-mit-scientists-have-developed-new; Singh, Snehal, “Herbicides Market Summary,” Market Future Research, July 2025, https://www.marketresearchfuture.com/reports/herbicides-market-4853.

17 Almulhim, Abdulaziz, Gabriela Nagle Alverio, Ayyoob Sharifi, Rejib Shaw, Saleemul Huq, Md Juel Mahmud, Shakil Ahmed, and Ismaila Rimi Abibakar, “Climate-Induced Migration in the Global South: An in Depth Analysis,” Nature, June 2024, https://www.nature.com/articles/s44168-024-00133-1.

18 “Planetary Boundaries,” Stockholm Resilience Centre, accessed: July 23, 2025, https://www.stockholmresilience.org/research/planetary-boundaries.html.

19 “Living Planet Report 2024,” World Wildlife Fund, accessed: July 23, 2025, https://livingplanet.panda.org/en-US/.

20 “Well … This Is the Million Dollar Question,” World Wildlife Fund, accessed: July 23, 2025, https://wwf.panda.org/discover/our_focus/biodiversity/biodiversity/.

21 “Media Release: Nature’s Dangerous Decline ‘Unprecedented’; Species Extinction Rates ‘Accelerating,’ ” ipbes, May 2019, https://www.ipbes.net/news/Media-Release-Global-Assessment.

22 “Media Release: Nature’s Dangerous Decline ‘Unprecedented’; Species Extinction Rates ‘Accelerating,’ ” ipbes, May 2019, https://www.ipbes.net/news/Media-Release-Global-Assessment.

23 Carrington, Damian, “Global Pollinator Losses Causing 500,000 Early Deaths a Year—Study,” *The Guardian*, January 2023, https://www.theguardian.com/environment/2023/jan/09/global-pollinator-losses-causing-500000-early-deaths-a-year-study#:~:text=Global%20pollinator%20losses%20causing%20500%2C000%20early%20deaths%20a%20year%20–%20study,-This%20article%20is&text=The%20global%20loss%20of%20pollinators,insects%20are%20in%20sharp%20decline.

24 “The State of Knowledge of Soil Biodiversity,” Food and Agriculture Organization of the United Nations, accessed: July 23, 2025, https://www.fao.org/interactive/soil-biodiversity/en/.

25 “Is Farmed Fish from Aquaculture a Healthy Source of Protein?” EIT Food, 2023, https://www.eitfood.eu/blog/is-farmed-fish-a-healthy-source-of-protein.

26 “Fish Consumption by Country 2025,” World Population Review, 2025, https://worldpopulationreview.com/country-rankings/fish-consumption-by-country.

27 Collins-Haynes, Lisa, “The Fearless Lionfish Huntresses of the Caribbean,” BBC, 2025, https://www.bbc.co.uk/travel/article/20250214-the-fearless-lionfish-huntresses-of-the-caribbean.

28 “Impacts of Invasive Lionfish,” NOAA Fisheries, https://www.fisheries.noaa.gov/southeast/ecosystems/impacts-invasive-lionfish.

29 "Is Farmed Fish from Aquaculture a Healthy Source of Protein?" EIT Food, 2023, https://www.eitfood.eu/blog/is-farmed-fish-a-healthy-source-of-protein.

30 Collins-Haynes, Lisa, "The Fearless Lionfish Huntresses of the Caribbean," BBC, 2025, https://www.bbc.co.uk/travel/article/20250214-the-fearless-lionfish-huntresses-of-the-caribbean.

31 Levien, Simon, Keep an Invasive Species in Check: Eat a Big Rat-Like Rodent, U.S. Says, *New York Times*, 2025, https://www.nytimes.com/2025/03/01/us/nutria-rodent-invasive.html?smid=nytcore-ios-share&referringSource=articleShare

32 Levien, Simon, "Keep an Invasive Species in Check: Eat a Big Rat-Like Rodent, U.S. Says," *New York Times*, 2025, https://www.nytimes.com/2025/03/01/us/nutria-rodent-invasive.html?smid=nytcore-ios-share&referringSource=articleShare.

33 Gravem, S. A, W. N. Heady, V. R. Saccomanno, K. F. Alvstad, A. L. M. Gehman, T. N. Frierson, and S. L. Hamilton, "Sunflower Sea Star," IUCN Red List, 2020, https://www.iucnredlist.org/species/178290276/178341498.

34 "Global Consumption of Agricultural Fertilizer from 1965 to 2022, by Nutrient," Statista, November 2024, https://www.statista.com/statistics/438967/fertilizer-consumption-globally-by-nutrient/; "Fertilizer Market Size to Worth around USD 276.92 Billion by 2034," Globe Newswire, October 2024, https://www.globenewswire.com/news-release/2024/10/08/2959512/0/en/Fertilizer-Market-Size-to-Worth-Around-USD-276-92-Billion-by-2034.html.

35 Jenssen, T. K., and G. Kongshaug, *Energy Consumption and Greenhouse Gas Emissions in Fertiliser Production*, International Fertiliser Society, 2003, https://fertiliser-society.org/store/energy-consumption-and-greenhouse-gas-emissions-in-fertiliser-production/#:~:text=Today%2C%20fertiliser%20production%20consumes%20approximately,as%20flue%20gas%20CO2; Menegat, S., A. Ledo, and R. Tirado, "Greenhouse Gas Emissions from Global Production and Use of Nitrogen Synthetic Fertilisers in Agriculture," Scientific Reports 12 (1) (August 25, 2022): 14490. doi: 10.1038/s41598-022-18773 w.

36 Hoxha, Antione, and Bjarne Christensen, "The Carbon Footprint of Fertiliser Production: Regional Reference Values," International Fertiliser Society, 2018, https://www.fertilizerseurope.com/wp-content/uploads/2020/01/The-carbon-footprint-of-fertilizer-production_Regional-reference-values.pdf.

37 "Nitrogen Driving Force and Pressure Relationships at Contrasting Scales Implications for Catchment Management | Publication Review," Stockholm Resilience Centre, accessed: July 23, 2025, https://www.stockholmresilience.org/publications/publications/2010-03-13-nitrogen-driving-force-and-pressure-relationships-at-contrasting-scales-implications-for-catchment-management.html.

38 Core Writing Team, R. K. Pachauri, and L. A. Meyer, "AR5 Climate Change 2014: Mitigation of Climate Change," Intergovernmental Panel on Climate Change, 2014, https://www.ipcc.ch/report/ar5/wg3/.

39 Diaz, Robert, and Rutger Rosenberg, "Spreading Dead Zones and Consequences for Marine Ecosystems," Science, August 2008, https://www.science.org/doi/10.1126/science.1156401.

40 Diaz, Robert, and Rutger Rosenberg, "Spreading Dead Zones and Consequences for Marine Ecosystems," Science, August 2008, https://www.science.org/doi/10.1126/science.1156401.

41 Ward, Mary, Rena Jones, Jean Brender, Theo de Kok, Peter Weyer, Bernard Nolan, Cristina Villanueva, and Simone Breda, "Drinking Water Nitrate and Human Health: An Updated Review," National Library of Medicine, July 2018, https://pmc.ncbi.nlm.nih.gov/articles/PMC6068531/#:~:text=Considering%20all%20studies%2C%20the%20strongest,disease%2C%20and%20neural%20tube%20defects.

42 Deniliquin, Mathare, "The Poisonous Global Politics of Water," *The Economist*, August 2024, https://www.economist.com/international/2024/08/26/the-poisonous-global-politics-of-water.

43 Bhagat, Sudheer, et al., "The Water Footprint of Livestock Production System and Livestock Products: A Dark Area: A Review," *International Journal of Fauna and Biological Studies*, 2020, https://www.faunajournal.com/archives/2020/vol7issue1/PartB/7-1-18-591.pdf.

44 Mekonnen, M., and A. Hoekstra, "The Green, Blue and Grey Water Footprint of Crops and Derived Crop Products," European Geosciences Union, May 2011, https://hess.copernicus.org/articles/15/1577/2011/.

45 Kuzma, Samantha, Liz Saccoccia, and Marlena Chertock, "25 Countries, Housing One-Quarter of the Population, Face Extremely High Water Stress," World Resources Institute, August 2023, https://www.wri.org/insights/highest-water-stressed-countries#:~:text=And%20at%20least%2050%25%20of,jobs%2C%20food%20and%20energy%20security.

46 Macpherson, E., C. Clark, P. W. Salazar, N. Baird, A. Akhtar-Khavari, and E. Challies, "Evolving Rights to (and of) Water in Chile: A Case for Relationship-Based Water Law and Governance," *International Journal of Human Rights*, 2023, 1–26. https://doi.org/10.1080/13642987.2023.2266719.

47 Richter, Brian, Dominique Bartak, Peter Caldwell, Kyle Frankel Davis, Peter Debaere, Arjen Y. Hoekstra, Tianshu Li, Landon Marston, Ryan McManamay, Mesfin M. Mekonnen, Benjamin L. Ruddell, Richard R. Rushforth, and J. Tara. Troy, "Water Scarcity and Fish Imperilment Driven by Beef Production," United States Department of Agriculture," 2020, https://research.fs.usda.gov/treesearch/59918.

48 Wada, Yoshihide, L. P. H. van Beek, and Marc Bierkens, "Nonsustainable Groundwater Sustaining Irrigation: A Global Assessment," American Geophysical Union, January 2012, https://agupubs.onlinelibrary.wiley.com/doi/10.1029/2011wr010562.

49 Mekonnen, M. and A. Hoekstra, "The Green, Blue and Grey Water Footprint of Crops and Derived Crop Products," European Geosciences Union, May 2011, https://hess.copernicus.org/articles/15/1577/2011/.

50 "Water War: How a Life-Sustaining Resource Goes Geopolitical," Carnegie Endowment for International Peace, March 2022, https://carnegieendowment.org/events/2022/03/water-war-how-a-life-sustaining-resource-goes-geopolitical?lang=en.

51 "Water War: How a Life-Sustaining Resource Goes Geopolitical," Carnegie Endowment for International Peace, March 2022, https://carnegieendowment.org/events/2022/03/water-war-how-a-life-sustaining-resource-goes-geopolitical?lang=en.

52 Baffour Frimpong, Osei, "Climate Change and Violent Extremism in the Lake Chad Basin: Key Issues and Way Forward," Wilson Center, July 2020, https://www.wilsoncenter.org/publication/climate-change-and-violent-extremism-lake-chad-basin-key-issues-and-way-forward.

53 Frankopan, Peter, *The New Silk Roads*, 2015, https://www.audible.co.uk/pd/The-New-Silk-Roads-Audiobook/B07HRRWC1B?source_code=M2M30DFT1BkSH11221601A7&&ipRedirectOverride=true&overrideBaseCountry=true&gclid=Cj0KCQjwkt6aBhDKARIsAAyeLJ3PrNDo6MdGTc8ENTjjFongHQPtxO6qcsb0MonsOS8AycqfnaZ2mucaAgdqEALw_wcB&gclsrc=aw.ds.

54 Frankopan, Peter, *The New Silk Roads*, 2015, https://www.audible.co.uk/pd/The-New-Silk-Roads-Audiobook/B07HRRWC1B?source_code=M2M30DFT1BkSH11221601A7&&ipRedirectOverride=true&overrideBaseCountry=true&gclid=Cj0KCQjwkt6aBhDKARIsAAyeLJ3PrNDo6MdGTc8ENTjjFongHQPtxO6qcsb0MonsOS8AycqfnaZ2mucaAgdqEALw_wcB&gclsrc=aw.ds.

55 Elaine,Watson, "Honey … without Bees? MeliBio Rolls Out Plant-Based Alternative in Austria and Switzerland with ALDI," AFN, 2024, https://agfundernews.com/honey-without-bees-melibio-rolls-out-plant-based-alternative-in-austria-and-switzerland-with-aldi.

56 Elaine,Watson, "Honey … without Bees? MeliBio Rolls Out Plant-Based Alternative in Austria and Switzerland with ALDI," AFN, 2024, https://agfundernews.com/honey-without-bees-melibio-rolls-out-plant-based-alternative-in-austria-and-switzerland-with-aldi.

57 Canaday, Autumn, "DNA Research Finds Low Genetic Diversity Among U.S. Honey Bees," USDA Agricultural Research Service, 2023, https://www.ars.usda.gov/news-events/news/research-news/2023/dna-research-finds-low-genetic-diversity-among-us-honey-bees/.

58 "Global Land Outlook, United Nations Convention to Combat Desertification," 2022 https://www.unccd.int/resources/global-land-outlook/glo2.

59 "Soil Erosion Threatens Environment and Human Health, Study Reports," Science Daily, March 2006, https://www.sciencedaily.com/releases/2006/03/060322141021.htm.

60 "Soil Erosion," United Nations Office for Disaster Risk Reduction, accessed: July 24, 2025, https://www.undrr.org/understanding-disaster-risk/terminology/hips/en0019#:~:text=A%20total%20of%2075%20billion,FAO%20and%20IAEA,%202017.

61 Boehm, Sophie, and Clea Schumer, "10 Big Findings from the 2023 IPCC Report on Climate Change," World Resources Institute, March 2023, https://www.wri.org/insights/2023-ipcc-ar6-synthesis-report-climate-change-findings; "Living Planet Report 2020," World Wildlife Fund, September 2020, https://www.worldwildlife.org/publications/living-planet-report-2020.

62 Lang, Tim, *Feeding Britain*, Pelican Books, April 2021, https://www.penguin.co.uk/books/308380/feeding-britain-by-lang-tim/9780241404805.

63 Reynolds, Matthew, *The Future of Food (WIRED Guides): How to Feed the Planet without Destroying It*, Random House Business, September 2021, https://www.amazon.co.uk/Future-Food-WIRED-guides-Destroying/dp/1847943284/ref=asc_df_1847943284/?tag=googshopuk-21&linkCode=df0&hvadid=499252476918&hvpos=&hvnetw=g&hvrand=2780991607230015613&hvpone=&hvptwo=&hvqmt=&hvdev=c&hvdvcmdl=&hvlocint=&hvlocphy=1006567&hvtargid=pla-1168223889095&psc=1&mcid=92693168bea031d7b05b05bc9a644964&th=1&psc=1.

2 Public Health in Crisis

1 "Obesity and Overweight," World Health Organization, accessed: August 10, 2025, https://www.who.int/news-room/fact-sheets/detail/obesity-and-overweight.

2 "Unhealthy Diet," NCD Alliance, accessed: August 10, 2025, https://ncdalliance.org/explore-ncds/risk-factors/unhealthy-diets.

3 "Rising obesity rates across the world reveal massive gaps in healthcare and nutrition, with poorest populations most adversely affected," World Obesity Atlas 2024, 2024 https://www.worldobesity.org/news/world-obesity-atlas-2024.

4 "Rising Obesity Rates across the World Reveal Massive Gaps in Healthcare and Nutrition, with Poorest Populations Most Adversely Affected," World Obesity Atlas 2024, 2024, https://www.worldobesity.org/news/world-obesity-atlas-2024.

5 Alamirie Ejigu, Bedilu and Fentanesh Nibret Tiruneh, "The Link between Overweight/Obesity and Noncommunicable Diseases in Ethiopia: Evidences from Nationwide WHO STEPS Survey 2015," National Library of Medicine, 2023, https://pmc.ncbi.nlm.nih.gov/articles/PMC10667048/.

6 Hayes, Tara, and Rakeb Asres, "The Economic Costs of Poor Nutrition," American Action Forum, March 2022, https://www.americanactionforum.org/research/the-economic-costs-of-poor-nutrition/.

7 Hayes, Tara, and Rakeb Asres, "The Economic Costs of Poor Nutrition," American Action Forum, March 2022, https://www.americanactionforum.org/research/the-economic-costs-of-poor-nutrition/.

8 Merschel, Michael, "How Much Harm Can a Little Excess Salt Do? Plenty," American Heart Association, May 2021, https://www.heart.org/en/news/2021/05/26/how-much-harm-can-a-little-excess-salt-do-plenty.

9 "Effects of Excess Sodium Infographic," American Heart Association, accessed: August 10, 2025, https://www.heart.org/en/healthy-living/healthy-eating/eat-smart/sodium/effects-of-excess-sodium-infographic.

10 "The Sweet Danger of Sugar," Harvard Health Publishing, January 2022, https://www.health.harvard.edu/heart-health/the-sweet-danger-of-sugar.

11 "Fat: The Facts," National Health Service, accessed: August 10, 2025, https://www.nhs.uk/live-well/eat-well/food-types/different-fats-nutrition/.

12 "Indicator Metadata Registry List," World Health Organization, accessed: August 10, 2025, https://www.who.int/data/gho/indicator-metadata-registry/imr-details/3418#:~:text=Excessive%20dietary%20fat%20intake%20has,the%20shift%20in%20consumption%20patterns.

13 Pietrangelo, Ann, "What's the Difference between Saturated and Unsaturated Fat?," Healthline, July 2023, https://www.healthline.com/health/food-nutrition/saturated-and-unsaturated-fat.

14 "Taking Aim at Belly Fat," Harvard Health Publishing, March 2024, https://www.health.harvard.edu/newsletter_article/taking-aim-at-belly-fat.

15 "Five Days of Eating Fatty Foods Can Alter How Your Body's Muscle Processes Food, Researchers Find," Virginia Tech News, April 2015, https://news.vt.edu/articles/2015/04/041315-cals-skeletal.html.

16 Herb Neff, Kirstie, Angela Fay, and Karen Saules, "Foods and Nutritional Characteristics Associated with Addictive-Like Eating," National Library of Medicine, May 2021, https://pubmed.ncbi.nlm.nih.gov/33934668/.

17 Vogler, Pen, *Stuffed*, Atlantic Books, October 2024, https://atlantic-books.co.uk/book/stuffed/.

18 Elizabeth, Leonie, Priscila Machado, Marit Zinocker, Philip Baker, and Mark Lawrence, "Ultra-Processed Foods and Health Outcomes: A Narrative Review," National Library of Medicine, June 2020, https://pmc.ncbi.nlm.nih.gov/articles/PMC7399967/.

19 Van Tulleken, Chris, *Ultra-Processed People: Why Do We All Eat Stuff That Isn't Food … and Why Can't We Stop*, Penguin, 2024.

20 "The Broken Plate 2025," Food Foundation, 2025, https://foodfoundation.org.uk/sites/default/files/2025-01/TFF_The%20Broken%20Plate%202005%20FINAL%20DIGITAL.pdf.

21 Lane, Melissa, Elizabeth Garnage, Shutong Du, Deborah Ashtree, Amelia McGuinness, Sarah Gauci, Philip Baker, Mark Lawrence, Casey Rebholz, Bernard Srour, Mathilde Touvier, Felice Jacka, Adrienne O'Neil, Toby Segasby, and Wolfgang Marx, "Ultra-Processed Food Exposure and Adverse Health Outcomes: Umbrella Review of Epidemiological Meta-Analyses," The BMJ, February 2024, https://www.bmj.com/content/384/bmj-2023-077310.

22 Hall, Kevin, Alexis Ayuketah, Robert Brychta, Hongyi Cai, Thomas Cassimatis, Kong Chen, Stephanie Chung, Elise Costa, AmberCourville, Valerie Darcey, Laura Fletcher, Ciaran Forde, Ahmed Gharib, Juen Guo, Rebecca Howard, Paule Joseph, Suzanne McGehee, Ronald Ouwrkerk, Klaudia Rasinher, Irene Rozga, Michael Stagliano, Mary Walter, Peter Walter, Shanna Yong, and Megan Zhou, "Ultra-Processed Diets Cause Excess Calorie Intake and Weight Gain: An Inpatient Randomized Controlled Trial of Ad Libitum Food Intake," National Library of Medicine, May 2019, https://pmc.ncbi.nlm.nih.gov/articles/PMC7946062/#:~:text=Interestingly%2C%20Figure%202F%20illustrates%20that,that%20are%20typically%20eaten%20slowly.

23 Van Tulleken, Chris, *Ultra-Processed People: Why Do We All Eat Stuff That Isn't Food … and Why Can't We Stop*, Penguin, 2024.

24 Morea, Nicola, "Ultra-Processed Foods: Nova Classification," Mérieux NutriSciences Italy, March 2021, https://regulatory.mxns.com/en/ultra-processed-foods-nova-classification; Salge Blake, John, "Examining the Nova Food Classification System and the Healthfulness of Ultra-Processed Foods," Eat Right Pro, August 2024, https://www.eatrightpro.org/news-center/practice-trends/examining-the-nova-food-classification-system-and-healthfulness-of-ultra-processed-foods.

25 Martínez Steele, Eurídice, Larisssa Baraldi, Maria Louzada, jean-Claude Moubarac, Dariush Mozaffarian, and Carlos Monteiro, "Ultra-Processed Foods and Added Sugars in the US Diet: Evidence from a Nationally Representative Cross-Sectional Study," National Library of Medicine, January 2016, https://pmc.ncbi.nlm.nih.gov/articles/PMC4785287/.

26 Martínez Steele, Eurídice, et al., "Ultra-Processed Foods and Added Sugars in the US Diet: Evidence from a Nationally Representative Cross-Sectional Study," National Library of Medicine, January 2016, https://pmc.ncbi.nlm.nih.gov/articles/PMC4785287/.

27 Monteiro, Carlos, Geoffrey Cannon, Renata Levy, Jean-Claude Moubarac, Maria Louzada, Fernanda Rauber, Neha Khandpur, Gustavo Cediel, Daniela Neri, Euridice Martinez-Steele, Larissa Baraldi, and Patricia Jaime, "Ultra-Processed Foods: What

They Are and How to Identify Them," National Library of Medicine, February 2019, https://pmc.ncbi.nlm.nih.gov/articles/PMC10260459/.
28 Federal Register, "Ultra-Processed Foods; Request for Information," National Archives, July 2025, https://www.federalregister.gov/documents/2025/07/25/2025-14089/ultra-processed-foods-request-for-information.
29 Gibney, Michael, "Ultra-Processed Foods: Definitions and Policy Issues," National Library of Medicine, September 2018, https://pmc.ncbi.nlm.nih.gov/articles/PMC6389637/#sec6.
30 Gibney, Michael, "Ultra-Processed Foods: Definitions and Policy Issues," National Library of Medicine, September 2018, https://pmc.ncbi.nlm.nih.gov/articles/PMC6389637/#sec6.
31 Gibney, Michael, "Ultra-Processed Foods: Definitions and Policy Issues," National Library of Medicine, September 2018, https://pmc.ncbi.nlm.nih.gov/articles/PMC6389637/#sec6.
32 "Titanium Dioxide as a Color Additive in Foods," FDA, 2024, https://www.fda.gov/industry/color-additives/titanium-dioxide-color-additive-foods.
33 EFSA Panel on Additives and Products or Substances Used in Animal Feed, "Safety and Efficacy of a Feed Additive Consisting of Titanium Dioxide for All Animal Species," *EFSA Journal* 19 (6) (2021), https://efsa.onlinelibrary.wiley.com/doi/10.2903/j.efsa.2021.6630.
34 "Titanium Dioxide: E171 No Longer Considered Safe When Used as a Food Additive," EFSA, 2021, https://www.efsa.europa.eu/en/news/titanium-dioxide-e171-no-longer-considered-safe-when-used-food-additive.
35 "Guidance for Industry Considering Whether an FDA-Regulated Product Involves the Application of Nanotechnology," FDA, 2011, https://www.liebertpub.com/doi/10.1089/blr.2011.9814.
36 "EU-In Horizon Scanning Report, Nanotechnology-Based Medicinal Products for Human Use," Euopean Medicines Agency, 2025, https://www.ema.europa.eu/en/documents/report/nanotechnology-based-medicinal-products-human-use-eu-horizon-scanning-report_en.pdf.
37 "Titanium Dioxide: E171 No Longer Considered Safe When Used as a Food Additive," EFSA, 2021, https://www.efsa.europa.eu/en/news/titanium-dioxide-e171-no-longer-considered-safe-when-used-food-additive.
38 Kaluc, Nur, Sara Bertorello, Oguz Kaan Tombul. Simone Baldi, Guila Nannini, Gianluca Bartolucci, Elena Niccolai, and Amedeo Amedei, "Gut-Lung Microbiota Dynamics in Mice Exposed to Nanoplastics," NanoImpact, 36, 2024, https://www.sciencedirect.com/science/article/pii/S2452074824000417.
39 State of the Union, "Nanomaterials in Food," EU Food, Farming and Fisheries, https://food.ec.europa.eu/food-safety/novel-food/nanomaterials_en.

40 Kern, David, Amy Auchincloss, Mark Stehr, Ana Diez Roux, Latetia Moore, Genevieve Kater, and Lucy Robinson, "Neighborhood Prices of Healthier and Unhealthier Foods and Associations with Diet Quality: Evidence from the Multi-Ethnic Study of Atherosclerosis," National Library of Medicine, November 2017, https://pmc.ncbi.nlm.nih.gov/articles/PMC5708033/.

41 "How to Tax Sugary Drinks," *The Economist*, May 2019, https://www.economist.com/finance-and-economics/2019/05/23/how-to-tax-sugary-drinks.

42 Vogler, Pen, *Stuffed*, Atlantic Books, October 2024, https://atlantic-books.co.uk/book/stuffed/.

43 "Food Insecurity," Office of Disease Prevention and Health Promotion, accessed: August 10, 2025, https://odphp.health.gov/healthypeople/priority-areas/social-determinants-health/literature-summaries/food-insecurity#cit1.

44 Food and Agriculture Organization, Rome Declaration on World Food Security and World Food Summit Plan of Action. World Food Summit, Rome, 1996. http://www.fao.org/3/w3613e/w3613e00.htm.

45 Rabbitt, Matthew, Laura Hales, and Madeline Reed-Jones, "Food Security in the U.S.—Key Statistics & Graphics," USDA Economic Research Service, January 2025, https://www.ers.usda.gov/topics/food-nutrition-assistance/food-security-in-the-us/key-statistics-graphics#:~:text=13.5%20percent%20(18.0%20million)%20of,(17.0%20million)%20in%202022; "Hunger Numbers Stubbornly High for Three Consecutive Years as Global Crises Deepen: UN Report," World Health Organization, July 2024, https://www.who.int/news/item/24-07-2024-hunger-numbers-stubbornly-high-for-three-consecutive-years-as-global-crises-deepen--un-report.

46 Ritchie, Hannah, Pablo Rosado, and Max Roser, "Hunger and Undernourishment," Our World in Data, 2023, https://ourworldindata.org/hunger-and-undernourishment.

47 "World Hunger Facts," Action against Hunger, accessed: August 10, 2025, https://www.actionagainsthunger.org.uk/why-hunger/world-hunger-facts.

48 Reynolds, Matt, *The Future of Food: How to Feed the Planet without Destroying It*, Penguin Random House, 2021, https://www.amazon.co.uk/Future-Food-WIRED-Guides-Destroying/dp/B09B3B2312/ref=sr_1_1?crid=32GRM2O3O5F7P&dib=eyJ2IjoiMSJ9.6l9EOgsuXSEiVPLXlBK7S2zlS2Zcr9cQKQF30efGAAjGjHj071QN20LucGBJIEps.umuLpCO0C4iVDiQqvrcu2BT4jwbadcIXD1Pnxc2c818&dib_tag=se&keywords=Wired+future+of+food&qid=1716055661&s=books&sprefix=wired+future+of+food,stripbooks,126&sr=1-1.

49 "Obesity: Health and Economic Consequences of an Impending Global Challenge," World Bank Group, January 2020, https://www.worldbank.org/en/topic/nutrition/publication/obesity-health-and-economic-consequences-of-an-impending-global-challenge.

50 "Small Investments in Nutrition Could Make the World Brainier," *The Economist*, July 2024, https://www.economist.com/briefing/2024/07/11/small-investments-in-nutrition-could-make-the-world-brainier.

51 Jun Kim, Tae, and Olaf von dem Knesebeck, "Income and Obesity: What Is the Direction of the Relationship? A Systematic Review and Meta-Analysis," National Library of Medicine, January 2018, https://pmc.ncbi.nlm.nih.gov/articles/PMC5781054/#:~:text=Studies%20testing%20reverse%20causality,relationship%20that%20was%20statistically%20insignificant.

52 Datar, Ashlesha, Roland Sturm, and Jennifer Magnabosco, "Childhood Overweight and Academic Performance: National Study of Kindergartners and First-Graders," National Library of Medicine, January 2004, https://pubmed.ncbi.nlm.nih.gov/14742843/.

53 Lang, Tim, *Feeding Britain*, Pelican Books, April 2021, https://www.penguin.co.uk/books/308380/feeding-britain-by-lang-tim/9780241404805.

54 Banerjee, Kajori, and Laxmi Kant Dwivedi, "Disparity in Childhood Stunting in India: Relative Importance of Community-Level Nutrition and Sanitary Practices," National Library of Medicine, September 2020, https://pmc.ncbi.nlm.nih.gov/articles/PMC7462311/#:~:text=Mean%20HAZ%20in%202015–16%20in%20India%20and%20selected%20states.&text=The%20distribution%20graphs%20in%20Fig,in%20comparison%20to%20Tamil%20Nadu..

55 Balarajan, Yarlin, and Michael Reich, "Political Economy of Child Nutrition Policy: A Qualitative Study of India's Integrated Child Development Services (ICDS) Scheme," Science Direct, July 2016, https://www.sciencedirect.com/science/article/pii/S0306919216300252.i.

56 "The Whole Truth| Positioning Whole Foods at the Centre of Government Food Strategy," Soil Association, 2025, https://www.soilassociation.org/media/28184/the-whole-truth-final.pdf.

57 "Lifelong Healthy Eating," Global Child Nutrition Forum, 2024, https://gcnf.org/how-japan-leverages-school-meals-as-a-living-textbook-for-lifelong-healthy-eating/.

58 Narayan, Swati, "India Feeds 100 Million Children a Day—So Why Can't Britain Have Free School Meals?," *Prospect Magazine*, June 2017, https://www.prospectmagazine.co.uk/politics/44593/india-feeds-100-million-children-a-dayso-why-cant-britain-have-free-school-meals.

59 "Global Food Market Set to Reach US$10.5 Trillion in 2024, with VC Trends Pointing to Need for More Real Estate," Savills, accessed: August 10, 2025, https://www.savills.com/insight-and-opinion/savills-news/369075-0/global-food-market-set-to-reach-us%2410.5-trillion-in-2024--with-vc-trends-pointing-to-need-for-more-real-estate?utm_source=chatgpt.com.

60 Benton, Tim, Rob Bailey, and Bernice Lee, "Breaking the Vicious Circle: Food, Climate & Nutrition," Sustainability Accelerator, June 2018, https://accelerator.chathamhouse.org/article/breaking-the-vicious-cycle-food-climate-nutrition.

61 Lang, Tim, *Feeding Britain*, Pelican Books, April 2021, https://www.penguin.co.uk/books/308380/feeding-britain-by-lang-tim/9780241404805.

62 Bagekari, Aarti, "Pharmaceutical Market Report 2025 (Global Edition)," Cognitive Market Research, June 2025, https://www.cognitivemarketresearch.com/pharmaceutical-market-report.

63 Purcell, Conor, "How the Food Industry Manipulates Nutrition Science: Five Questions for Marion Nestle," Undark, November 2018, https://undark.org/2018/11/30/marion-nestle-food-science-industry/.

64 "Interest Groups," Open Secrets, February 2025, https://www.opensecrets.org/industries.

65 Zimmer, Scott, "Veblen's Theory of Conspicuous Consumption," EBSCO, February 2017, https://www.ebsco.com/research-starters/political-science/veblens-theory-conspicuous-consumption.

66 Madyibi, Siphe, "The Relationship between Conspicuous Consumption and Poverty in Developing Countries: Evidence from South Africa," University of the Western Cape, 2017, https://uwcscholar.uwc.ac.za/items/29c9aacf-13a3-4034-b1fb-c174affc369a.

67 Hawkes, Corinna, "Uneven Dietary Development: Linking the Policies and Processes of Globalization with the Nutrition Transition, Obesity and Diet-Related Chronic Diseases," Globalization and Health, March 2006, https://globalizationandhealth.biomedcentral.com/articles/10.1186/1744-8603-2-4.

68 Barquera, Simón, and Juan Rivera, "Obesity in Mexico: Rapid Epidemiological Transition and Food Industry Interference in Health Policies," National Library of Medicine, August 2020, https://pmc.ncbi.nlm.nih.gov/articles/PMC7434327/#:~:text=In%20the%20latest%20national%20survey,%25%20vs%2030·5%25).&text=Results%20also%20showed%20only%2023,group%20(15·4%25).

69 Hari, Johann, "The Land That Doesn't Need Ozempic," Time, May 2024, https://time.com/6974579/japan-food-culture-low-obesity/.

70 Windarti, Nanik, Suwahim Hlaing, and Makoto Kakinaka, "Obesity Kuznets Curve: International Evidence," Science Direct, April 2019, https://www.sciencedirect.com/science/article/pii/S0033350619300046.

71 Dunbar, Marina, "RFK Jr Calls Sugar 'Poison' but Says Government Probably Can't Eliminate It," *The Guardian*, April 2025, https://www.theguardian.com/us-news/2025/apr/22/rfk-jr-sugar-poison-food-dyes; "Fact-Checking RFK Jr's Views on Health Policy," BBC Verify, November 2024, https://www.bbc.com/news/articles/c0mzk2y41zvo; Snape, Joel, "Robert F Kennedy Jr Claims Seed Oils Are 'Poisoning' Us. Here's Why He's Wrong," *The Guardian*, January 2025, https://www.theguardian.

com/lifeandstyle/2025/jan/13/robert-f-kennedy-jr-claims-seed-oils-are-poisoning-us-heres-why-hes-wrong.

72 Speed, Madeleine, and Gregory Meyer, "Robert F Kennedy Jr's Battle with Big Food," Financial Post, May 2025, https://financialpost.com/financial-times/robert-f-kennedy-battle-with-big-food.

73 Aleccia, JoNel, and Matthew Perrone, "WATCH: RFK Jr and Health Officials Move to Phase Out Artificial Dyes from Food Supply," PBS News, April 2025, https://www.pbs.org/newshour/politics/watch-live-rfk-jr-to-announce-phasing-out-of-artificial-food-dyes#:~:text="We%20don't%20have%20an%20agreement%2C%20we%20have,hyperactivity%20and%20attention%20issues%2C%20in%20some%20children.

74 Sridhar, Devi, "The UK and EU Cracked Down on Food Dyes Long Ago—Maga Is Way behind the Curve," *The Guardian*, April 2025, https://www.theguardian.com/commentisfree/2025/apr/30/britain-us-food-safety-robert-f-kennedy-jr.

75 Speed, Madeline, and Gregory Meyerm, "Robert F Kennedt Jr's Battle with Big Food," *Financial Times*, 2025, https://www.ft.com/content/7520dfbf-49b2-4d9b-8247-a0b80c5f1314

76 "USDA Approves First State Waivers to Restrict SNAP Purchases," Hogan Lovells, May 2025, https://www.hoganlovells.com/en/publications/usda-approves-first-state-waivers-to-restrict-snap-purchases#:~:text=On%20May%2019%2C%202025%2C%20U.S.,eating%20and%20improve%20lifestyle%20habits.

77 Hobbs, Katie, "RE: House Bill 2165: SNAP; Prohibited Purchases; Waiver," State of Arizona Office of the Governor, April 2025, https://www.azleg.gov/govlettr/57leg/1r/hb2165.pdf.

78 "WHO Launches New Guideline on Fiscal Policies to Promote Healthy Diets," World Health Organization, June 2024, https://www.who.int/news/item/14-06-2024-who-launches-new-guideline-on-fiscal-policies-to-promote-healthy-diets#:~:text=This%20in%20turn%20encourages%20their,variety%20of%20unhealthy%20food%20categories.

79 Muhammad, Andrew, and Birgit Meade, "Low-Income Countries Are Most Responsive to Income and Food Price Changes," USDA Economic Research Service, September 2011, https://www.ers.usda.gov/amber-waves/2011/september/low-income-countries#:~:text=Consumers%20in%20low%2Dincome%20countries%2C%20such%20as%20the,United%20States%20)%20%2C%20Canada%2C%20and%20Germany.

80 Carlos Salgado Hernández, Juan, Ana Basto-Abreu, Isabel Junquera-Badilla, Luis Moreno-Aguilar, Tonatiuh Barrientos-Gutierrez, and Arantxa Colchero, "Building upon the Sugar Beverage Tax in Mexico: A Modelling Study of Tax Alternatives to Increase Benefits," National Library of Medicine, November 2023, https://pmc.ncbi.nlm.nih.gov/articles/PMC10649495/#:~:text=In%202014%2C%20Mexico%20impl

emented%20a%20one%20peso%2Dper%2Dlitre%20tax%20to%20sugar%2Dsweetened%20beverage%20(SSB).&text=In%20January%202014%2C%20as%20part%20of%20a,to%20all%20non%2Dalcoholic%20beverages%20with%20added%20sugar.

81 "Soda Sales in Mexico Continue to Decline since Sugary Beverage Tax in First Two Years," Mike Bloomberg, February 2017, https://www.mikebloomberg.com/news/soda-sales-mexico-continue-decline-since-sugary-beverage-tax-began-two-years-ago/#:~:text=Mexico%20implemented%20a%201%20peso,the%20burden%20of%20chronic%20disease.

82 Colchero, Arantxa, Barry M. Popkin, Juan A. Rivera, and Shu Wen Ng, "Beverage Purchases from Stores in Mexico under the Excise Tax on Sugar Sweetened Beverages: Observational Study," BMJ, January 2016, https://www.bmj.com/content/352/bmj.h6704#:~:text=Discussion,increased%20their%20purchases%20the%20most.

83 "South African Beverage Tax Has Reduced Purchases of Sugar-Sweetened Beverages," University of North Carolina Gillings School of Global Public Health, April 2021, https://sph.unc.edu/sph-news/south-african-beverage-tax-has-reduced-purchases-of-sugar-sweetened-beverages/.

84 "South African Beverage Tax Has Reduced Purchases of Sugar-Sweetened Beverages," University of North Carolina Gillings School of Global Public Health, April 2021, https://sph.unc.edu/sph-news/south-african-beverage-tax-has-reduced-purchases-of-sugar-sweetened-beverages/.

85 Basu, Sanjay, Euridice Martinez-Steele, Neha Khandpur, Gustavo Cediel, Maria Zapata, Fernanda Rauber, Joaquin Marron-Ponce, Priscila Machado, Maria Laura Louzada, Giovanna Andrade, Carolina Batis Nancy Babio, Jordi Salas-Salvado, Christopher Millett, Carlos Monteiro, and Renata Levy, "Averting Obesity and Type 2 Diabetes in India through Sugar-Sweetened Beverage Taxation: An Economic-Epidemiologic Modeling Study," National Library of Medicine, January 2014, https://pubmed.ncbi.nlm.nih.gov/24409102/#:~:text=The%2020%25%20SSB%20tax%20was,used%20to%20inform%20our%20calculations.

86 Scarborough, Peter, Vyas Adhikari, Richard Harrington, Ahmend Eluhussein, Adam Briggs, Mike Rayner, Jean Adams, Steven cummins, Tarra Penney, and Martin White, "Impact of the Announcement and Implementation of the UK Soft Drinks Industry Levy on Sugar Content, Price, Product Size and Number of Available Soft Drinks in the UK, 2015–19: A Controlled Interrupted Time Series Analysis," National Library of Medicine, February 2020, https://pmc.ncbi.nlm.nih.gov/articles/PMC7012398/#:~:text=In%20March%202016%2C%20the%20United,15%25%20over%20the%20time%20period.

87 "New Report Shows Further Sugar Reduction Progress by Food Industry," Public Health England, September 2019, https://www.gov.uk/government/news/new-report-shows-further-sugar-reduction-progress-by-food-industry-1#:~:text=Soft%20

Drinks%20Industry%20Levy%20data,brand%20and%20manufacturer%2Dbranded%20products.&text=Public%20Health%20England%20(%20PHE%20)%20has,simple%20average%20sugar%20per%20100g.

88 Sasse, Tom, and Sophie Metcalfe, "Sugar Tax," Institute for Government, November 2022, https://www.instituteforgovernment.org.uk/explainer/sugar-tax.

89 "The False Economy of Big Food," Food, Farming & Countryside Commission, November 2024, https://ffcc.co.uk/publications/the-false-economy-of-big-food.

90 "Official Statistics: Soft Drinks Industry Levy Statistics Commentary 2024," HM Revenue & Customs, September 2024, https://www.gov.uk/government/statistics/soft-drinks-industry-levy-statistics/soft-drinks-industry-levy-statistics-commentary-2021.

91 Smith, Eileen, "Chile Battles Obesity with Stop Signs on Packaged Foods," National Public Radio, August 2016, https://www.npr.org/sections/thesalt/2016/08/12/486898630/chile-battles-obesity-with-stop-signs-on-packaged-foods.

92 Smith, Eileen, "Chile Battles Obesity with Stop Signs on Packaged Foods," National Public Radio, August 2016, https://www.npr.org/sections/thesalt/2016/08/12/486898630/chile-battles-obesity-with-stop-signs-on-packaged-foods.

93 Wilson, Bee, "Ultra-Processed Babies: Are Toddler Snacks One of the Great Food Scandals of Our Time?," *The Guardian*, 2025, https://www.theguardian.com/society/2025/mar/15/ultra-processed-babies-are-toddler-snacks-one-of-the-great-food-scandals-of-our-time.

94 Wilson, Bee, "Ultra-Processed Babies: Are Toddler Snacks One of the Great Food Scandals of Our Time?," *The Guardian*, 2025, https://www.theguardian.com/society/2025/mar/15/ultra-processed-babies-are-toddler-snacks-one-of-the-great-food-scandals-of-our-time.

95 Wilson, Bee, "Ultra-Processed Babies: Are Toddler Snacks One of the Great Food Scandals of Our Time?," *The Guardian*, 2025, https://www.theguardian.com/society/2025/mar/15/ultra-processed-babies-are-toddler-snacks-one-of-the-great-food-scandals-of-our-time.

96 Cosslett, Rhiannon Lucy, "Baby Food Pouch Hysteria? It's Just Another Way of Making Mothers Feel Guilty," *The Guardian*, 2025, https://www.theguardian.com/commentisfree/2025/may/04/baby-food-pouch-parents-society-choices.

97 Daniela, Neri, Euridice Martinez-Steele, Neha Khandpur, Gustavo Cediel, Maria Zapata, Fernanda Rauber, Joaquin Marron-Ponce, Priscila Machado, Maria Laura Louzada, Giovanna Andrade, Carolina Batis Nancy Babio, Jordi Salas-Salvado, Christopher Millett, Carlos Monteiro, and Renata Levy, "Ultraprocessed Food Consumption and Dietary Nutrient Profiles Associated with Obesity: A Multicountry Study of Children and Adolescents," Obesity Reviews,

23: Determinants of Childhood Obesity, 2022, https://onlinelibrary.wiley.com/doi/epdf/10.1111/obr.13387

98 Daniela, Neri, et al. "Ultraprocessed Food Consumption and Dietary Nutrient Profiles Associated with Obesity: A Multicountry Study of Children and Adolescents," Obesity Reviews, 23: Determinants of Childhood Obesity, 2022, https://onlinelibrary.wiley.com/doi/epdf/10.1111/obr.13387.

99 Daniela, Neri, et al. "Ultraprocessed Food Consumption and Dietary Nutrient Profiles Associated with Obesity: A Multicountry Study of Children and Adolescents," Obesity Reviews, 23: Determinants of Childhood Obesity, 2022, https://onlinelibrary.wiley.com/doi/epdf/10.1111/obr.13387.

100 Wilson, Bee, "Ultra-Processed Babies: Are Toddler Snacks One of the Great Food Scandals of Our Time?," *The Guardian*, 2025, https://www.theguardian.com/society/2025/mar/15/ultra-processed-babies-are-toddler-snacks-one-of-the-great-food-scandals-of-our-time.

101 Beasley, Deena, "Weight-Loss Drug Forecasts Jump to $150 Billion as Supply Grows," *Reuters*, May 2024, https://www.reuters.com/business/healthcare-pharmaceuticals/weight-loss-drug-forecasts-jump-150-billion-supply-grows-2024-05-28/.

102 Garvey, Timothy, Rachel Batterham, Meena Bhatta, Silvio Buscemi, Louise Christensen, Juan Frias, Esteban Jodar, Kristian Kandler, Georgia Rigas, Thomas wadden, Sean Wharton, and the STEP 5 Study Group, "Two-Year Effects of Semaglutide in Adults with Overweight or Obesity: The STEP 5 Trial," Nature Medicine, October 2022, https://www.nature.com/articles/s41591-022-02026-4#Sec2.

103 Davis, Nicola, "Weight-Loss Jabs Linked to Reduced Risk of 42 Conditions Including Dementia," *The Guardian*, January 2025, https://www.theguardian.com/society/2025/jan/20/weight-loss-jabs-linked-to-reduced-risk-of-42-conditions-including-dementia.

104 Sloan, Erica, "People with PCOS Are Taking GLP-1s for Their Symptoms. Here's What to Know," SELF, May 2025, https://www.self.com/story/glp1-for-pcos-treatment-ozempic-mounjaro-wegovy?utm_source=chatgpt.com.

105 Cantor, Carla, "The Rise of Ozempic for Weight Loss Sparks Ethical Concerns," Columbia University Department of Psychiatry, October 2024, https://www.columbiapsychiatry.org/news/rise-ozempic-sparks-ethical-concerns.

106 Wilson, Amelia, "How Much Does Ozempic Cost without Insurance?," Ro, March 2023, https://ro.co/weight-loss/ozempic-cost-without-insurance/.

107 Grimmelt, Anne, Jessica Moulton, Chirag Pandya, and Nadya Snezhkova, "Hungry and Confused: The Winding Road to Conscious Eating," McKinsey & Company, October 2022, https://www.mckinsey.com/industries/consumer-packaged-goods/

our-insights/hungry-and-confused-the-winding-road-to-conscious-eating?utm_source=chatgpt.com.

108 Grimmelt, Anne, et al., "Hungry and Confused: The Winding Road to Conscious Eating," McKinsey & Company, October 2022, https://www.mckinsey.com/industries/consumer-packaged-goods/our-insights/hungry-and-confused-the-winding-road-to-conscious-eating?utm_source=chatgpt.com.

109 Grimmelt, Anne, et al., "Hungry and Confused: The Winding Road to Conscious Eating," McKinsey & Company, October 2022, https://www.mckinsey.com/industries/consumer-packaged-goods/our-insights/hungry-and-confused-the-winding-road-to-conscious-eating?utm_source=chatgpt.com.

110 "Dietary Supplements Market Size, Share & Trends Analysis Report By Ingredients (Vitamin, Botanicals, Minerals), by Type, by Form (Tablets, Capsules, Powder), by Application, by End User, by Distribution Channel, by Region, and Segment Forecasts, 2025–2033," Grand View Research, 2024, https://www.grandviewresearch.com/industry-analysis/dietary-supplements-market-report.

111 Sender, Ron, Shai Fuchs, and Ron Milo, "Revised Estimates for the Number of Human and Bacteria Cells in the Body," National Library of Medicine, August 2016, https://pmc.ncbi.nlm.nih.gov/articles/PMC4991899/.

112 "FAQ: Human Microbiome," American Society for Microbiology, 2013, https://www.ncbi.nlm.nih.gov/books/NBK562894/.

113 Thursby, Elizabeth, and Nathalie Juge, "Introduction to the Human Gut Microbiota," National Library of Medicine, May 2017, https://pmc.ncbi.nlm.nih.gov/articles/PMC5433529/#BCJ-2016-0510CC9.

114 Tung Choy, Chi Chi, Un Kei Chan, Pui Ling Kella Siu, Junwei Zhou, Chi Ho Wong, Yuk Wai Lee, Ho Wang Chan, Joseph Chi Ching Tsui, Steven King Fan Loo, and Stephen Kwok Wing Tsui, "A Novel E3 Probiotics Formula Restored Gut Dysbiosis and Remodelled Gut Microbial Network and Microbiome Dysbiosis Index (MDI) in Southern Chinese Adult Psoriasis Patients," MDPI, March 2023, https://www.mdpi.com/1422-0067/24/7/6571.

115 Davis, Donald, Melvin Epp, and Hugh Riordan, "Changes in USDA Food Composition Data for 43 Garden Crops, 1950 to 1999," National Library of Medicine, 2004, https://pubmed.ncbi.nlm.nih.gov/15637215/.

116 Fan, Ming-Sheng, Fang-Jie Zhao, Susan Fairweather-Tait, Paul Pulton, Sarah Dunham, and Steve McGrath, "Evidence of Decreasing Mineral Density in Wheat Grain Over the Last 160 Years," National Library of Medicine, 2008, https://pubmed.ncbi.nlm.nih.gov/19013359/; Dirt Poor: Have Fruits and Vegetables Become Less Nutritious?," Scientific American, 2011, https://www.scientificamerican.com/article/soil-depletion-and-nutrition-loss/.

117 Pollan, Michael, "Unhappy Meals," *New York Times*, 2007, https://www.nytimes.com/2007/01/28/magazine/28nutritionism.t.html.

118 "Personalized Nutrition and Supplements Market Size, Share & Trends Analysis Report by Ingredient (Proteins & Amino Acid, Vitamins), by Dosage Form (Tablets/Capsules. Liquids), by Distribution Channel, by Age Group, by Region, and Segment Forecasts, 2025–2030," Grand View Research, 2024, https://www.grandviewresearch.com/industry-analysis/personalized-nutrition-supplements-market-report.

119 Julian McClements, David, *Future Foods: How Modern Science Is Transforming the Way We Eat*, Springer, May 2019, https://www.amazon.co.uk/Future-Foods-Modern-Science-Transforming/dp/3030129942?dplnkId=4fffdcfc-0dd1-463b-8202-9e77d5da1e21&nodl=1.

120 Rutzerveld, Chloé, *Food Futures*, BIS Publishers, September 2018, https://www.bispublishers.com/food-futures.html.

121 "Obesity and Overweight," World Health Organization, May 2025, https://www.who.int/news-room/fact-sheets/detail/obesity-and-overweight.

122 "Health at a Glance 2023," OECD, November 2023, https://www.oecd.org/en/publications/2023/11/health-at-a-glance-2023_e04f8239/full-report/overweight-and-obesity_590d3909.html.

123 "Obesity and Overweight," World Health Organization, May 2025, https://www.who.int/news-room/fact-sheets/detail/obesity-and-overweight.

3 Social and Political Challenges

1 Saik, Robert, *Food 5.0: How We Feed the Future*, Lioncrest Publishing, August 2019, https://www.amazon.co.uk/Food-5-0-How-Feed-Future/dp/1544504519/?_encoding=UTF8&pd_rd_w=GTpxz&content-id=amzn1.sym.3413293e-3815-4359-96ba-1ec5110e0b30&pf_rd_p=3413293e-3815-4359-96ba-1ec5110e0b30&pf_rd_r=259-7033390-3999948&pd_rd_wg=srAf7&pd_rd_r=9f8b4b09-add8-43ef-9263-cf6db6b665cb&ref_=aufs_ap_sc_dsk.

2 Gilliard, Alexandra, "Democracy in the Shadow of the Global Rise in Authoritarian Populism," Harvard Carr-Ryan Center for Human Rights, February 2025, https://www.hks.harvard.edu/centers/carr-ryan/our-work/carr-ryan-commentary/democracy-shadow-global-rise-authoritarian-populism.

3 "Goal 2: Zero Hunger," United Nations Sustainable Development Goals, accessed: August 18, 2025, https://www.un.org/sustainabledevelopment/hunger/.

4 "Department of Economic and Social Affairs," United Nations, accessed: August 18, 2025, https://www.un.org/en/desa/world-population-projected-reach-98-billion-2050-and-112-billion-2100.

5 Malthus, Thomas, "An Essay on the Principle of Population," University of Chicago, 1798, https://math.uchicago.edu/~shmuel/Modeling/Malthus,%20An%20essay%20on%20the%20principle%20of%20population.pdf.

6 "MapMaker: Transatlantic and Intra-Americas Slave Trade," National Geographic, accessed: August 18, 2025, https://education.nationalgeographic.org/resource/mapmaker-transatlantic-and-intra-americas-slave-trade/.

7 Ehrlich, Paul, "The Population Bomb," Ballantine Books, 1968, http://pinguet.free.fr/ehrlich68.pdf.

8 "Making Better Policies for Food Systems," OECD, January 2021, https://www.oecd.org/en/publications/making-better-policies-for-food-systems_ddfba4de-en.html.

9 "Global Average GDP Per Capita Over the Long Run," Our World in Data, June 2025, https://archive.ourworldindata.org/20250624-125417/grapher/global-average-gdp-per-capita-over-the-long-run.html?tab=table&time=1820; Rundle, Michael, "Good Data Will Make You an Economic Optimist," Wired, October 2015, https://www.wired.com/story/max-roser-data-visualisation-wired-2015/.

10 Kochhar, Rakesh, "Despite Poverty's Plunge, Middle-Class Status Remains Out of Reach for Many," Pew Research Center, July 2015, https://www.pewresearch.org/global/2015/07/08/despite-povertys-plunge-middle-class-status-remains-out-of-reach-for-many/; "How Well-off Is China's Middle Class?," China Power, accessed: August 18, 2025, https://chinapower.csis.org/china-middle-class/#:~:text=the%20Middle%20Class-,The%20Rise%20of%20China's%20Middle%20Class,2000%2C%20reaching%20%2410%2C410%20in%202019.

11 "How Well-off Is China's Middle Class?," China Power, accessed: August 18, 2025, https://chinapower.csis.org/china-middle-class/#:~:text=the%20Middle%20Class-,The%20Rise%20of%20China's%20Middle%20Class,2000%2C%20reaching%20%2410%2C410%20in%202019.

12 Reynolds, Matthew, "The Future of Food (WIRED Guides): How to Feed the Planet Without Destroying It," Wired, September 2021, https://www.amazon.co.uk/Future-Food-WIRED-guides-Destroying/dp/1847943284/ref=asc_df_1847943284/?tag=googshopuk-21&linkCode=df0&hvadid=499252476918&hvpos=&hvnetw=g&hvrand=2780991607230015613&hvpone=&hvptwo=&hvqmt=&hvdev=c&hvdvcmdl=&hvlocint=&hvlocphy=1006567&hvtargid=pla-1168223889095&psc=1&mcid=92693168bea031d7b05b05bc9a644964&th=1&psc=1.

13 Fukuyama, Francis, *The End of History and the Last Man*, Free Press, 1992.

14 Hu, Zuliu, and Mohsin Khan, "Why Is China Growing So Fast?," International Monetary Fund, June 1997, https://www.imf.org/external/pubs/ft/issues8/index.htm#:~:text=While%20pre%2D1978%20China%20had,force%20behind%20the%20economic%20boom.

15 Hu, Zuliu, and Mohsin Khan, "Why Is China Growing So Fast?," International Monetary Fund, June 1997, https://www.imf.org/external/pubs/ft/issues8/index.htm#:~:text=While%20pre%2D1978%20China%20had,force%20behind%20the%20economic%20boom.
16 "When Will China's GDP Overtake America's?," *The Economist*, June 2023, https://www.economist.com/graphic-detail/2023/06/07/when-will-chinas-gdp-overtake-americas.
17 Shen, Simon, and Wilson Chen, "A Comparative Study of the Belt and Road Initiative and the Marshall plan," Nature, March 2018, https://www.nature.com/articles/s41599-018-0077-9.
18 Omelicheva, Mariya, "Repression Trap: The Mechanism of Escalating State Violence in Russia," Center for Strategic and International Studies, July 2021, https://www.csis.org/analysis/repression-trap-mechanism-escalating-state-violence-russia.
19 "Ukraine: Protection of Civilians in Armed Conflict December 2024 Update," United Nations Human Rights, January 2025, https://ukraine.ohchr.org/sites/default/files/2025-01/Ukraine%20-%20protection%20of%20civilians%20in%20armed%20conflict%20%28December%20%202024%29_ENG_1.pdf; "Needs Growing for Over 8 Million Internally Displaced in Ukraine," IOM UN Migration, May 2022, https://www.iom.int/news/needs-growing-over-8-million-internally-displaced-ukraine.
20 Welsh, Caitlin, and Joseph Glauber, "Food as the 'Silent Weapon': Russia's Gains and Ukraine's Losses," Center for Strategic and International Studies, February 2024, https://www.csis.org/analysis/food-silent-weapon-russias-gains-and-ukraines-losses.
21 "Ukraine Agricultural Production and Trade," USDA Foreign Agricultural Service, April 2022, https://www.fas.usda.gov/sites/default/files/2022-04/Ukraine-Factsheet-April2022.pdf.
22 Welsh, Caitlin, and Joseph Glauber, "Food as the 'Silent Weapon': Russia's Gains and Ukraine's Losses," Center for Strategic and International Studies, February 2024, https://www.csis.org/analysis/food-silent-weapon-russias-gains-and-ukraines-losses.
23 "Ukraine's Front-Line Farms Battle Russians and Weather," *The Economist*, July 2025, https://www.economist.com/europe/2025/07/13/ukraines-front-line-farms-battle-russians-and-weather.
24 "Ukraine's Front-Line Farms Battle Russians and Weather," *The Economist*, July 2025, https://www.economist.com/europe/2025/07/13/ukraines-front-line-farms-battle-russians-and-weather.
25 "Дмитрий Медведев," Telegram, March 2022, https://t.me/s/medvedev_telegram/29; Welsh, Caitlin, and Joseph Glauber, "Food as the 'Silent Weapon': Russia's Gains and Ukraine's Losses," Center for Strategic and International Studies,

February 2024, https://www.csis.org/analysis/food-silent-weapon-russ ias-gains-and-ukraines-losses; Banfield-Nwachi, Mabel, Yohannes Lowe, and Helen Sullivan, "Putin Tells African Nations Russia Can Take Ukraine's Place as Supplier of Grain—as It Happened," *The Guardian*, August 2023, https://www.theguardian.com/world/live/2023/aug/22/russia-ukraine-war-live-updates-ukraine-reportedly-destroys-russian-bomber-drones-damage-moscow-apartments.

26 "Ukraine's Front-Line Farms Battle Russians and Weather," *The Economist*, July 2025, https://www.economist.com/europe/2025/07/13/ukraines-front-line-farms-battle-russians-and-weather.

27 Welsh, Caitlin, and Joseph Glauber, "Food as the 'Silent Weapon,'" Center for Strategic and International Studies, February 2024, https://csis-website-prod.s3.amazonaws.com/s3fs-public/2024-02/240229_Welsh_Silent_Weapon_0.pdf?VersionId=Ss6dcxYB_dIk7i8oO9xkZhVGpdvP_h8R.

28 "Amid Increasingly Dire Humanitarian Situation in Gaza, Secretary-General Tells Security Council Hamas Attacks Cannot Justify Collective Punishment of Palestinian People," United Nations, United Nations Press, October 2023, https://press.un.org/en/2023/sc15462.doc.htm.

29 "The Scale of Destruction in Gaza," *New York Times*, October 2024, https://www.nytimes.com/issue/world/2024/10/21/the-scale-of-destruction-in-gaza; Shurafa, Wafaa, and Fatma Khaled, "More than 55,000 Palestinians Have Been Killed in the Israel-Hamas War, Gaza Health Officials Say," Associated Press, June 2025, https://www.ap.org/news-highlights/spotlights/2025/more-than-55000-palestini ans-have-been-killed-in-the-israel-hamas-war-gaza-health-officials-say/; Kingsley, Patrick, "A Glimpse Inside a Devasted Gaza," *New York Times*, 2024, https://www.nytimes.com/2024/01/09/world/middleeast/israel-military-gaza-ruins.html.

30 Iraqi, Amjad, "The Real Schism in the Israel–Hamas Ceasefire Talks Is about Who Decides Gaza's Future," Chatham House, August 2024, https://www.chathamhouse.org/2024/08/real-schism-israel-hamas-ceasefire-talks-about-who-decides-gazas-future.

31 "Over 200 Non-Profits Call for Shutting Down US-Backed Israeli supply distribution scheme in Gaza," Medecins Sans Frontieres | Doctors Without Borders, July 2025, https://prezly.msf.org.uk/over-200-non-profits-call-for-shutt ing-down-us-backed-israeli-supply-distribution-scheme-in-gaza.

32 "What Is Happening in Gaza? Children Suffering from Severe Malnutrition as 470,000 People Face Mass Starvation," British Red Cross, August 2025, https://www.redcross.org.uk/stories/disasters-and-emergencies/world/whats-happening-in-gaza-humanitarian-crisis-grows; "State of Palestine," World Food Programme, accessed: August 18, 2025, https://www.wfp.org/emergencies/palestine-emergency.

33 World Health Organization, "Famine Confirmed for the First Time in Gaza," WHO, 2025, https://www.who.int/news/item/22-08-2025-famine-confirmed-for-first-time-in-gaza.

34 "Analyzing Israel's Strategy in Gaza," Brookings Institution, January 2024, https://www.brookings.edu/events/analyzing-israels-strategy-in-gaza/.

35 "Israel Must End 'Cruel Collective Punishment' in Gaza, Urges UN Relief chief," United Nations News, May 2025, https://news.un.org/en/story/2025/05/1162806; Berman, Lazar, "IDF Says Documents Show Hamas Has Been Confiscating Aid as a Matter of Policy," *Times of Israel*, June 2025, https://www.timesofisrael.com/idf-says-documents-show-hamas-has-been-confiscating-aid-as-a-matter-of-policy/; "Gaza Aid Crisis Deepens as Border Closure Stretches into 50th Day," United Nations News, April 2025, https://news.un.org/en/story/2025/04/1162471.

36 Yehoshua, Yossi, "Israel Shifts Its Military Strategy—and Hamas Is Losing Its Grip," *Jewish Chronicle*, June 2025, https://www.thejc.com/opinion/israel-shifts-its-military-strategy-and-hamas-is-losing-its-grip-jynv0khv.

37 "People in Gaza Starving, Sick and Dying as Aid Blockade Continues," World Health Organization, May 2025, https://www.who.int/news/item/12-05-2025-people-in-gaza-starving--sick-and-dying-as-aid-blockade-continues.

38 "Gaza Malnutrition Crisis Reaches Catastrophic Levels as the Threat of Famine Looms," British Red Cross, August 2025, https://www.redcross.org.uk/stories/disasters-and-emergencies/world/gaza-malnutrition-crisis-starvation#:~:text=Published%2024%20July%202025&text=Gaza%20is%20now%20on%20the,the%20imminent%20threat%20of%20starvation; "Gaza Malnutrition Crisis Reaches Catastrophic Levels as the Threat of Famine Looms," British Red Cross, August 2025, https://www.redcross.org.uk/stories/disasters-and-emergencies/world/gaza-malnutrition-crisis-starvation#:~:text=Published%2024%20July%202025&text=Gaza%20is%20now%20on%20the,the%20imminent%20threat%20of%20starvation.

39 Ellis, Michael, "Executive Summary," UK Committees, 2025, https://committees.parliament.uk/writtenevidence/135245/html/.

40 "Israel Must End 'Cruel Collective Punishment' in Gaza, Urges UN Relief Chief," United Nations News, May 2025, https://news.un.org/en/story/2025/05/1162806.

41 "Figures at a Glance," UNHCR, UN Refugee Agency, 2024, https://www.unhcr.org/uk/node/5388.

42 Bowman, Emma, "Boycotts of Russian Products and Groups Spread, but the Effects May Be Limited," NPR, 2022, https://www.npr.org/2022/02/28/1083385057/boycotts-russian-effects.

43 Baldwin, Sian, "All the Brands to Have Boycotted Israel and Their Reasons behind It," Standard, 2025, https://www.standard.co.uk/news/uk/brands-boycott-israel-reasons-b1228771.html.

44 "Myanmar Earthquake 2025: Recovery Efforts Continue as Thousands Confirmed Dead," British Red Cross, 2025, https://www.redcross.org.uk/stories/disasters-and-emergencies/world/myanmar-earthquake-2025.
45 Gamboa, Glenn, "Chef Jose Andres Urges Donors to Give More," AP, 2025, https://www.arkansasonline.com/news/2025/jun/15/chef-jose-andres-urges-donors-to-give-more/.
46 "Myanmar Rohingya: What You Need to Know about the Crisis," BBC News, 2020, https://www.bbc.co.uk/news/world-asia-41566561.
47 Bourne, Joel, "Eating the Earth," Science, November 2024, https://www.science.org/content/article/burgeoning-global-food-trade-lifeline-billions-is-it-breaking-planet#:~:text=About%20one%2Dquarter%20of%20all,business%2C%20nearly%20quadrupling%20since%202000; "Future of Farming: Will We Grow Enough Food?," UBS Investment Bank, April 2023, https://www.ubs.com/global/en/investment-bank/insights-and-data/2023/will-we-grow-enough-food.html#:~:text=Our%20analysis%20suggests%20that%20rising,occurring%20as%20early%20as%202030.
48 D'Odorico, Paolo, Joel A. Carr, Francesco Laio, Luca Ridolfi, and Stefano Vandoni, "Feeding Humanity through Global Food Trade," AGU Publications, August 2014, https://agupubs.onlinelibrary.wiley.com/doi/epdf/10.1002/2014EF000250; Barrett, Mike, "How Far Does Your Food Travel? 4 Reasons to Choose Local," Natural Society, December 2013, https://naturalsociety.com/how-far-food-travel-food-miles-1500-average/.
49 Reidy, Susan, "IGC: Global Grain Production Increases 2% to New Record," World-Grain, January 2024, https://www.world-grain.com/articles/19548-igc-global-grain-production-increases-2-to-new-record.
50 "Report Name: Grain and Feed Annual," USDA Foreign Agricultural Service, April 2024, https://apps.fas.usda.gov/newgainapi/api/Report/DownloadReportByFileName?fileName=Grain%20and%20Feed%20Annual_Beijing_China%20-%20People%27s%20Republic%20of_CH2024-0054.pdf; "Data and Analysis," USDA Foreign Agricultural Service, accessed: August 18, 2025, https://www.fas.usda.gov/data.
51 "General Agreement on Tariffs and Trade," World Trade Organization, July 1986, https://www.wto.org/english/docs_e/legal_e/gatt47.pdf; "Codex Alimentarius," Food and Agriculture Organization of the United Nations, accessed: August 18, 2025, https://www.fao.org/fao-who-codexalimentarius/en/.
52 "Current Status of Disputes," World Trade Organization, 2025, https://www.wto.org/english/tratop_e/dispu_e/dispu_current_status_e.htm; "Dispute Settlement," World Trade Organization, 2025, https://www.wto.org/english/tratop_e/dispu_e/dispu_e.htm.

53 "WTO Dispute Settlement Body Developments in 2012," World Trade Organization, February 2013, https://www.wto.org/english/tratop_e/dispu_e/bashir_13_e.htm.

54 Zahniser, Steven, Sahar Angadjivand, and Thomas Hertz, "NAFTA at 20: With Regional Trade Liberalization Complete, Focus Shifts to Other Methods of Deepening Economic Integration," USDA Economic Research Service, April 2015, https://www.ers.usda.gov/amber-waves/2015/april/nafta-at-20-with-regional-trade-liberalization-complete-focus-shifts-to-other-methods-of-deepening-economic-integration.

55 Lang, Tim, *Feeding Britain*, Pelican Books, April 2021, https://www.penguin.co.uk/books/308380/feeding-britain-by-lang-tim/9780241404805.

56 Gilliard, Alexandra, "Democracy in the Shadow of the Global Rise in Authoritarian Populism," Harvard Carr-Ryan Center for Human Rights, February 2025, https://www.hks.harvard.edu/centers/carr-ryan/our-work/carr-ryan-commentary/democracy-shadow-global-rise-authoritarian-populism.

57 Jones, Kent, "The Populist Damage to the Trading System," VoxEU, January 2022, https://cepr.org/voxeu/columns/populist-damage-trading-system; Tilin, Louise, "The Political Economy of Populism in India," Kings College London, 2024, https://www.kcl.ac.uk/the-political-economy-of-populism-in-india; Tilin, Louise, "The Political Economy of Populism in India," Kings College London, 2024, https://www.kcl.ac.uk/the-political-economy-of-populism-in-india; Kohlebberg, Nathan, "Turkey's Election: Mostly Free, Anything but Fair," Alliance for Securing Democracy, May 2023, https://securingdemocracy.gmfus.org/turkeys-election-mostly-free-anything-but-fair/.

58 "Giorgia Meloni," Italian Government Presidency of the Council of Ministers, accessed: August 18, 2025, https://www.governo.it/en/governo/meloni/president-council-ministers/giorgia-meloni.

59 Cash, Joe, Mei Mei Chu, and Xiuhao Chen, "China Targets US Agriculture Over Trump Tariff Threat, Global Times Says," *Reuters*, March 2025, https://www.reuters.com/world/china/china-preparing-countermeasures-us-tariff-threat-global-times-says-2025-03-03/.

60 "Pork Production in China in 2023 Jumps to Record 57.94 Million Tons," Tridge, January 2024, https://www.tridge.com/news/pork-production-in-china-in-2023-jumps-to-record-5.

61 "Chart of Note: U.S. Exports of Feed Grains in All Forms (GIAF) Decline in 2018/2019, but Still Represent Third Largest Year on Record," US Grains & Byproducts, October 2019, https://grains.org/chart-of-note-u-s-exports-of-feed-grains-in-all-forms-giaf-decline-in-2018-2019-but-still-represent-third-largest-year-on-record/#:~:text=The%20final%20numbers%20for%20the,according%20to%20data%20from%20the.

62 McCrimmon, Ryan, "'Here's Your Check': Trump's Massive Payouts to Farmers Will Be Hard to Pull Back," Politico, July 2020, https://www.politico.com/news/2020/07/14/donald-trump-coronavirus-farmer-bailouts-359932.

63 "Soya Beans Exports to Brazil in 2018," WTO World Integrated Trade Solution, 2018, https://wits.worldbank.org/trade/comtrade/en/country/All/year/2018/tradeflow/Exports/partner/BRA/product/120100.

64 Minsberg, Talya, "A Timeline of Trump's On-Again, Off-Again Tariffs," *New York Times*, August 2025, https://www.nytimes.com/2025/03/13/business/economy/trump-tariff-timeline.html.

65 Minsberg, Talya, "A Timeline of Trump's On-Again, Off-Again Tariffs," *New York Times*, August 2025, https://www.nytimes.com/2025/03/13/business/economy/trump-tariff-timeline.html.

66 Harithas, Barath, Kyle Meng, Evan Brown, and Catharine Mouradian, "'Liberation Day' Tariffs Explained," Center for Strategic and International Studies, April 2025, https://www.csis.org/analysis/liberation-day-tariffs-explained; Feingold, Spencer, "Tracking Tariffs: Key Moments in the US-China Trade Dispute," World Economic Forum, June 2025, https://www.weforum.org/stories/2025/06/trumps-us-china-trade-tariffs-timeline/.

67 Feingold, Spencer, "Tracking Tariffs: Key Moments in the US-China Trade Dispute," World Economic Forum, June 2025, https://www.weforum.org/stories/2025/06/trumps-us-china-trade-tariffs-timeline/.

68 Held, Lisa, "How Trump's Tariffs Will Affect Farmers and Food Prices," Civil Eats, March 2025, https://civileats.com/2025/03/05/how-trumps-tariffs-will-affect-farmers-and-food-prices/.

69 Cash, Joe, Mei Mei Chu, and Xiuhao Chen, "China Targets US Agriculture Over Trump Tariff Threat, Global Times Says," *Reuters*, March 2025, https://www.reuters.com/world/china/china-preparing-countermeasures-us-tariff-threat-global-times-says-2025-03-03/.

70 Boak, Josh, and Rob Gillies, "Trump Grants One-Month Exemption for US Automakers from New Tariffs on Imports from Mexico, Canada," Associated Press, March 2025, https://apnews.com/article/trump-tariffs-canada-mexico-china-lutnick-2b269614084027a4894aa14f3dc16227.

71 Neuman, Scott, "Farmers Strongly Back Trump. A New Trade War Could Test Their Loyalty," National Public Radio, March 2025, https://www.npr.org/2025/03/07/g-s1-52362/tariffs-farmers-trade-china-mexico-canada.

72 "U.S. Trade by Industry Sectors and Selected Trading Partners," United States International Trade Commission, 2021, https://www.usitc.gov/research_and_analysis/tradeshifts/2021/trade_by_industry_sectors.

73 "Trump Pauses Some Canada and Mexico Tariffs until Next Month," British Broadcasting Corporation, March 2025, https://www.bbc.com/news/live/cy4v0x27jrpt.

74 "Modifying Reciprocal Tariff Rates to Reflect Trading Partner Retaliation and Alignment," The White House, April 2025, https://www.whitehouse.gov/presidential-actions/2025/04/modifying-reciprocal-tariff-rates-to-reflect-trading-partner-retaliation-and-alignment/.

75 "How a Steel Box Changed the World: A Brief History of Shipping," *Wall Street Journal*, January 2018, https://www.youtube.com/watch?v=0MUkgDIQdcM.

76 Durisin Albery, Megan, "Global Food Import Bill Set for Record, Taking Toll on Poorest," Bloomberg UK, 2022, https://www.bloomberg.com/news/articles/2022-06-09/global-food-import-bill-set-for-record-taking-toll-on-poorest?embedded-checkout=true.

77 Baraniuk, Chris, "Why Container Ships Probably Won't Get Bigger," British Broadcasting Corporation, July 2022, https://www.bbc.co.uk/future/article/20220629-why-container-ships-probably-wont-get-bigger.

78 "Boom Times Are Back for Container Shipping," *The Economist*, June 2024, https://www.economist.com/business/2024/06/27/boom-times-are-back-for-container-shipping.

79 Ritchie, Hannah, "Very Little of Global Food Is Transported by Air; This Greatly Reduces the Climate Benefits of Eating Local," Our World in Data, January 2020, https://ourworldindata.org/food-transport-by-mode.

80 "Suez Canal," Encyclopedia Britannica, 2025, https://www.britannica.com/topic/Suez-Canal.

81 "Boom Times Are Back for Container Shipping," *The Economist*, 2025, https://www.economist.com/business/2024/06/27/boom-times-are-back-for-container-shipping.

82 "The Price of Passage: Red Sea Tensions and the Ripple Effect," Indian Aerospace and Defence Bulletin, 2025, https://www.iadb.in/2025/01/31/the-price-of-passage-red-sea-tensions-the-ripple-effect/.

83 Blackburn, Gavin, "US Defence Secretary Claims China Poses an Ongoing Threat to the Panama Canal," Euro News, August 2025, https://www.euronews.com/2025/04/08/us-defence-secretary-claims-china-poses-an-ongoing-threat-to-the-panama-canal.

84 "The Panama Canal and the Torrijos-Carter Treaties," Office of the Historian, accessed: August 18, 2025, https://history.state.gov/milestones/1977-1980/panama-canal#:~:text=The%20second%2C%20called%20The%20Panama,Panamanians%20on%20December%2031%2C%201999.

85 Roy, Diana, "Who Controls the Panama Canal?," Council for Foreign Relations, January 2025, https://www.cfr.org/article/who-controls-panama-canal; "Expert Context: What's Going on with Trump and the Panama Canal?," Atlantic Council,

January 2025, https://www.atlanticcouncil.org/blogs/new-atlanticist/expert-context-whats-going-on-with-trump-and-the-panama-canal/.

86 Roy, Diana, “Who Controls the Panama Canal?,” Council for Foreign Relations, January 2025, https://www.cfr.org/article/who-controls-panama-canal. “Expert Context: What’s Going on with Trump and the Panama Canal?,” Atlantic Council, January 2025, https://www.atlanticcouncil.org/blogs/new-atlanticist/expert-context-whats-going-on-with-trump-and-the-panama-canal/.

87 “Cargo Volume in the Panama Canal in 2024, by Country of Destination,” Statista, 2025, https://www.statista.com/statistics/710491/cargo-volume-panama-canal-country-of-destination/.

88 Tatananni, Mackenzie, “BlackRock Buys Majority Stake in Ports on Both Sides of Panama Canal Following Trump Complaints,” Barron’s, March 2025, https://www.barrons.com/articles/blackrock-panama-canal-ports-trump-5fbfdc5f.

89 Burk, Ana, “Panama Faces Water Crisis as Drought Impacts Canal, Tourism, Agriculture,” Cronkite News, May 2025, https://cronkitenews.azpbs.org/2025/05/13/panama-water-scarcity-amid-climate-change-canal-demands/.

90 “Arctic Marine Transport Workshop,” Institute of the North, September 2004, https://web.archive.org/web/20070808145951/http:/www.arctic.gov/files/AMTW_book.pdf.

91 Raleigh, Clionadh, and Katayoun Kishi, “Conflict Index,” ACLED, December 2024, https://acleddata.com/series/acled-conflict-index.

4 Farming in Crisis

1 “Agricultural Production Statistics 2010–2023,” Food and Agriculture Organization of the United Nations, accessed: August 1, 2025, https://www.fao.org/statistics/highlights-archive/highlights-detail/agricultural-production-statistics-2010-2023/en?utm_source=chatgpt.com; “Paying the Price | Cheap Food, Big Business and the Cost to Farming and Food Security,” Food, Farming & Countryside Commission, March 2025, https://ffcc-uk.files.svdcdn.com/production/assets/downloads/FFCC_False-Economy_report_part2_v5-1.pdf?dm=1741863510.

2 Root, Skye, “Agriculture in the Red: Net Farm Income Drops Again in 2024 Forecast,” Root Agricultural Advisory, November 2024, https://rootagadvisory.com/agriculture-in-the-red-net-farm-income-drops-again-in-2024-forecast/.

3 “Paying the Price | Cheap Food, Big Business and the Cost to Farming and Food Security,” Food, Farming & Countryside Commission, March 2025, https://ffcc-uk.files.svdcdn.com/production/assets/downloads/FFCC_False-Economy_report_part2_v5-1.pdf?dm=1741863510.

4 "U.S. Farm Production Expenditures, 2023," USDA National Agricultural Statistics Service, July 2024, https://www.nass.usda.gov/Publications/Highlights/2024/2023_FarmExpenditures_Highlights.pdf.
5 "Agrochemicals & Commercial Seeds," ETC Group, September 2022, https://www.etcgroup.org/files/files/01_agrochemicals.pdf.
6 "Farm Tractors Market Size, Share, Growth, and Industry Analysis, by Types (Harvesting,Haying,Planting and Fertilizing,Plowing and Cultivating,Spraying,Others,), Applications (CNH Industrial,Deere,AGCO Corporation,Same Deutz-Fahr,Claas,Kubota,Agri Argo,JCB,ITMCO,MTZ (Minsk),Mahindra,China YTO,) and Regional Insights and Forecast to 2033," Global Growth Insights, 2023 https://www.globalgrowthinsights.com/market-reports/farm-tractors-market-112688.
7 Agriculture Concentration Data," Farm Action, July 2024, https://farmaction.us/concentrationdata/.
8 Paulson, Nick, Carl Zulauf, and Gary Schnitkey, "Trends in Input Costs for Corn and Soybean Production in Illinois," farmdoc daily, April 2024, https://farmdocdaily.illinois.edu/2024/04/trends-in-input-costs-for-corn-and-soybean-production-in-illinois.html.
9 Wise, Timothy, *Eating Tomorrow: Agribusiness, Family Farmers, and the Battle for the Future of Food*, New Press, March 2019.
10 Held, Lisa, "Trump Revokes Biden Order Targeting Food System Consolidation," Civil Eats, 2025, www.civileats.com/2025/08/26/trummp-revokes-biden-order-targeting-food-system-consolidation/.
11 "Paying the Price | Cheap Food, Big Business and the Cost to Farming and Food Security," Food, Farming & Countryside Commission, March 2025, https://ffcc-uk.files.svdcdn.com/production/assets/downloads/FFCC_False-Economy_report_part2_v5-1.pdf?dm=1741863510.
12 Ari, Anil, Nicolas Arregui, Simon Black, Oya Celasun, Dora M. Iakova, Aiko Mineshima, Victor Mylonas, Ian W. H. Parry, Iulia Teodoru, and Karlygash Zhunussova, "Surging Energy Prices in Europe in the Aftermath of the War: How to Support the Vulnerable and Speed up the Transition away from Fossil Fuels," International Monetary Fund, July 2022, https://www.imf.org/en/Publications/WP/Issues/2022/07/28/Surging-Energy-Prices-in-Europe-in-the-Aftermath-of-the-War-How-to-Support-the-Vulnerable-521457.
13 Brown, Oli, and Christia Sander, *Supermarket Buying Power: Global Supply Chains and Smallholder Farmers*, IISD, 2007.
14 Inklebarger, Timothy, "Walmart, Kroger, Costco Make Top Three Grocery Retailers List," Supermarket News, March 2024, https://www.supermarketnews.com/independents-regional-grocers/walmart-kroger-costco-make-top-three-grocery-retailers-list.

15 Maddren, Erin, "The Grocery Retail Market in Germany," Europe Market Research, October 2022, https://assets.ctfassets.net/pn8wbiqtnzw9/eeyTMpaRI0dHLXem5f4rN/420522bb353b0ba16a37388e02419a7c/The_German_grocery_retail_market.pdf.

16 "Paying the Price | Cheap Food, Big Business and the Cost to Farming and Food Security," Food, Farming & Countryside Commission, March 2025, https://ffcc-uk.files.svdcdn.com/production/assets/downloads/FFCC_False-Economy_report_part2_v5-1.pdf?dm=1741863510.

17 "Social Cooperatives," CONFCOOPERATIVE FEDERSOLIDARIETÀ, accessed: August 1, 2025, https://www.federsolidarieta.confcooperative.it/EN/Social-Cooperatives-and-Social-Enterprises.

18 Das, Dayeeta, "Arla Foods Reports Revenue Of €13.8bn in FY 2024," European Supermarket Magazine, February 2025, https://www.esmmagazine.com/fresh-produce/arla-foods-reports-revenue-of-e13-8bn-in-fy-2024-282949; "Fonterra," Companies Market Cap, accessed: August 1, 2025, https://companiesmarketcap.com/gbp/fonterra/revenue/; "Land O'Lakes Reports Strength in Branded Butter and Foodservice Businesses," Dairy Processor News, April 2024, https://www.dairyfoods.com/articles/97272-land-olakes-reports-strength-in-branded-butter-and-foodservice-businesses#:~:text=Land%20O'Lakes%20reports%20strength%20in%20Branded%20Butter%20and%20Foodservice%20businesses,-April%2030%2C%202024&text=Land%20O'Lakes%20in%202023,based%20company%2C%20which%20placed%20No.

19 "Kuapa Kokoo," Slow Food, September 2007, https://www.slowfood.com/blog-and-news/kuapa-kokoo/.

20 "JBS," CompaniesMarketcap, 2025, https://companiesmarketcap.com/gbp/jbs/revenue/#:~:text=According%20to%20JBS's%20latest%20financial,were%20of%20%C2%A348.20%20Billion; "Brazil Marks 50 Years as Top Global Chicken Exporter," The Poultry Site, 2025, https://www.thepoultrysite.com/news/2025/08/brazil-marks-50-years-as-top-global-chicken-exporter.

21 "Brazilian Agribusiness Exports Fall 1.3% and Reach US$164.4 Billion in 2024," Agropages, 2025, https://news.agropages.com/news/NewsDetail---52646.htm.

22 "SENAR's Technical and Management Assistance: Changing Lives in the Countryside," Brazilian Farmers, 2024, https://brazilianfarmers.com/news/senars-technical-and-management-assistance-changing-lives-in-the-countryside/.

23 "SENAR's Technical and Management Assistance: Changing Lives in the Countryside, Brazilian Farmers," 2024, https://brazilianfarmers.com/news/senars-technical-and-management-assistance-changing-lives-in-the-countryside/.

24 "SENAR's Technical and Management Assistance: Changing Lives in the Countryside," Brazilian Farmers, 2024, https://brazilianfarmers.com/news/senars-technical-and-management-assistance-changing-lives-in-the-countryside/.

25 Schechinger, Anne, "Triple Dipping: House Farm Bill Increases Likelihood of Wealthy Farmers Raking in Billions Each Year," Environmental Working Group, May 2024, https://www.ewg.org/research/triple-dipping-house-farm-bill-increases-likelihood-wealthy-farmers-raking-billions-each.

26 "Farm Bill Home," USDA Farm Service Agency, accessed: August 1, 2025, https://www.fsa.usda.gov/tools/informational/farm-bill#:~:text=The%20American%20Relief%20Act%2C%202025,30%2C%202025.

27 Brown, Marcia, "USDA Cancels $1B in Local Food Purchasing for Schools, Food Banks," Politico, March 2025, https://www.politico.com/news/2025/03/10/usda-cancels-local-food-purchasing-for-schools-food-banks-00222796.

28 Brown, Marcia, "USDA Cancels $1B in Local Food Purchasing for Schools, Food Banks," Politico, March 2025, https://www.politico.com/news/2025/03/10/usda-cancels-local-food-purchasing-for-schools-food-banks-00222796.

29 Douglas, Leah, "U.S. Black Farmers Lost $326 bln Worth of Land in 20th Century -Study," *Reuters*, May 2022, https://www.reuters.com/world/us/us-black-farmers-lost-326-bln-worth-land-20th-century-study-2022-05-02/.

30 Costa, Daniel, "How Many Farmworkers Are Employed in the United States?," Economic Policy Institute, October 2023, https://www.epi.org/blog/how-many-farmworkers-are-employed-in-the-united-states/#:~:text=There%20are%20roughly%202.4%20million%20farmworkers%20nationwide&text=But%20QCEW%20also%20estimated%20that,of%201.6%20million%20in%202022.

31 "Agriculture," American Immigration Council, accessed: August 1, 2025, https://www.americanimmigrationcouncil.org/about-immigration/agriculture/.

32 "Agriculture," American Immigration Council, accessed: August 1, 2025, https://www.americanimmigrationcouncil.org/about-immigration/agriculture/.

33 "Agriculture," American Immigration Council, accessed: August 1, 2025, https://www.americanimmigrationcouncil.org/about-immigration/agriculture/.

34 Nigh, Veronica, "H-2A Processing Timeliness Falls as Positions Climb," Market Intel, June 2019, https://www.fb.org/market-intel/h-2a-processing-timeliness-falls-as-positions-climb.

35 Held, Lisa, "H-2A Guestworker Program Impacted by Stalled USDA Payments, Cuts to USAID," Civil Eats, March 2025, https://civileats.com/2025/03/21/federal-cuts-and-funding-pauses-add-to-farmers-labor-woes/.

36 McGuinness, Terry, and Gabrielle Garton Grimwood, "Migrant Workers in Agriculture," UK House of Commons Library, July 2017, https://researchbriefings.files.parliament.uk/documents/CBP-7987/CBP-7987.pdf.

37 "Are There Shortages of Labour in the Agriculture and Horticulture Sector?," UK Parliamentary Business Publications, April 2017, https://publications.parliament.uk/pa/cm201617/cmselect/cmenvfru/1009/100904.htm.

38 McGuinness, Terry, and Gabrielle Garton Grimwood, "Migrant Workers in Agriculture," UK House of Commons Library, July 2017, https://researchbriefings.files.parliament.uk/documents/CBP-7987/CBP-7987.pdf.

39 "Labour Shortages in the Food and Farming Sector," House of Commons Environment, Food and Rural Affairs Committee, March 2022, https://committees.parliament.uk/publications/9580/documents/162177/default/.

40 "Summary of Conclusions and Recommendations," UK Parliamentary Business Publications, 2017, https://publications.parliament.uk/pa/ld201617/ldselect/ldeucom/169/16911.htm.

41 Moreda, Tsegaye, "Cross-Border Seasonal Migrant Labour and Agricultural Commodity Production in the Ethiopia–Sudan Borderlands," *Agric Hum Values*, February 2025, https://doi.org/10.1007/s10460-025-10716-1.

42 Rogero, Tiago, "'Morally Repugnant': Brazilian Workers Sue Coffee Supplier to Starbucks Over 'Slavery-Like Conditions,'" *The Guardian*, April 2025, https://www.theguardian.com/world/2025/apr/24/starbucks-brazil-coffee-forced-labour.

43 "Ending Child Labour and Promoting Sustainable Cocoa Production in Cote D'ivoire and Ghana," European Commission, 2021, https://international-partnerships.ec.europa.eu/system/files/2021-07/ending-child-labour-cote-ivoire-ghana-2021-executive-summary_en.pdf.

44 "Convention on the Rights of the Child," United Nations General Assembly Resolution 44/25, November 1989, https://www.ohchr.org/en/instruments-mechanisms/instruments/convention-rights-child; "Brazil," United States Department of Labor's Bureau of International Labor Affairs, 2014, https://www.dol.gov/sites/dolgov/files/ILAB/child_labor_reports/tda2014/Brazil.pdf.

45 Kay, Abi, "Neurodiversity in Farming May be Double Rate of General Public," *Farmers Weekly*, March 2025, https://www.fwi.co.uk/news/neurodiversity-in-farming-may-be-double-rate-of-general-public.

46 Bjornestad, Andrea, Courtney Cuthbertson, and Jessie Hendricks, "An Analysis of Suicide Risk Factors among Farmers in the Midwestern United States," International Journal of Environmental Research and Public Health, March 2021.

47 "India's Farmer Suicides: Are Deaths Linked to GM Cotton?—in Pictures," *The Guardian*, May 2014, https://www.theguardian.com/global-development/gallery/2014/may/05/india-cotton-suicides-farmer-deaths-gm-seeds.

48 "Agritourism Market Size, Share and COVID-19 Impact Analysis," Fortune Business Insight, 2025, https://www.fortunebusinessinsights.com/agritourism-market-103297.

49 "Agritourism Market Size, Share and COVID-19 Impact Analysis," Fortune Business Insight, 2025, https://www.fortunebusinessinsights.com/agritourism-market-103297.

50 "A Swiss Farm in a Digital World," Microsoft News, 2015, https://news.microsoft.com/europe/features/a-swiss-farm-in-a-digital-world/.

51 Wilson, Katherine, "When Dairy Farmers Leave, What Happens to the Farm?," *The Guardian*, March 2025, https://www.theguardian.com/australia-news/2025/mar/02/when-dairy-farmers-leave-what-happens-to-the-farm?; Eckelkamp, Elizabeth, "America's Dairy Farms Have Vanished," Wired, September 2024, https://www.wired.com/story/americas-dairy-farms-have-vanished/.

52 "National Agricultural Statistics Service, Farm Producers," Census of Agriculture 2022 Highlights, USDA, 2024, https://www.nass.usda.gov/Publications/Highlights/2024/Census22_HL_FarmProducers_FINAL.pdf; "Exchange on Advancing Young and Historically Underserved Farmers and Addressing Intergenerational Farm Issues," Collaboration Platform on Agriculture, 2023, https://agriculture.ec.europa.eu/system/files/2023-12/event-report-useu-exchange-advancing-young-farmers-addressing-intergenerational-farm-issues_en.pdf.

53 MacDonald, D., J. R. Crabtree, G. Wiesinger, T. Dax, N. Stamou, P. Fleury, J. Gutierrez Lazpita, A. Gibon, "Agricultural Abandonment in Mountain Areas of Europe: Environmental Consequences and Policy Response," *Journal of Environmental Management*, May 2002, https://www.sciencedirect.com/science/article/abs/pii/S0301479799903353.

54 McGreal, Chris, "How America's Food Giants Swallowed the Family Farms," *The Guardian*, March 2019, https://www.theguardian.com/environment/2019/mar/09/american-food-giants-swallow-the-family-farms-iowa.

55 "Huge Quantities of Productive Land Lost to Development, Our Research Shows," The Countryside Charity, July 2025, https://www.cpre.org.uk/about-us/cpre-media/huge-quantities-of-productive-land-lost-to-development-our-research-shows/#:~:text=New%20research%20by%20CPRE,%20the,lost%20to%20development%20since%202010.

56 "Urban-Rural Europe—Labour Market," Europa Statistics Explained, accessed: August 1, 2025 https://ec.europa.eu/eurostat/statistics-explained/index.php?oldid=587838#Unemployment.

57 O'Trakoun, John, "Rural Populations Face Higher and More Volatile Unemployment Rates," Federal Reserve Bank of Richmond, November 2024, https://www.richmondfed.org/research/national_economy/macro_minute/2024/rural_unemployment_rates_20241119.

58 Abbott, Chuck, "Rural Poverty Rate Dips to Pre-pandemic Levels," Successful Farming, September 2024, https://www.agriculture.com/rural-poverty-rate-dips-to-pre-pandemic-levels-8710132.

59 Monnat, Shannon, "The Contributions of Socioeconomic and Opioid Supply Factors to U.S. Drug Mortality Rates: Urban-Rural and Within-Rural Differences,"

Journal of Rural Studies 68 (2019): 319–35, https://doi.org/10.1016/j.jrurstud.2018.12.004.

60 Treene, Alayna, "Opioid-Related Deaths Linked to High Unemployment Rates," Axios, March 2017, https://www.axios.com/2017/12/15/opioid-related-deaths-linked-to-high-unemployment-rates-1513300733.

61 Kayode Oladapo, Rasaq, and Mariam Olaitan Oladapo, "Tackling the Rural Opioid Crisis in the United States (U.S): Strategies for Comprehensive Intervention and Resilience Building," ClinMed International Library, 2024, https://clinmedjournals.org/articles/iaarm/international-archives-of-addiction-research-and-medicine-iaarm-9-042.php?jid=iaarm.

62 "EIG Applauds U.S. Conference of Mayors' Heartland Visa Endorsement," Economic Innovation Group, June 2024, https://eig.org/mayors-skilled-immigration-2024/.

63 "Policy for Establishing New Farmers and Transferring Farms," Cour Des Comptes, December 2023, https://www.ccomptes.fr/en/publications/policy-establishing-new-farmers-and-transferring-farms.

64 Norwood, Graham, "Rural Tenants Hit Hardest by Rent Rises Caused by Stock Shortage," Letting Agent Today, December 2023, https://www.lettingagenttoday.co.uk/breaking-news/2023/12/rural-tenants-hit-hardest-by-rent-rises-caused-by-stock-shortage/.

Part 2

1 "Dietary Guidelines for Americans," USDA, 2020, https://www.dietaryguidelines.gov/sites/default/files/2020-12/Dietary_Guidelines_for_Americans_2020-2025.pdf.

2 "Animal Production," Food and Agriculture Organization of the United Nations, accessed: March 10, 2025, https://www.fao.org/animal-production/en/.

3 Luk, Johnny, "Is This the End of the Road for Meat Consumption?," Aljazeera, 2021, https://www.aljazeera.com/opinions/2021/8/12/is-this-the-end-of-the-road-for-meat#:~:text=Similarly%2C%20numbers%20show%20that%20India,percent%20between%202005%20and%202025.

4 Quinton, Amy, "Cows and Climate Change," UC Davis, 2019, https://www.ucdavis.edu/food/news/making-cattle-more-sustainable#:~:text=Each%20year%2C%20a%20single%20cow,the%20Department%20of%20Animal%20Science.

5 Chandler, David, "PAOC Spotlights," Program in Atmospheres, Oceans, and Climate, 2017, http://paocweb.mit.edu/about/paoc-spotlights/explained-greenhouse-gases.

6 Richards, Peter, "What Drives Indirect Land Use Change? How Brazil's Agriculture Sector Influences Frontier Deforestation," National Library of Medicine, 2015, https://pmc.ncbi.nlm.nih.gov/articles/PMC4789281/#:~:text=Work%20in%20land%20change%20science,of%20the%20region's%20forest%20loss.

7 Trovo Garofalo, Danilo, Renan Milagres Novas, Ricardo Pazianotto, Vinicius Goncalves Maciel, Miguel Brandao, Julia Zanin Shimno, and Marila Folegatti-Matsurra, "Land-Use Change CO_2 Emissions Associated with Agricultural Products at Municipal Level in Brazil," Science Direct, 2022, https://www.sciencedirect.com/science/article/abs/pii/S0959652622021503#:~:text=Land%2Duse%20change%20(LUC),footprint%20of%20Brazilian%20agricultural%20products.

8 Denicoff, Josh, and Mia Nielsen, "How Cattle Ranching in Brazil Could Lead to the End of the Amazon," IISD, 2024, https://sdg.iisd.org/commentary/generation-2030/how-cattle-ranching-in-brazil-could-lead-to-the-end-of-the-amazon/#:~:text=In%20Brazil%2C%20deforestation%20is%20largely,current%20deforestation%20in%20the%20Amazon.

9 "Corporate Control of Our Food," Greenpeace, accessed: March 10, 2025, https://www.greenpeace.org/usa/sustainable-agriculture/issues/corporate-control/#:~:text=A%20small%20handful%20of%20corporations,executives%20putting%20profits%20before%20people.

10 "Inhumane Practices on Factory Farms," Animal Welfare Institute, accessed: March 10, 2025, https://awionline.org/content/inhumane-practices-factory-farms.

11 Macitelli, F., J. S. Braga, D. Gellatly, and M. J. R. Paranhos da Costa, "Reduced Space in Outdoor Feedlot Impacts b=Beef Cattle Welfare," National Library of Medicine, 2020, https://pubmed.ncbi.nlm.nih.gov/32669155/.

12 Dohoo, I. R., L. DesCoteaux, K. Leslie, A. Fredeen, W. Shewfelt, A. Preston, P. Dowling, "A Meta-Analysis Review of the Effects of Recombinant Bovine Somatotropin," National Library of Medicine, 2003, https://pmc.ncbi.nlm.nih.gov/articles/PMC280709/.

13 Anomaly, Jonathan, "What's Wrong with Factory Farming?," National Library of Medicine, 2014, https://pmc.ncbi.nlm.nih.gov/articles/PMC9757169/.

14 Anomaly, Jonathan, "What's Wrong with Factory Farming?," National Library of Medicine, 2014, https://pmc.ncbi.nlm.nih.gov/articles/PMC9757169/.

15 Kelly Tang, Ka Wah, Beverley Millar, and John Moore, "Antimicrobial Resistance (AMR)," National Library of Medicine, 2023, https://pmc.ncbi.nlm.nih.gov/articles/PMC10336207/#:~:text=According%20to%20estimates%20provided%20globally,to%20control%20AMR%20%5B5%5D.

16 "Annual Report on Antimicrobial Agents Intended for Use in Animals," World Organisation for Animal Health, 2023, https://www.woah.org/app/uploads/2024/05/woah-amu-report-2024-final.pdf.

17 "Annual Report on Antimicrobial Agents Intended for Use in Animals," World Organisation for Animal Health, 2023, https://www.woah.org/app/uploads/2024/05/woah-amu-report-2024-final.pdf.
18 Brown-Brandl, Tami, Elaine Berry, James wells, Terrance Arthur, and John Nienaber, "Impacts of Individual Animal Response to Heat and Handling Stresses on Escherichia Coli and E. Coli O157:H7 Fecal Shedding by Feedlot Cattle," National Library of Medicine, 2009, https://pubmed.ncbi.nlm.nih.gov/19459757/.
19 "African Swine Fever Fact Sheet," US Department of Homeland Security, 2023, https://www.dhs.gov/publication/st-african-swine-fever-factsheet#:~:text=African%20Swine%20Fever%20(ASF)%20is,to%20100%25%20mortality%20in%20swine.
20 "African Swine Fever," UK Department of Agriculture, Environment and Rural Affairs, accessed: March 10, 2025, https://www.daera-ni.gov.uk/articles/african-swine-fever#:~:text=African%20Swine%20Fever%20can%20be,pig%20meat%20or%20meat%20products.
21 "African Swine Fever," USDA, accessed: March 10, 2025, https://www.aphis.usda.gov/livestock-poultry-disease/swine/african-swine-fever.
22 "Production—Pork," USDA, accessed: March 10, 2025, https://www.fas.usda.gov/data/production/commodity/0113000.
23 Ma, Meilin, Holly Wang, Yizhou Hua, Fei Qin, and Jing Yang, "African Swine Fever in China: Impacts, Responses, and Policy implications," Science Direct, 2021, https://www.sciencedirect.com/science/article/abs/pii/S0306919221000427.
24 Anomaly, Jonathan, "What's Wrong with Factory Farming?," National Library of Medicine, 2014, https://pmc.ncbi.nlm.nih.gov/articles/PMC9757169/.
25 "Avian Influenza A(H5N1)—United States of America," World Health Organization, 2024, https://www.who.int/emergencies/disease-outbreak-news/item/2024-DON512#:~:text=From%202003%20to%201%20April,in%20Viet%20Nam%20(11).
26 "Cancer: Carcinogenicity of the Consumption of Red Meat and Processed Meat," World Health Organization, 2015, https://www.who.int/news-room/questions-and-answers/item/cancer-carcinogenicity-of-the-consumption-of-red-meat-and-processed-meat.
27 Goodier, Michael, and Viktor Sunnemark, "UK Meat Consumption at Lowest Level since Records Began, Data Reveals," *The Guardian*, 2023, https://www.theguardian.com/environment/2023/oct/24/uk-meat-consumption-lowest-level-since-record-began-data-reveal.
28 "Chicken Consumption Rockets Ahead When Compared with Other Proteins," Poultry Network, 2024, https://poultry.network/chicken-consumption-rockets-ahead-when-compared-with-other-proteins/#:~:text=Wit

hin%20this%2C%20poultry%20consumption%20at,growth%20averaging%20 6.7%25%20a%20year.

29 "Interest in Veganism Is Surging," *The Economist*, 2020, https://www.economist.com/graphic-detail/2020/01/29/interest-in-veganism-is-surging.

30 Niranjan, Ajit, "Belching Livestock to Incur Green Levy in Denmark from 2030," *The Guardian*, 2024, https://www.theguardian.com/world/article/2024/jun/26/flatulent-livestock-to-incur-green-levy-in-denmark-from-2035.

5 Reconceptualizing Alternative Protein Sources

1 Ritchie, Hannah, Pablo Rosado, and Max Roser, "Meat and Dairy Production," Our World in Data, 2019, https://ourworldindata.org/meat-production.

2 Willett, Walter, Johan Rockstrom, Brent Loken, Marco Springmann, Tim Lang, Sonja Vermeulen, Tara garnett, David Tilman, Fabrice DeClerk, Amanda Wood, Malin Jonell, Michael Clark, Line Gordon, Jessica Fanzo, Corinna Hawkes, Rami Zurayk, Juan Rivera, Wim De Vries, Lindiwe Majele Sibanda, Ashkan Afshin, Abhishek Chaudhary, Mario Herrero, Rina Agustina, Francesco Branca, Anna Lartey, Shenggen Fan, Beatrice Crona, Elizabeth Fox, Victoria Bignet, Max Troell, Therese Lindahl, Sudhvir Singh, Sarah Cornell, Srinath Reddy, Sunita Narain, Sania Nistar, and Christopher Murray, "Food in the Anthropocene: The EAT-*Lancet* Commission on Healthy Diets from Sustainable Food Systems," The Lancet Commissions, 2019, https://www.thelancet.com/action/showPdf?pii=S0140-6736(18)31788-4.

3 Willett, Walter, et al., "Food in the Anthropocene: The EAT-*Lancet* Commission on Healthy Diets from Sustainable Food Systems," The Lancet Commissions, 2019, https://www.thelancet.com/action/showPdf?pii=S0140-6736(18)31788-4.

4 Troell, Max, Malin Jonell, and Beatrice Crona, "Scoping Report: The Role of Seafood in Sustainable and Healthy Diets," Stockholm Resilience Centre, 2019, https://eatforum.org/content/uploads/2019/11/Seafood_Scoping_Report_EAT-Lancet.pdf.

5 Ritchie, Hannah, "The World Now Produces More Seafood from Fish Farms than Wild Catch," Our World in Data, 2019, https://ourworldindata.org/rise-of-aquaculture.

6 Ritchie, Hannah, "The World Now Produces More Seafood from Fish Farms than Wild Catch," Our World in Data, 2019, https://ourworldindata.org/rise-of-aquaculture.

7 Osterloff, Emily, "The Problem of Sea Lice in Salmon Farms," Natural History Museum, https://www.nhm.ac.uk/discover/the-problem-of-sea-lice-in-salmon-farms.html.

8 Osterloff, Emily, "The Problem of Sea Lice in Salmon Farms," Natural History Museum, https://www.nhm.ac.uk/discover/the-problem-of-sea-lice-in-sal mon-farms.html.

9 "Antibiotics in Seafood Farming," Agriculture Stewardship Council, accessed: January 26, 2025, https://asc-aqua.org/.

10 "Environmental impacts," Sustain, accessed: January 26, 2025, https://www.sustain web.org/goodcatch/environmental_impacts/#:~:text=Waste%2520from%2520f ish%252Dfeed%2520and,and%2520impact%2520surrounding%2520mar ine%2520life.

11 De Augustinis, Francesco, "Antarctic Krill Fishing Boosted by Fish Farming," The Ferret, 2022, https://theferret.scot/antarctic-krill-fishing-boosted-fish-farming/.

12 "Fish Feed: Why We Need Sustainable Alternatives," European Institute of Innovation and Technology (EIT) Food, 2023, https://www.eitfood.eu/blog/ fish-feed-why-we-need-sustainable-alternatives#:~:text=Why%2520are%2520com mercial%2520fish%2520feeds,are%2520caught%2520for%2520feed%2520pro duction.

13 "Hima," Hima Seafood, accessed: January 28, 2025, https://himaseafood.com/; De Lorenzo, Daniela, "A Norwegian Company Set Up a Land-Based Aquaculture System, But It's Not Salmon They Bet On," Forbes, 2024, https://www.forbes.com/ sites/danieladelorenzo/2024/01/07/a-norwegian-company-set-up-a-land-based-aquaculture-farm-but-its-not-salmon-they-bet-on/.

14 "Norway's Fish Health Report 2023: What Killed Nearly 17% of Farmed Salmon," Fish Farming Expert, 2024, accessed: January 28, 2025, https://www.fishfarmingexp ert.com/about-us.

15 "Building Smart Land-Based Fish Farms," The Conservation Fund, accessed: January 26, 2025, https://www.conservationfund.org/our-impact/projects/ building-smart-land-based-fish-farms/.

16 Holland, Jason, "Next Tuna Leveling Up Its Land-Based Atlantic BluefinFarming Concept," Seafood Source, 2023, https://www.seafoodsource.com/news/premium/ aquaculture/startup-next-tuna-ready-to-level-up-its-novel-atlantic-bluefin-farm ing-concept.

17 Johnson, Kurt, "Cultural Tabu: How an Ancient Ocean Custom Is Saving Fiji's Reefs," *The Guardian*, 2020, https://www.theguardian.com/world/2020/jun/20/cultu ral-tabu-how-an-ancient-ocean-custom-is-saving-fijis-reefs.

18 Johnson, Kurt, "Cultural Tabu: How an Ancient Ocean Custom Is Saving Fiji's Reefs," *The Guardian*, 2020, https://www.theguardian.com/world/2020/jun/20/cultu ral-tabu-how-an-ancient-ocean-custom-is-saving-fijis-reefs.

19 "Sustainable Clams, Mussels, Oysters, and Scallops Guide," Monterey Bay Aquarium Seafood Watch, accessed: January 24, 2025, https://www.seafoodwatch.org/reco mmendations/download-consumer-guides/sustainable-bivalves-guide.

20 "Global Study Benefits of Shellfish and Seaweed Aquaculture," National Oceanic and Atmospheric Administration (NOAA) Fisheries, 2022, accessed: January 24, 2025, https://www.fisheries.noaa.gov/feature-story/global-study-sheds-light-valuable-benefits-shellfish-and-seaweed-aquaculture.

21 "Global Study Benefits of Shellfish and Seaweed Aquaculture," National Oceanic and Atmospheric Administration (NOAA) Fisheries, 2022, accessed: January 24, 2025, https://www.fisheries.noaa.gov/feature-story/global-study-sheds-light-valuable-benefits-shellfish-and-seaweed-aquaculture.

22 De Souza Celente, Gleison, Yixing Sui, and Parag Acharya, "Seaweed as an Alternative Protein Source: Prospective Protein Extraction Technologies," Science Direct, 2023, https://www.sciencedirect.com/science/article/pii/S146685642300108X.

23 De Souza Celente, Gleison, Yixing Sui, and Parag Acharya, "Seaweed as an Alternative Protein Source: Prospective Protein Extraction Technologies," Science Direct, 2023, https://www.sciencedirect.com/science/article/pii/S146685642300108X.

24 Espinosa-Ramírez, Johanan, Alicia Mondrago-Portocarrero, Jose Rodriguez, Jose Lorenzo, and Eva Santos, "Algae as a Potential Source of Protein Meat Alternatives," National Library of Medicine, 2023, https://pmc.ncbi.nlm.nih.gov/articles/PMC10513374/.

25 Brown, Rob, "Algae: Can Pond Slime Feed a Growing Population?," The Grocer, 2019, https://www.thegrocer.co.uk/analysis-and-features/algae-can-pond-slime-feed-a-growing-global-population/599244.article.

26 Oonincx, Dennis, Joost van Itterbeeck, Marcel J. W. Heetkamp, Henry van den Brand, Joop J. A. van Loon, and Arnold van Huis, "An Exploration on Greenhouse Gas and Ammonia Production by Insect Species Suitable for Animal or Human Consumption," *PLOS One*, 2010, https://journals.plos.org/plosone/article?id=10.1371/journal.pone.0014445.

27 "FAO Knowledge Repository," Food and Agriculture Organization of the United Nations, accessed: January 25, 2025, https://openknowledge.fao.org/home.

28 Liceaga, Andrea, "Edible Insects, a Valuable Protein Source from Ancient to Modern Times," National Library of Medicine, 2022, https://pmc.ncbi.nlm.nih.gov/articles/PMC9107018/.

29 Smith, Matthew, Valerie J. Stull, Jonathan A. Patz, and Samuel S. Myers, "Nutritional and Environmental Benefits of Increasing Insect Consumption in Africa and Asia," IOP Science, 2021, https://iopscience.iop.org/article/10.1088/1748-9326/abf06c.

30 Severson, Kim, "Kennedy Plans to Phase Out 8 Commonly Used Food Dyes," *New York Times*, 2025, https://www.nytimes.com/2025/04/21/dining/rfk-jr-food-dyes.html.

31 "Food Defect Levels Handbook," US Food and Drug Administration, 2018, accessed: January 24, 2025, https://www.fda.gov/food/current-good-manufacturing-practices-cgmps-food-and-dietary-supplements/food-defect-levels-handbook.

32 Matthews, Dylan, "The Biggest Problem with Eating Insects Isn't the 'ew' Factor," Vox, 2021, https://www.vox.com/future-perfect/22445822/insect-farming-crickets-mealworm-ethics.
33 "Approval of Another Insect/Insect-Derived Food as a Novel Food," European Commission, 2025, accessed: January 28, 2025, https://food.ec.europa.eu/food-safety/novel-food/authorisations/approval-insect-novel-food_en.
34 "Insect Protein Market by Application (Food & Beverages, Animal Nutrition, Pharmaceutical & Cosmetics), Insect Type (Cricket, Grasshoppers, Ants, Mealworms, Black Soldier Flies, and Others),Distribution Channel and Region—Global Forecast to 2027," Markets and Markets, 2022, accessed: January 28, 2025, https://www.marketsandmarkets.com/Market-Reports/insect-protein-market-150067243.html.
35 Godwin, Richard, "If We Want to Save the Planet, the Future of Food is Insects," *The Guardian*, 2021, https://www.theguardian.com/food/2021/may/08/if-we-want-to-save-the-planet-the-future-of-food-is-insects.
36 Kępińska-Pacelik, Jagoda, and Alessia Giannetto, "Insects in Pet Food Industry—Hope or Threat?," National Library of Medicine, 2022, https://pmc.ncbi.nlm.nih.gov/articles/PMC9219536/.
37 Wall, Tim, "Pet Food Growth May Outstrip Animal Protein Supply," Pet Food Industry, 2022, https://www.petfoodindustry.com/news-newsletters/pet-food-news/article/15469008/pet-food-growth-may-outstrip-animal-protein-supply#:~:text=In%2520their%2520study%252C%2520the%2520researchers,1.5%2520million%2520tons%2520(16.85%2525.
38 "Insects for Human Consumption," Federal Agency for the Safety of the Food Chain, 2022, https://www.fasfc.be/foodthemesnovel-food/insects-human-consumption.
39 Matandirotya N. R, W. L. Filho, G. Mahed, B. Maseko, C. V. Murandu, "Edible Insects Consumption in Africa towards Environmental Health and Sustainable Food Systems: A Bibliometric Study," *International Journal of Environmental Research and Public Health* 19 (22) (2022): 14823, 10.3390/ijerph192214823.
40 "Application of 'Novel Food' Regulation Regarding Insects for Human Consumption in Belgium," Health Food Chain Safety, 2023, https://www.health.belgium.be/sites/default/files/uploads/fields/fpshealth_theme_file/2023_11_20_qna_novel_foods_insects_final_en.pdf.
41 Fletcher, Jenna, "39 High-Protein Foods," *Medical News Today*, 2023, https://www.medicalnewstoday.com/articles/321522#high-protein-foods.
42 "Soybeans," Union of Concerned Scientists, 2015, accessed: January 28, 2025, https://www.ucsusa.org/resources/soybeans#:~:text=But%2520in%2520reality%252C%2520a%2520typical,food%2520products%2520for%2520human%2520consumption.
43 "Tofu Market Size, Share, Growth Analysis, by Application (Hotels, Restaurants and Catering, Food Processing, Household), by Product (Processed Tofu, Unprocessed

Tofu), by Type (Organic, Conventional, Firm Tofu, Soft (or silken) Tofu), by Distribution Channel (Dealer/Retailer, Supermarkets and Hypermarkets, Direct Selling), by Region—Industry Forecast 2025–2032," Skyquest, 2024, accessed: January 28, 2025, https://www.skyquestt.com/report/tofu-market#:~:text=Tofu%2520Market%2520Regional%2520Insights&text=Particularly%2520in%2520the%2520United%2520States,regional%2520demand%2520for%2520the%2520product.

6 Lab-Grown "Meat"

1 Chriki, Sghaier, Marie-Pierre Ellies-Oury, and Jean-François Hocquette, "Is 'Cultured Meat' a Viable Alternative to Slaughtering Animals and a Good Comprise between Animal Welfare and Human Expectations?," National Library of Medicine, 2022, https://pmc.ncbi.nlm.nih.gov/articles/PMC8929989/.
2 Kyriakopoulou, Konstantina, Julia K. Keppler, and Atze Jan van der Goot, "Functionality of Ingredients and Additives in Plant-Based Meat Analogues," MDPI, 2021, https://www.mdpi.com/2304-8158/10/3/600.
3 Albuquerque Penna Franca, Paula, Patricia Duque-Estrada, Breno Fraga da Fonseca e Sa, Atze Jan van der Goot, and Anna Poala Trinidade Rocha Pierucci, "Meat Substitutes—Past, Present, and Future of Products Available in Brazil: Changes in the Nutritional Profile," Science Direct, 2022, https://www.sciencedirect.com/science/article/pii/S2666833522000211.
4 Featherstone, Susan, "A Complete Course in Canning and Related Processes," Science Direct, 2015, https://www.sciencedirect.com/topics/agricultural-and-biological-sciences/textured-vegetable-protein.
5 Wang, Yi, Long Chen, Tianyi Yang, Yun Ma, David Julian McClements, Fei Ren, Yaoqu Tian, and Zhengyu Jin, "A Review of Structural Transformations and Properties Changes in Starch During Thermal Processing of Foods," Science Direct, 2021, https://www.sciencedirect.com/topics/agricultural-and-biological-sciences/food-extrusion#:~:text=Extrusion%20is%20a%20high%2Dtemperature,forward%20along%20the%20screw%20direction.
6 Plattner, Brian, "Extrusion Techniques for Meat Analogues," Cereal Foods World, 2020, https://www.cerealsgrains.org/publications/cfw/2020/Documents/CFW-65-4-0043.pdf.
7 Kyriakopoulou, Konstantina, Julia K. Keppler, and Atze Jan van der Goot, "Functionality of Ingredients and Additives in Plant-Based Meat Analogues," MDPI, 2021, https://www.mdpi.com/2304-8158/10/3/600.
8 Wolf, Julie, "The Microbial Reasons Why the Impossible Burger Tastes So Good," The American Society for Microbiology, 2019, https://asm.org/articles/2019/may/the-microbial-reasons-why-the-impossible-burger-ta.

9 Zidarič, Tanja, Mark Milojevic, Jernej Vajda, Bostjan Vihar, and Uros Maver, "Cultured Meat: Meat Industry Hand in Hand with Biomedical Production Methods," Springer Nature, 2020, https://link.springer.com/article/10.1007/s12393-020-09253-w; Cao, Shanbo, Fang Wang, and Lin Liu, "Isolation and Culture of Bovine Embryonic Stem Cells," Springer Nature, 2013, https://link.springer.com/protocol/10.1007/978-1-62703-628-3_9.

10 Zidarič, Tanja et. al., "Cultured Meat: Meat Industry Hand in Hand with Biomedical Production Methods," Springer Nature, 2020, https://link.springer.com/article/10.1007/s12393-020-09253-w.

11 Soice, Emily, and Jeremiah Johnston, "Immortalizing Cells for Human Consumption," National Library of Medicine, 2021, https://pmc.ncbi.nlm.nih.gov/articles/PMC8584139/.

12 Zidarič, Tanja, et al., "Cultured Meat: Meat Industry Hand in Hand with Biomedical Production Methods," Springer Nature, 2020, https://link.springer.com/article/10.1007/s12393-020-09253-w.

13 Zidarič, Tanja, et al., "Cultured Meat: Meat Industry Hand in Hand with Biomedical Production Methods," Springer Nature, 2020, https://link.springer.com/article/10.1007/s12393-020-09253-w.

14 Zidarič, Tanja, et al., "Cultured Meat: Meat Industry Hand in Hand with Biomedical Production Methods," Springer Nature, 2020, https://link.springer.com/article/10.1007/s12393-020-09253-w.

15 Zidarič, Tanja, et al., "Cultured Meat: Meat Industry Hand in Hand with Biomedical Production Methods," Springer Nature, 2020, https://link.springer.com/article/10.1007/s12393-020-09253-w.

16 Ben-Arye, Tom, and Shulamit Levenberg, "Tissue Engineering for Clean Meat Production," Frontiers, 2019, https://www.frontiersin.org/journals/sustainable-food-systems/articles/10.3389/fsufs.2019.00046/full.

17 Azhar, Asim, Md Zeyaullah, Shushruta Bhunia, Santhosh Kacham, Girish Patil, Khursheed Muzammil, Mohammad Suhail Khan, and Sandeep Sharma, "Cell-Based Meat: The Molecular Aspect," Frontiers, 2023, https://www.frontiersin.org/journals/food-science-and-technology/articles/10.3389/frfst.2023.1126455/full

18 Zidarič, Tanja, et al., "Cultured Meat: Meat Industry Hand in Hand with Biomedical Production Methods," Springer Nature, 2020, https://link.springer.com/article/10.1007/s12393-020-09253-w.

19 "Definitions of Bovine Serum Types," International Serum Industry Association, accessed: February 2, 2025, https://www.serumindustry.org/standardization/serum-definitions/.

20 Lee, Da Young, Seung Yun Lee, Seung Hyeon Yun, Jae Won Jeong, Jae Hyeon Kim, Hyun Woo Kim, Jung Seok Choi, Gap-Don Kim, Seon Tea Joo, Inho Choi, and Sun Jin Hur, "Review of the Current Research on Fetal Bovine Serum and the

Development of Cultured Meat," National Library of Medicine, 2022, https://pmc.ncbi.nlm.nih.gov/articles/PMC9478980/.

21 Zidarič, Tanja, et al., "Cultured Meat: Meat Industry Hand in Hand with Biomedical Production Methods," Springer Nature, 2020, https://link.springer.com/article/10.1007/s12393-020-09253-w.

22 Reiss, Jacob, Samantha Robertson, and Masatoshi Suzuki, "Cell Sources for Cultivated Meat: Applications and Considerations throughout the Production Workflow," National Library of Medicine, 2021, https://pmc.ncbi.nlm.nih.gov/articles/PMC8307620/.

23 Zidarič, Tanja, et al., "Cultured Meat: Meat Industry Hand in Hand with Biomedical Production Methods," Springer Nature, 2020, https://link.springer.com/article/10.1007/s12393-020-09253-w; Kulus, Magdalena, Maurycy Jankowski, Wieslawa Kranc, Afsaneh Golkar Narenji, Maryam Farzaneh, Piotr Dziegiel, Maciej Zabel, Pawel Antosik, Dorota Bukowska, Paul Mozdziak, and Bartosz Kempisty, "Bioreactors, Scaffolds and Microcarriers and In Vitro Meat Production—Current Obstacles and Potential Solutions," National Library of Medicine, 2023, https://pmc.ncbi.nlm.nih.gov/articles/PMC10513094/.

24 Zidarič, Tanja, et al., "Cultured Meat: Meat Industry Hand in Hand with Biomedical Production Methods," Springer Nature, 2020, https://link.springer.com/article/10.1007/s12393-020-09253-w.

25 "Cell-Cultivated Meat: An Overview," Congressional Research Service, 2023, https://crsreports.congress.gov/product/pdf/R/R47697.

26 "USDA and FDA Announce a Formal Agreement to Regulate Cell-Cultured Food Products from Cell Lines of Livestock and Poultry," USDA, 2019, https://www.usda.gov/about-usda/news/press-releases/2019/03/07/usda-and-fda-announce-formal-agreement-regulate-cell-cultured-food-products-cell-lines-livestock-and.

27 "Cell-Cultivated Meat: An Overview," Congressional Research Service, 2023, https://crsreports.congress.gov/product/pdf/R/R47697.

28 "Cell-Cultivated Meat: An Overview," Congressional Research Service, 2023, https://crsreports.congress.gov/product/pdf/R/R47697.

29 Poinski, Megan, "Cultivated Meat to Enter US Market with Full USDA Approval for Eat Just and Upside Foods," Food Dive, 2023, https://www.fooddive.com/news/us-approves-cultivated-meat-eat-just-upside-foods/653509/.

30 "Master Plan Exercise to Transform Lim Chu Kang into a High Tech Agri Food Cluster," Singapore Food Agency, 2020, https://www.ourfoodfuture.gov.sg/speeches-and-media-releases/lckmasterplan/.

31 "From Precision Fermentation to Aquaculture, How Singapore Is Driving Innovation in Food Production,' EDB Singapore, 2024, https://www.edb.gov.sg/en/business-insights/insights/from-precision-fermentation-to-aquaculture-how-singapore-is-driving-innovation-in-food-production.html.

32 "From Precision Fermentation to Aquaculture, How Singapore Is Driving Innovation in Food Production," EDB Singapore, 2024, https://www.edb.gov.sg/en/business-insights/insights/from-precision-fermentation-to-aquaculture-how-singapore-is-driving-innovation-in-food-production.html.

33 "ScaleUp Bio Secures Food Manufacturing License from Singapore Food Agency," FnBNews, 2025, https://www.fnbnews.com/Top-News/scaleup-bio-secures-food-manufacturing-licence-from-singapore-food-agency-77586.

34 "Dairy Alternatives Market by Source (Soy, Almond, Coconut, Oats, Hemp), Application (Milk, Yogurt, Ice Creams, Cheese, Creamers), Distribution Channel (Retail, Online Stores, Foodservice), Formulation and Region—Global Forecast to 2028," Markets and Markets, 2023, https://www.marketsandmarkets.com/Market-Reports/dairy-alternatives-market-677.html.

35 "Baby Milk & Infant Formula," Statista, accessed: February 1, 2025, https://www.statista.com/outlook/cmo/food/baby-food/baby-milk-infant-formula/worldwide.

36 "Regulation (EU) No 1308/2013 of the European Parliament and of the Council," EU Legislation, 2013. https://www.legislation.gov.uk/eur/2013/1308/annex/VII/part/III; "FDA Provides Draft Labeling Recommendations for Plant-based Milk Alternatives to Inform Consumers," US Food and Drug Administration, 2023, https://www.fda.gov/news-events/press-announcements/fda-provides-draft-labeling-recommendations-plant-based-milk-alternatives-inform-consumers#:~:text=The%20draft%20guidance%2C%20%E2%80%9CLabeling%20of,that%20is%20different%20than%20milk%2C.

37 Turrell, Claire, "How a Microbe from Yellowstone's Hot Springs Could Help Feed the World," *Smithsonian Magazine*, 2024, https://www.smithsonianmag.com/innovation/how-a-microbe-from-yellowstones-hot-springs-could-help-feed-the-world-180983872/.

38 Holt, Roberta, John Munafo Jr., Julie Salmen, Carl Keen, Behroze Mistry, Justin Whiteley, and Harold Schmitz, "Mycelium: A Nutrient-Dense Food to Help Address World Hunger, Promote Health, and Support a Regenerative Food System," ACS Publications, 2024, https://pubs.acs.org/doi/epdf/10.1021/acs.jafc.3c03307.

39 Turrell, Claire, "How a Microbe from Yellowstone's Hot Springs Could Help Feed the World," *Smithsonian Magazine*, 2024, https://www.smithsonianmag.com/innovation/how-a-microbe-from-yellowstones-hot-springs-could-help-feed-the-world-180983872/.

40 Jolly, Jasper, "Eating Light: Finnish Startup Begins Making Food 'from Air and Solar Power,'" *The Guardian*, 2024, https://www.theguardian.com/business/2024/apr/19/finnish-startup-food-air-solar-power-solein.

41 Wurgaft, Benjamin, *Meat Planet: Artificial Flesh and the Future of Food*, University of California Press, 2019.

42 "US Cultured Meat Market Size, Trends and Insights by Type (Red Meat, Poultry Meat, Seafood), by End User (Household, Food Service), and by Region—Industry Overview, Statistical Data, Competitive Analysis, Share, Outlook, and Forecast 2024–2033," Custom Market Insights, accessed: January 31, 2025, https://www.custommarketinsights.com/report/us-cultured-meat-market/#:~:text=The%20U.S.%20Cultured%20Meat%20market,alternative%20to%20conventional%20meat%20production.

43 "Bell Food Group CEO Expects Cultivated Meat Burgers to Hit Swiss Restaurants in 3 Years," Cultivated X, 2024, https://cultivated-x.com/company-news/people/bell-food-group-ceo-cultivated-meat-burgers-swiss-restaurants-3-years/.

44 "Consider a Future in Which Lab-Grown Meat Alternatives Are Indistinguishable from Animal Meat in Terms of Taste, Nutrition, and Cost. In this Scenario, Would You Prefer to Eat …?," YouGov, 2024, https://today.yougov.com/topics/politics/survey-results/daily/2024/05/29/9b1cc/2.

45 "Study Finds Singaporean Meat Consumers Prefer the Term 'Cultivated Meat,'" Cultivated X, 2023, https://cultivated-x.com/studies-numbers/singaporean-meat-consumers-prefer-term-cultivated-meat/.

46 Smith, Matthew, "One in Five Britons Would Try Lab-Grown Human Meat," YouGov, 2024, https://yougov.co.uk/consumer/articles/50195-one-in-twenty-britons-would-try-lab-grown-human-meat.

47 "French Parliamentarians Propose Law to Ban Cultivated Meat, Calling it 'Junk Food,'" Cultivated X, 2023, https://cultivated-x.com/politics-law/french-parliamentarians-propose-law-ban-cultivated-meat-junk-food/.

48 "Nearly 70% of Americans Say Animal Wellness Plays an Important Role in Purchasing Decisions," National Science Foundation, 2024, https://www.nsf.org/news/nsf-reveals-americans-say-animal-wellness-important-role-purchasing-decisions.

49 Narawad, Aniket, "Global Surveys Show People's Growing Concern about Climate Change," Clean Energy Wire, 2024, https://www.cleanenergywire.org/factsheets/global-surveys-show-peoples-growing-concern-about-climate-change#:~:text=Overall%2C%20a%20majority%20of%20people,country%20should%20fight%20climate%20change.

50 Chriki, Sghaier, Marie-Pierre Ellies-Oury, and Jean-François Hocquette, "Is 'Cultured Meat' a Viable Alternative to Slaughtering Animals and a Good Comprise between Animal Welfare and Human Expectations?," National Library of Medicine, 2022, https://pmc.ncbi.nlm.nih.gov/articles/PMC8929989/.

51 Risner, Derrick, Yoobin Kim, Cuong Nguyen, Justin Siegel, and Edward Spang, "Environmental Impacts of Cultured Meat: A Cradle-to-Gate Life Cycle Assessment," bioRxiv, 2023, https://www.biorxiv.org/content/10.1101/2023.04.21.537778v1.full.

52 Sinke, Pelle, "Ex-ante Life Cycle Assessment of Commercial-Scale Cultivated Meat Production in 2030," Springer Nature, 2023, https://link.springer.com/article/10.1007/s11367-022-02128-8.

53 Chriki, Sghaier, and Jean-François Hocquette, "The Myth of Cultured Meat: A Review," National Library of Medicine, 2020, https://pmc.ncbi.nlm.nih.gov/articles/PMC7020248/.
54 "Cost-Effective Biotechnology Solutions for Cultivated Meat," Wacker, accessed: February 2, 2025, https://www.wacker.com/cms/en-us/products/insights/biotechnology-cultivated-meat.html.
55 Zidarič, Tanja, et al., "Cultured Meat: Meat Industry Hand in Hand with Biomedical Production Methods," Springer Nature, 2020, https://link.springer.com/article/10.1007/s12393-020-09253-w.
56 Wurgaft, Benjamin, *Meat Planet: Artificial Flesh and the Future of Food*, University of California Press, 2019.
57 Rutzerveld, Chloe, "Food Futures," BIS Publishers, 2018, https://www.bispublishers.com/food-futures.html.
58 "Fatwa on Cultivated Meat," Majlis Ugama Islam Singapura, 2024, https://www.muis.gov.sg/Media/Media-Releases/2024/2/3-Feb-24-Fatwa-on-Cultivated-Meat.
59 "Fatwa on Cultivated Meat," Majlis Ugama Islam Singapura, 2024, https://www.muis.gov.sg/Media/Media-Releases/2024/2/3-Feb-24-Fatwa-on-Cultivated-Meat.
60 "Kera Pitch Deck," Unpublished Internal Company Document, Kera, 2024.
61 Ritchie, Hannah, Pablo Rosado, and Max Roser, "Meat and Dairy Production," Our World in Data, 2023, https://ourworldindata.org/meat-production.
62 "The Rise of Flexitarianism," Cardiff Metropolitan University, accessed: January 30, 2025, https://www.cardiffmet.ac.uk/health/zero2five/news/Pages/The-rise-of-flexitarianism.aspx#:~:text=According%20to%20Kantar%2C%20flexitarians%20account,identify%20as%20flexitarian%20by%202025.
63 "The Rise of Flexitarianism," Cardiff Metropolitan University, accessed: January 30, 2025, https://www.cardiffmet.ac.uk/health/zero2five/news/Pages/The-rise-of-flexitarianism.aspx#:~:text=According%20to%20Kantar%2C%20flexitarians%20account,identify%20as%20flexitarian%20by%202025.
64 "U.S. Retail Market Insights for the Plant-Based Industry," Good Food Institute, accessed: February 2, 2025, https://gfi.org/marketresearch/.
65 Mascaraque, Maria, "Coronavirus Accelerates Shift towards Plant-Based Food," Euromonitor International, 2020, https://www.euromonitor.com/article/coronavirus-accelerates-shift-towards-plant-based-food; Ho, Shirley, "Exploring the General Public's and Experts' Risk and Benefit Perceptions of Cultured Meat in Singapore: A Mental Models Approach," National Library of Medicine, 2023, https://pmc.ncbi.nlm.nih.gov/articles/PMC10688898/.
66 Zidarič, Tanja, et al., "Cultured Meat: Meat Industry Hand in Hand with Biomedical Production Methods," Springer Nature, 2020, https://link.springer.com/article/10.1007/s12393-020-09253-w.

67 Whitehead, Dalton, and Yuan H Brad Kim, "The Impact of COVID 19 on the Meat Supply Chain in the USA: A Review," National Library of Medicine, 2022, https://pmc.ncbi.nlm.nih.gov/articles/PMC9478983/#B5.

68 Whitehead, Dalton, and Yuan H Brad Kim, "The Impact of COVID 19 on the Meat Supply Chain in the USA: A Review," National Library of Medicine, 2022, https://pmc.ncbi.nlm.nih.gov/articles/PMC9478983/#B5.

69 Kee, Jennifer, Fred Gale, and Joshua Huang, "Pork Exports to China Surged as African Swine Fever Curtailed China's Pork Output," USDA, 2023, https://www.ers.usda.gov/data-products/charts-of-note/chart-detail?chartId=107941.

70 Bryant, Christopher, Keri Szejda, Nishant Parekh, Varun Deshpende, and Brian Tse, "A Survey of Consumer Perceptions of Plant-Based and Clean Meat in the USA, India, and China," Frontiers, 2019, https://www.frontiersin.org/journals/sustainable-food-systems/articles/10.3389/fsufs.2019.00011/full.

71 Interesse, Giulia, *Investing in China's Meat Industry: Trends and Opportunities*, China Briefing, 2023.

72 "Workplace Safety and Health: Safety in the Meat and Poultry Industry, While Improving, Could Be Further Strengthened," US Government Accountability Office, 2005, https://www.gao.gov/products/gao-05-96.

73 "India to Set Regulatory Framework to Move Forward Cultivated Meat & Seafood," Cultivated X, 2024, https://cultivated-x.com/politics-law/india-regulatory-framework-cultivated-meat-seafood/.

74 "Consumer Insights," Statista, accessed: February 2, 2025, https://www.statista.com/global-consumer-survey/tool/10/gcs_esp_202203?bars=1&missing=1&heatmap=0&tgeditor=0&token=0&index=0&absolute=1&population=1&rows%5B0%5D=v0230e_ecom_fooddiet.

75 Bhalla, Surjit, and Karan Bhasin, "India Eliminates Extreme Poverty," Brookings, 2024, https://www.brookings.edu/articles/india-eliminates-extreme-poverty/.

76 "Cell-Cultivated Meat: An Overview," Congressional Research Service, 2023, https://crsreports.congress.gov/product/pdf/R/R47697.

77 "Cell-Cultivated Meat: An Overview," Congressional Research Service, 2023, https://crsreports.congress.gov/product/pdf/R/R47697.

78 Fassler, Joe, "Lab-Grown Meat Is Supposed to Be Inevitable. The Science Tells a Different Story," *The Counter*, 2021, https://thecounter.org/lab-grown-cultivated-meat-cost-at-scale/.

79 "Cell-Cultivated Meat: An Overview," Congressional Research Service, 2023, https://crsreports.congress.gov/product/pdf/R/R47697.

80 Garwood, Grace, "Analysis: Could Hybridge Meat be a Plant-Based Breakthrough?," The Food Institute, 2022, https://foodinstitute.com/focus/analysis-could-hybrid-meat-be-a-plant-based-breakthrough/.

81 Humbird, D., Scale-up Economics for Cultured Meat," *Biotechnology Bioengineering* 118 (2021): 3239–50, https://doi.org/10.1002/bit.27848.

82 "Cell-Cultivated Meat: An Overview," Congressional Research Service, 2023, https://crsreports.congress.gov/product/pdf/R/R47697.

83 "Cell-Cultivated Meat: An Overview," Congressional Research Service, 2023, https://crsreports.congress.gov/product/pdf/R/R47697.

84 "Cell-Cultivated Meat: An Overview," Congressional Research Service, 2023, https://crsreports.congress.gov/product/pdf/R/R47697.

85 "Cultivated Meat LCA and TEA: Policy Recommendations," The Good Food Institute, accessed: February 1, 2025, https://gfi.org/resource/cultivated-meat-lca-and-tea-policy-recommendations/.

86 Carlsson, Isabelle, "Denmark Will Be the First to Impose CO2 Tax on Farms, government says," Reuters, 2024, https://www.reuters.com/sustainability/denmark-will-be-first-impose-co2-tax-farms-government-says-2024-06-25/.

87 Battaglia Richi, Evelyne, Beatrice Baumer, Beatrice Conrad, Roger Darioli, Alexandra Schmid, and Ulrich Keller, "Health Risks Associated with Meat Consumption: A Review of Epidemiological Studies," National Library of Medicine, 2015, https://pubmed.ncbi.nlm.nih.gov/26780279/.

88 "Meat in Your Diet," National Health Service, accessed: February 2, 2025, https://www.nhs.uk/live-well/eat-well/food-types/meat-nutrition/.

89 Gibney M. J., "Ultra-Processed Foods: Definitions and Policy Issues," Current Developments in Nutrition 2018 Sep 14;3(2):nzy077. doi: 10.1093/cdn/nzy077. PMID: 30820487; PMCID: PMC6389637.

90 Colombet, Zoé, Ellen Schwaller, Anna Head, Chris Kypridemos, Simon Capewell, and Martin O'Flaherty, "OP12 Social Inequalities in Ultra-Processed Food Intakes in the United Kingdom: A time Trend Analysis (2008–2018)," *BMJ Journals*, 2022, https://jech.bmj.com/content/76/Suppl_1/A6.2.

91 Bryant, Christopher, et al., "A Survey of Consumer Perceptions of Plant-Based and Clean Meat in the USA, India, and China," Frontiers, 2019, https://www.frontiersin.org/journals/sustainable-food-systems/articles/10.3389/fsufs.2019.00011/full.

7 Sustainably Raised Meat

1 "Sustainable Livestock Farming Practices for Resilience," Alliance Biodiversity & CIAT, October 2024, https://alliancebioversityciat.org/stories/sustainable-livestock-farming-practices-resilience.

2 "Defining Regenerative Agriculture for State Policies and Programs," Recommendation from the California State Board of Food and Agriculture to CDFA Secretary, January 2025, https://www.cdfa.ca.gov/RegenerativeAg/docs/2025/

board_recommendation_defining_regenerative_ag_for_state_policies_and_progra ms_jan_2025.pdf.

3 “The 10 Elements of Agroecology Guiding the Transition to Sustainable Food a Agricultural Systems,” Food and Agriculture Organization of the United Nations, accessed: August 10, 2025, https://openknowledge.fao.org/server/api/core/bitstre ams/3d7778b3-8fba-4a32-8d13-f21dd5ef31cf/content.

4 Hickey, Louise, “Main Barriers for Methane Reduction Research Outlined,” Agriland, https://www.agriland.co.uk/farming-news/main-barriers-for-methane-reduction-research-outlined/.

5 Keenan, Rachel, “Conspiracy Theory on Methane-Cutting Cow Feed a ‘Wake-up Call’ Say Scientists,” *The Guardian*, 2025, https://www.theguardian.com/envi ronment/2025/feb/11/conspiracy-theory-on-methane-cutting-cow-feed-bova era-wake-up-call-say-scientists; Khan, Yusuf, “A Milk Brand Wanted to Tout Its Sustainability Push. Then It Got Swept Up in a Conspiracy Theory,” *Wall Street Journal*, 2025, https://www.wsj.com/articles/a-milk-brand-wanted-to-tout-its-sus tainability-push-then-it-got-swept-up-in-a-conspiracy-theory-f0c23ee4.

6 Keenan, Rachel, “Conspiracy Theory on Methane-Cutting Cow Feed a ‘Wake-up Call’ Say Scientists,” *The Guardian*, 2025, https://www.theguardian.com/envi ronment/2025/feb/11/conspiracy-theory-on-methane-cutting-cow-feed-bova era-wake-up-call-say-scientists.

7 “Background Information on Bovaer, the Feed Additive that Enables Farmers to Achieve Consistent Methane Emission Reductions by on Average 30% for Dairy Cows and on Average 45% for Beef Cattle,” DSM, 2022, https://www.dsm.com/cont ent/dam/dsm/corporate/en_US/documents/media-backgrounder_introduction_b ovaer_update_september_2022.pdf.

8 Keenan, Rachel, “Conspiracy Theory on Methane-Cutting Cow Feed a ‘Aake-up Call’ Say Scientists,” *The Guardian*, 2025, https://www.theguardian.com/envi ronment/2025/feb/11/conspiracy-theory-on-methane-cutting-cow-feed-bova era-wake-up-call-say-scientists.

9 Marsden, Jen, “Burping Cows, Bovaer and Boycotts: The Anti-Methane Additive That’s Taking Social Media by Storm,” EuroNews, 2024, https://www.euronews. com/green/2024/12/03/burping-cows-bovaer-and-boycotts-the-anti-methane-addit ive-thats-taking-social-media-by-st.

10 Gibson, Eloise, “Fonterra Caught up in Misinformation around Use of Feed Additive Bovaer,” RNZ, 2025, https://www.rnz.co.nz/news/country/540043/fonte rra-caught-up-in-misinformation-around-use-of-feed-additive-bovaer.

11 Keenan, Rachel, “Conspiracy Theory on Methane-Cutting Cow Feed a ‘Wake-up Call’ Say Scientists,” *The Guardian*, 2025, https://www.theguardian.com/envi ronment/2025/feb/11/conspiracy-theory-on-methane-cutting-cow-feed-bova era-wake-up-call-say-scientists.

12 Keenan, Rachel, "Conspiracy Theory on Methane-Cutting Cow Feed a 'Wake-up Call' Say Scientists," *The Guardian*, 2025, https://www.theguardian.com/environment/2025/feb/11/conspiracy-theory-on-methane-cutting-cow-feed-bovaera-wake-up-call-say-scientists.

13 Khan, Yusuf, "A Milk Brand Wanted to Tout Its Sustainability Push. Then It Got Swept Up in a Conspiracy Theory," *Wall Street Journal*, 2025, https://www.wsj.com/articles/a-milk-brand-wanted-to-tout-its-sustainability-push-then-it-got-swept-up-in-a-conspiracy-theory-f0c23ee4.

14 Gov.uk, "Bovaer Cow Feed Additive Explained," Food Standards Agency, 2024, https://food.blog.gov.uk/2024/12/05/bovaer-cow-feed-additive-explained/

15 "How Climate Change Is Stressing the Global Food Supply and Public Health," Columbia Mailman School of Public Health, September 2024, https://www.publichealth.columbia.edu/news/how-climate-change-stressing-global-food-supply-public-health; Ritchie, Hannah, "Food Production Is Responsible for One-Quarter of the World's Greenhouse Gas Emissions," Our World in Data, November 2019, https://ourworldindata.org/food-ghg-emissions.

16 Bai, Yongfei, and Francesca Cotrufo, "Grassland Soil Carbon Sequestration: Current Understanding, Challenges, and Solutions," Science, August 2022, https://www.science.org/doi/10.1126/science.abo2380.

17 "Global Symposium on Soil Erosion," Food and Agriculture Organization of the United Nations, accessed: August 10, 2025, https://www.fao.org/about/meetings/soil-erosion-symposium/key-messages/en/.

18 "Grasslands as a Carbon Store," Plantlife, July 2023, https://www.plantlife.org.uk/wp-content/uploads/2023/08/Grasslands-as-a-Carbon-Store.pdf.

19 "Land Use: Policies for a Net Zero UK," Climate Change Committee, accessed: August 10, 2025, https://www.theccc.org.uk/publication/land-use-policies-for-a-net-zero-uk/.

20 "Nature-Based Solutions," The International Union for Conservation of Nature, accessed: August 10, 2025, https://iucn.org/our-work/nature-based-solutions.

21 Seitz, D., R. Dechow, D. Emde, F. Schneider, and A. Don, "Improved Broad-Scale Modelling of Soil Organic Carbon Dynamics Following Land-Use Changes," *European Journal of Soil Science* 76 (4): e70159. https://doi.org/10.1111/ejss.70159.

22 "Are Free-Range Chickens the Best Choice for the Environment?," British Broadcasting Cooperation, accessed: August 10, 2025, https://www.bbc.co.uk/programmes/articles/3zL4mc4RmF0CjRYLmSxZvrs/are-free-range-chickens-the-best-choice-for-the-environment#:~:text=Carbon%2Dwise%20the%20answer%20is,chicken%20(around%2013%20weeks).

23 "*Chicken v. Wildlife: The Environmental Costs of Eating Poultry*," Take Extinction Off Your Plate, accessed: August 10, 2025, https://www.biologicaldiversity.org/takeextinctionoffyourplate/pdfs/ChickenFactsheet.pdf.

24 "SFT Welcomes Support for Small Abattoirs," Sustainable Food Trust, accessed: August 10, 2025, https://sustainablefoodtrust.org/news-views/sft-welcomes-support-for-small-abattoirs/.

25 Weber, Christopher, and Scott Matthews, "Food-Miles and the Relative Climate Impacts of Food Choices in the United States," ACS Publications, April 2008, https://pubs.acs.org/doi/10.1021/es702969f.

26 Bell, Ryan, "Temple Grandin, Killing Them Softly at Slaughterhouses for 30 Years," National Geograhpic, 2015, https://www.nationalgeographic.com/culture/article/temple-grandin-killing-them-softly-at-slaughterhouses-for-30-years.

27 Leasca, Stacey, "'Pasture Raised' Has a Brand-New Definition—Here's Why That Matters at the Meat Counter," Food & Wine, 2025, https://www.foodandwine.com/pasture-raised-vs-free-range-usda-definition-11754667

28 "Pasture Raised Is Not Synonymous With Free Range," American Pastured Poultry Producers Association, 2023, https://www.apppa.org/news/13410329.

29 McAfee, A. J., E. M. McSorley, G. J. Cuskelly, A. M. Fearon, B. W. Moss, J. A. M. Beattie, J. M. W. Wallace, M. P. Bonham, J. J. Strain, "Red Meat from Animals Offered a Grass Diet Increases Plasma and Platelet *n*-3 PUFA in Healthy Consumers," *British Journal of Nutrition*, September 2010, https://www.cambridge.org/core/journals/british-journal-of-nutrition/article/red-meat-from-animals-offered-a-grass-diet-increases-plasma-and-platelet-n3-pufa-in-healthy-consumers/B52D258D6025E6D7FC76FEDA49B4032D.

30 Dhiman, T. R., G. R. Anand, L. D. Satter, M. W. Pariza., "Conjugated Linoleic Acid Content of Milk from Cows Fed Different Diets," National Library of Medicine, October 1999, https://pubmed.ncbi.nlm.nih.gov/10531600/.

31 Dunne, Daisy, "Eating More Offal Could Help Cut Meat Emissions by '14%,'" Carbon Brief, April 2019, https://www.carbonbrief.org/eating-more-offal-could-help-cut-meat-emissions-by-14-per-cent/#:~:text=Choosing%20more%20meat%20by%2Dproducts,emissions%20from%20Germany's%20meat%20industry.

32 Biel, Wioletta, Ewa Czerniawska-Piątkowska, and Alicja Kowalczyk, "Offal Chemical Composition from Veal, Beef, and Lamb Maintained in Organic Production Systems," National Library of Medicine, July 2019, https://pmc.ncbi.nlm.nih.gov/articles/PMC6721235/.

Part 3

1 "Chemical Fertilizers Market Size, Global Forecast Report 2037," Research Nester, 2024, https://www.researchnester.com/reports/chemical-fertilizers-market/329.

2 "Glyphosate General Fact Sheet," National Pesticide Information Center, 2010, http://npic.orst.edu/factsheets/glyphogen.html?ref=hir.harvard.edu.

3 "What is the Internet of Things (IoT)?," IBM, accessed: February 28, 2025. https://www.ibm.com/think/topics/internet-of-things.

4 Wise, Timothy, *Eating Tomorrow: Agribusiness, Family Farmers, and the Battle for the Future of Food*, New Press, 2019, https://thenewpress.com/books/eating-tomorrow.

5 Saik, Robert, *Food 5.0: How We Feed the Future*, Lioncrest Publishing, 2019, https://www.amazon.co.uk/Food-5-0-How-Feed-Future/dp/1544504519/?_encoding=UTF8&pd_rd_w=GTpxz&content-id=amzn1.sym.3413293e-3815-4359-96ba-1ec5110e0b30&pf_rd_p=3413293e-3815-4359-96ba-1ec5110e0b30&pf_rd_r=259-7033390-3999948&pd_rd_wg=srAf7&pd_rd_r=9f8b4b09-add8-43ef-9263-cf6db6b665cb&ref_=aufs_ap_sc_dsk.

6 Nye, Caroline, and Matt Lobley, "Farm Labour in the U.K. Accessing the workforce the Industry Needs," University of Exeter, 2021, https://www.exeter.ac.uk/v8media/research/crpr/documents/Farm_labour_in_the_UK._Accessing_the_workforce_the_industry_needs_.pdf.

7 Ritchie, Hannah, Veronika Samborska, and Max Roser, "Urbanization," Our World in Data, 2018, https://ourworldindata.org/urbanization.

8 Ritchie, Hannah, Veronika Samborska, and Max Roser, "Urbanization," Our World in Data, 2018, https://ourworldindata.org/urbanization.

9 Rukmana, Deden ed., *The Routledge Handbook of Planning Megacities in the Global South*, Routledge, 2020, https://www.routledge.com/The-Routledge-Handbook-of-Planning-Megacities-in-the-Global-South/Rukmana/p/book/9781032400020#:~:text=In%202016%20there%20were%2031,located%20in%20the%20Global%20South.

10 Nakamura, Shohei, Mark Roberts, and Benjamin Stewart, "Has Extreme Poverty Been Urbanized?," World Bank Blogs, 2024, https://blogs.worldbank.org/en/developmenttalk/has-extreme-poverty-been-urbanized.

11 Ellis-Petersen, Hannah, "Fighting Giants: Eco-Activist Vandana Shiva on her Battle against GM Multinationals," *The Guardian*, 2023, https://www.theguardian.com/global-development/2023/apr/28/fighting-giants-eco-activist-vandana-shiva-on-her-battle-against-gm-multinationals.

12 "Agroecology in Cuba: For the Farmer, Seeing Is Believing," MACAC, accessed: Februrary 28, 2025, https://openknowledge.fao.org/server/api/core/bitstreams/2dc8a59f-61ca-47e0-880b-67811b167585/content.

13 Miley-Smith, Ben, "Everything You Need to Know about Hunger in Cuba," The Borgen Project, 2024, https://borgenproject.org/hunger-in-cuba/#:~:text=A%202023%20Cuban%20Observatory%20for,struggling%20to%20buy%20essential%20goods.

14 "AgroEcoTech: How Can Technology Accelerate a Transition to Agroecology?," Report for the Soil Association, 2021, https://www.soilassociation.org/media/22821/agroecotech-soil-association-report.pdf.

8 The Future of Breeding

1 Turnbull, Crystal, Morten Lillemo, and Trine A. K. Hvoslef-Eide, "Global Regulation of Genetically Modified Crops Amid the Gene Edited Crop Boom—a Review," Frontiers, 2021, https://www.frontiersin.org/journals/plant-science/articles/10.3389/fpls.2021.630396/full#B18.

2 "Modernizing the Regulatory System for Biotechnology Products: An Update to the Coordinated Framework for the Regulation of Biotechnology," Obama White House Archives, 2015, https://obamawhitehouse.archives.gov/sites/default/files/microsites/ostp/biotech_coordinated_framework.pdf.

3 "Update to the List of Bioengineered Foods," USDA, 2023, https://www.ams.usda.gov/rules-regulations/be.

4 "HHS Secretary Kennedy Directs FDA to Explore Rulemaking to Eliminate Pathway for Companies to Self-Affirm Food Ingredients Are Safe," Department of Health and Human Services, 2025, https://www.hhs.gov/press-room/revising-gras-pathway.html.

5 "Food safety in the EU," European Union, accessed: February 14, 2025, https://european-union.europa.eu/priorities-and-actions/actions-topic/food-safety_en#:~:text=The%2520European%2520food%2520safety%2520authority,the%2520EU's%2520food%2520safety%2520standards.

6 Williams, Holly, "U.S. Food Additives Banned in Europe: Expert Says What Americans Eat Is 'Almost Certainly' Making Them Sick," CBS News, 2023, https://www.cbsnews.com/news/us-food-additives-banned-europe-making-americans-sick-expert-says/.

7 "U.S. vs. EU: An Examination of the Trade Issues Surrounding Genetically Modified Food," PEW Trusts, 2002, https://www.pewtrusts.org/-/media/legacy/uploadedfiles/wwwpewtrustsorg/reports/food_and_biotechnology/vfbiotechtrade0602pdf.pdf.

8 "European Commission Authorizes 10 GM Crops for Food and Feed," ISAAA, 2021, https://www.isaaa.org/kc/cropbiotechupdate/article/default.asp?ID=18969#:~:text=The%2520European%2520Commission%2520has%2520authorized,published%2520on%2520August%252017%252C%25202021.

9 News, "New Genomic Techniques: MEPs Back Rules to Support Green Transition of Farmers," European Parliament, 2024, https://www.europarl.europa.eu/news/en/press-room/20240202IPR17320/new-genomic-techniques-meps-back-rules-to-support-green-transition-of-farmers.

10 "The Agreements on the Application of Sanitary and Phytosanitary Measures and Technical Barriers to Trade," FAO, accessed: February 15, 2025, https://www.fao.org/4/X3452E/x3452e06.htm#:~:text=The%20Agreement%20on%20the%20Application%20of%20Sanitary%20and,which%20may%20directly%20or%20indirectly%20affect%20international%20trade.

11 Cooper, Charlie, and Adam Behsudi, "Food Fight Looms over Liam Fox's Trade Mission to US," POLITICO, 2017, https://www.politico.eu/article/uk-us-trade-deal-gmo-food-fight-looms-over-liam-foxs-trade-mission-to-us/.

12 "Cartagena Protocol on Biosafety to the Convention on Biological Diversity," Convention on Biological Diversity, 2000, https://www.cbd.int/doc/legal/cartagena-protocol-en.pdf.

13 "Global Agricultural Information Network, MARA Announces 51 Genetically Engineered Cord and Soybean Varieties Registered," USDA Foreign Agricultural Service, 2023, https://apps.fas.usda.gov/newgainapi/api/Report/DownloadReportByFileName?fileName=MARA+Announces+51+Genetically+Engineered+Corn+and+Soybean+Varieties+Registered_Beijing_China+-+People%27s+Republic+of_CH2023-0185&.

14 "China's Genetically Modified Dilemma," Asia Society, 2024, https://asiasociety.org/policy-institute/chinas-genetically-modified-dilemma.

15 Jin, Yan, Simon Schaub, Jale Tosun, and Justus Wesseler, "Does China Have a Public Debate on Genetically Modified Organisms? A Discourse Network Analysis of Public Debate on Weibo," National Library of Medicine, 2022, https://pmc.ncbi.nlm.nih.gov/articles/PMC9344491/#bibr17-09636625211070150.

16 Little, Daniel, "Small Farmers in Indian Agriculture," Foundation for Agrarian Studies, 2017, https://fas.org.in/small-farmers-in-indian-agriculture/; Nagesh Kumar, Yarasi Bharathi, and T. Pradeep, "Indian Seed Sector—Challenges and Opportunities," Journal of Research PJTSAU 46 (4) (2018): 1–12, https://www.researchgate.net/publication/348382076_INDIAN_SEED_SECTOR_-CHALLENGES_AND_OPPORTUNITIES.

17 Thomas Abraham, Supreme Court to Begin Hearing on GM Mustard Release in April, *Hindustan Times*, 2025, https://www.hindustantimes.com/india-news/supreme-court-to-begin-hearing-on-gm-mustard-release-in-april-101741259331902.html.

18 Green, Jerry, "The Benefits of Herbicide-Resistant Crops," Society of Chemical Industry, 2012, http://www.ask-force.org/web/HerbizideTol/Green-Benefits-Herbicide-Resistant-Crops-2012.pdf?ref=hir.harvard.edu.

19 "Food Waste FAQs," USDA, https://www.usda.gov/about-food/food-safety/food-loss-and-waste/food-waste-faqs; National Institute of Food and Agriculture, "Importance of Addressing Food Loss and Waste," USDA, https://www.nifa.usda.gov/topics/food-loss-waste.

20 Pollock, Andrew, "Gene-Altered Apples Gets U.S. Approval," *New York Times*, 2015, https://www.nytimes.com/2015/02/14/business/gmo-apples-are-approved-for-growing-in-us.html.

21 McPherson, Malia, "Three Reasons I'm Excited about Arctic Apples," Stanford Law Society, 2015, https://law.stanford.edu/2015/02/21/three-reasons-im-excited-about-arctic-apples/.

22 Leu, Andre, "The Myth of No-Till: The Future is Regenerative Organic Agriculture," Regeneration International, 2022, https://regenerationinternational.org/2022/12/12/the-myth-of-no-till-the-future-is-regenerative-organic-agriculture/.

23 Leu, Andre, "The Myth of No-Till: The Future Is Regenerative Organic Agriculture," Regeneration International, 2022, https://regenerationinternational.org/2022/12/12/the-myth-of-no-till-the-future-is-regenerative-organic-agriculture/.

24 Bloch, Sarah, Rosemary Clark, Shayin Gottlieb, Kent Wood, Neal Shah, San-Ming Mak, James Lorigan, Jenny Johnson, Austin Davis-Richardson, Lorna Williams, Megan McKellar, Dominic Soriano, Max Petersen, Alana Horton, Olivia Smith, Leslie Wu, Emily Tung, Richard Broglie, Alvin Tamsir, and Karsten Temme, "Biological Nitrogen Fixation in Maize: Optimizing Nitrogenase Expression in a Root-Associated Diazotroph," National Library of Medicine, 2020, https://pmc.ncbi.nlm.nih.gov/articles/PMC7382387/.

25 Bloch, Sarah, et al., "Biological Nitrogen Fixation in Maize: Optimizing Nitrogenase Expression in a Root-Associated Diazotroph," National Library of Medicine, 2020, https://pmc.ncbi.nlm.nih.gov/articles/PMC7382387/.

26 "Chemical Fertilizers Market Size, Global Forecast Report 2037," Research Nester, 2024, https://www.researchnester.com/reports/chemical-fertilizers-market/329.

27 Passarelli, Simone, et al., "Global Estimation of Dietary Micronutrient Inadequacies: A Modelling Analysis," *Lancet Global Health*, 12 (10): e1590–9.

28 "Vitamin A Deficiency," World Health Organization, accessed: February 15, 2025, https://www.who.int/data/nutrition/nlis/info/vitamin-a-deficiency.

29 Solh, Mahmoud, "Foreword," FAO, https://www.fao.org/4/y4347e/y4347e01.htm.

30 Solh, Mahmoud, "Foreword," FAO, https://www.fao.org/4/y4347e/y4347e01.htm.

31 E. McLean, M. Cogswell, I. Egli, D. Wojdyla, and B. de Benoist, "Worldwide Prevalence of Anaemia, WHO Vitamin and Mineral Nutrition Information System, 1993–2005," Public Health Nutrition 12 (4) (April 2009): 444–54. doi: 10.1017/S1368980008002401. Epub May 23, 2008. PMID: 18498676.

32 "Dirt Poor: Have Fruits and Vegetables Become Less Nutritious?," Scientific American, April 2011, https://www.scientificamerican.com/article/soil-depletion-and-nutrition-loss/#:~:text=A%20landmark%20study%20on%20the,over%20the%20past%20half%20century.

33 Schmitt, Kellie, "Less Nutritious Crops: Another Result of Rising CO2," *Public Health Magazine*, 2024, https://magazine.publichealth.jhu.edu/2024/less-nutritious-crops-another-result-rising-co2#:~:text=The%2520amount%2520of%2520nutrient%2520decline,and%2520magnesium%2520by%2520about%25209%2525.
34 Turnbull, Crystal, Morten Lillemo, and Trine A. K. Hvoslef-Eide, "Global Regulation of Genetically Modified Crops Amid the Gene Edited Crop Boom—a Review," Frontiers, 2021, https://www.frontiersin.org/journals/plant-science/articles/10.3389/fpls.2021.630396/full#B18.
35 "Tangerine," Wikipedia, accessed: February 15, 2025, https://en.wikipedia.org/wiki/Tangerine.
36 Ritchie, Hannah, "Smallholders Produce One-Third of the World's Food, Less Than Half of What Many Headlines Claim," Our World in Data, 2021, https://ourworldindata.org/smallholder-food-production; Ritchie, Hannah, and Max Roser, "Farm Size and Productivity," Our World in Data, 2022, https://ourworldindata.org/farm-size#article-citation.
37 Loft, Phillip, "UK Aid: FCDO Merger and the Move to 0.5% Four Years on," House of Commons Library, 2024, https://commonslibrary.parliament.uk/uk-aid-fcdo-merger-and-the-move-to-0-5-four-years-on/.
38 "Annual Report 2023," Gates Foundation, 2023, https://www.gatesfoundation.org/about/financials/annual-reports/annual-report-2023; "Future of Food," Bezos Earth Fund, accessed: February 15, 2025, https://www.bezosearthfund.org/our-programs/future-of-food.
39 Fuglie, Keith, "Prices for Genetically Modified Seeds Have Risen Much Faster Than Non-GM Seeds," USDA, 2023, https://www.ers.usda.gov/data-products/charts-of-note/chart-detail?chartId=106785.
40 "Cost of Developing a GM Crop Demystified," ISAAA, 2016, https://www.isaaa.org/kc/cropbiotechupdate/article/default.asp?ID=14638#:~:text=Such%2520costs%2520are%2520not%2520far,donating%2520as%2520little%2520as%2520$10.
41 Fuglie, Keith, "Prices for Genetically Modified Seeds Have Risen Much Faster Than Non-GM Seeds," USDA, 2023, https://www.ers.usda.gov/data-products/charts-of-note/chart-detail?chartId=106785.
42 Conway, Hannah, "Farmers, Food and Freedom," Transform Trade, 2024, https://static1.squarespace.com/static/62a067e5deea9a028a3eae4b/t/662a4b9884540d37b4f5cad4/1714047898547/Farmers_food_freedom_seeds_report_Transform_Trade_FINAL.pdf.
43 Grubb, Norton, "From Seeds of Suicide to Seeds of Hope: Why Are Indian Farmers Committing Suicide and How Can We Stop This Tragedy?," Huffpost, 2009, https://www.huffpost.com/entry/from-seeds-of-suicide-to_b_192419.

44 Klein, Daniel, "Farmer Suicide, GMOs and Seed Saving," Huffpost, 2013, https://www.huffpost.com/entry/farmer-suicide-gmos-seed-saving_b_2662604?utm_hp_ref=food.

45 Grubb, Norton, "From Seeds of Suicide to Seeds of Hope: Why Are Indian Farmers Committing Suicide and How Can We Stop This Tragedy?," Huffpost, 2009, https://www.huffpost.com/entry/from-seeds-of-suicide-to_b_192419.

46 Bayable, M., A. Tsunekawa, N. Haregeweyn, T. Ishii, G. Alemayehu, M. Tsubo, E. Adgo, A. Tassew, W. Tsuji, F. Asaregew, and T. Masunaga, "Biomechanical Properties and Agro-Morphological Traits for Improved Lodging Resistance in Ethiopian Teff (*Eragrostis tef* (Zucc.) Trottor) Accessions," *Agronomy* 10 (2020): 1012. https://doi.org/10.3390/agronomy10071012.

47 Beyene G., R. D. Chauhan, J. Vilmer, N. Music, N. Wang, E. Gebre, D. Girma, S. Chanyalew, K. Assefa, G. Tabor, M. Gehan, M. McGrone, M. Yang, B. Lenderts, C. Schwartz, H. Gao, W. Gordon-Kamm, N. Taylor, and D. MacKenzie, "CRISPR/Cas9-Mediated Tetra-Allelic Mutation of the 'Green Revolution' SEMIDWARF-1 (SD-1) Gene Confers Lodging Resistance in Tef (Eragrostis Tef)," *Plant Biotechnology Journal* 20 (9) (September 2022): 1716–29. doi: 10.1111/pbi.13842.

48 "USDA Clears Danforth Center's Genome Edited Teff," Donald Danforth Plant Science Center, 2023, https://www.danforthcenter.org/news/usda-clears-danforth-centers-genome-edited-teff/.

49 Abraham Adane, "Towards a Workable Biosafety System for Regulating Genetically Modified Organisms in Ethiopia: Balancing Conservation and Competitiveness," GM Crops & Food, 2013, https://www.researchgate.net/publication/236196750_Toward_a_workable_biosafety_system_for_regulating_genetically_modified_organisms_in_Ethiopia_Balancing_conservation_and_competitiveness.

50 Gudeta, Bedane, Endale Gabre-Kedisso, Donia Gurnessa, Dawit Tesfaye, Samuel Damtew, Workishet Taye, Arkebe Gebre-Egziabher, Merdasa Balcha, Tadessa Daba, Alehegn Workie, and Karim Maredia, "Adaptability of Genetically Engineered Bt Cotton Varieties in Different Growing Regions of Ehtiopia," *Advances in Agriculture* 2023(1) (2023), https://doi.org/10.1155/2023/8224053.

51 "IARC Monograph on Glyphosate," International Agency for Research on Cancer, 2018, https://www.iarc.who.int/featured-news/media-centre-iarc-news-glyphosate/.

52 Aga, Aniket, "Genetically Modified Crops: Twenty Years of Debate in India," Regional Office New Delhi, 2023, https://in.boell.org/en/gm-crops-india#:~:text=As%2520of%25202019%252C%2520over%252099,with%2520which%2520they%2520are%2520modified.

53 "Study Confirms GM Crops Lead to Increased Pesticide Use," Science in Society Archive, 2012, https://www.i-sis.org.uk/Study_Confirms_GM_crops_Increased_Pesticide_Use.php.

54 "Friend of the Earth Europe, Introducing Glyphosate, the World's Biggest Selling Herbicide," 2013, https://friendsoftheearth.eu/wp-content/uploads/2013/06/foee_1_introducing_glyphosate.pdf.

55 Kanissery, Ramdas, Biwek Gairhe, Davie Kadyampakeni, Ozgur Batuman, and Fernando Alferez, "Glyphosate: Its Environmental Persistence and Impact on Crop Health and Nutrition," National Library of Medicine, 2019, https://pmc.ncbi.nlm.nih.gov/articles/PMC6918143/.

56 Costas-Ferreira, Carmen, Rafael Durán, and Lilian Faro, "Toxic Effects of Glyphosate on the Nervous System: A Systematic Review," National Library of Medicine, 2022, https://pmc.ncbi.nlm.nih.gov/articles/PMC9101768/.

57 Pleasants, J., "Milkweed Restoration in the Midwest for Monarch Butterfly Recovery: Estimates of Milkweeds Lost, Milkweeds Remaining and Milkweeds That Must Be Added to Increase the Monarch Population. Insect Conververtion and Diversity 10 (2017): 42–53. https://doi.org/10.1111/icad.12198.

58 Gandhi, Kavita, Shaila Khan, Mansi Patrikar, Aishwarya Markad, Niraj Kumar, Aaditee Choudhari, Pallavi Sagar, and Shreya Indurka, "Exposure Risk and Environmental Impacts of Glyphosate: Highlights on the Toxicity of Herbicide Co-Formulants," Science Direct, 2021, https://www.sciencedirect.com/science/article/pii/S2667010021001281#:~:text=Common%2520milkweed%2520plants%2520were%2520killed,(Monarch%2520Watch%252C%25202008).

59 "Monarchs in Decline," Xerces Society, accessed: February 15, 2025, https://xerces.org/monarchs/conservation-efforts.

60 Shushkova, Tatyana, Inna Ermakova, and Alexey Leontievsky, "Glyphosate Bioavailability in Soil," National Library of Medicine, 2010, https://pubmed.ncbi.nlm.nih.gov/19902365/.

61 Eker, Salim, Levent Ozturk, Atila Yazici, Bulent Erenoglu, Volker Romheld, and Ismail Cakmak, "Foliar-Applied Glyphosate Substantially Reduced Uptake and Transport of Iron and Manganese in Sunflower (Helianthus annuus L.) Plants," National Library of Medicine, 2006, https://pubmed.ncbi.nlm.nih.gov/17177536/.

62 Haney, R. L., S. Senseman, F. M Hons, and David Zuberer, "Effect of Glyphosate on Soil Microbial Activity and Biomass," Research Gate, 2009, https://www.researchgate.net/publication/232668121_Effect_of_Glyphosate_on_Soil_Microbial_Activity_and_Biomass; Wardle, D. A., and D. Parkinson, "Effects of Three Herbicides on Soil Microbial Biomass and Activity," Springer Nature, 1990, https://link.springer.com/article/10.1007/BF02851906.

63 Piola, Lucas, Julio Fuchs, Maria Luisa Oneto, Silvana Basack, Eva Kesten, and Norma Casabe, "Comparative Toxicity of Two Glyphosate-Based Formulations to Eisenia Andrei under Laboratory Conditions," National Library of Medicine, 2013, https://pubmed.ncbi.nlm.nih.gov/23332878/; Casabé, Norma, Lucas Piola, Julio Fuchs, Maria Luisa Oneto, Laura Pamparato, Silvana Basack, Rosana Gimenez,

Ruben Massaro, Juan Carlos Papa, and Eva Kesten, "Ecotoxicological Assessment of the Effects of Glyphosate and Chlorpyrifos in an Argentine Soya Field," Research Gate, 2007, https://www.researchgate.net/publication/226955680_Ecotoxicological_assessment_of_the_effects_of_glyphosate_and_chlorpyrifos_in_an_Argentine_soya_field; Springett, J. A., and R. A. J. Gray, "Effect of Repeated Low Doses of Biocides on the Earthworm Aporrectodea Caliginosa in Laboratory Culture," Scispace, 1992, https://typeset.io/papers/effect-of-repeated-low-doses-of-biocides-on-the-earthworm-47n8og2szk.

64 Kuklinsky-Sobral, Júlia, "Isolation and Characterization of Endophytic Bacteria from Soybean (Glycine Max) Grown in Soil Treated with Glyphosate Herbicide," Springer Nature, 2005, https://link.springer.com/article/10.1007/s11104-004-6894-1; Zobiole, L. H. S., "Glyphosate Affects Micro-Organisms in Rhizospheres of Glyphosate-Resistant Soybeans," National Library of Medicine, 2010, https://pubmed.ncbi.nlm.nih.gov/20880215/.

65 Kanissery R., B. Gairhe, D. Kadyampakeni, O. Batuman, and F. Alferez, "Glyphosate: Its Environmental Persistence and Impact on Crop Health and Nutrition," Plants (Basel). 13 8(11) (November 2019): 499. doi: 10.3390/plants8110499. PMID: 31766148; PMCID: PMC6918143.

66 Relyea, Rick, "The Lethal Impact of Roundup on Aquatic and Terrestrial Amphibians," The Nuclear Regulatory Commission, 2005, https://www.nrc.gov/docs/ml1434/ml14345a564.pdf.

67 Bringolf, Robert, Gregory Cope, Shad Mosher, Chris Barnhart, and Damian Shea, "Acute and Chronic Toxicity of Glyphosate Compounds to Glochidia and Juveniles of Lampsilis Siliquoidea (Unionidae)," National Library of Medicine, 2007, https://pubmed.ncbi.nlm.nih.gov/17867870/; Guilherme, S., "European Eel (Anguilla Anguilla) Genotoxic and Pro-Oxidant Responses Following Short-Term Exposure to Roundup—a Glyphosate-Based Herbicide," National Library of Medicine, 2010, https://pubmed.ncbi.nlm.nih.gov/20643706/.

68 Molyneaux, Paul, "The Rise and Fall of Genetically Engineered Salmon," National Fisherman, 2025, https://www.nationalfisherman.com/the-rise-and-fall-of-genetically-engineered-salmon.

69 Huffstutter, P. J, "Japan, South Korea Block Certain U.S. Wheat Varieties over GMO Concerns: USDA," Reuters, 2016, https://www.reuters.com/article/business/environment/japan-south-korea-block-certain-us-wheat-varieties-over-gmo-concerns-usda-idUSKCN10C3CG/.

70 Petrequin, Samuel, "Environmentalists Are Up in Arms after EU Relaxes Rules on Genetically Modified Plants," Fortune, 2024, https://fortune.com/europe/2024/01/24/environmentalists-furious-european-union-non-gmo-genetically-modified-plants-food/.

71 Mallory-Smith C., and M. Zapiola, "Gene Flow from Glyphosate-Resistant Crops," Pest Management Science 64 (4) (April 2008): 428–40, doi: 10.1002/ps.1517. PMID: 18181145.

72 Gurian-Sherman, Doug, and Margaret Mellon, "The Rise of Superweeds—and What to Do about It," Union of Concerned Scientists, 2013, https://navdanyainternational.org/wp-content/uploads/2017/09/rise-of-superweeds.pdf?ref=hir.harvard.edu; Baek, Yousoon, "Evolution of Glyphosate-Resistant Weeds," National Library of Medicine, 2021, https://pubmed.ncbi.nlm.nih.gov/33932185/.

73 Gurian-Sherman, Doug, and Margaret Mellon, "The Rise of Superweeds—and What to Do about It," Union of Concerned Scientists, 2013, https://navdanyainternational.org/wp-content/uploads/2017/09/rise-of-superweeds.pdf?ref=hir.harvard.edu.

74 "How Much Glyphosate Is Used Worldwide?," Global Glyphosate Study, accessed: February 16, 2025, https://glyphosatestudy.org/hrf_faq/how-much-glyphosate-is-used-worldwide/.

75 "Dicamba," US Environmental Protection Agency, accessed: February 17, 2025, https://www.epa.gov/pesticides/dicamba.

76 Mueller, Thomas, and Lawrence Steckel, "Changes in Dicamba Use Are Ahead," *Outlooks on Pest Management* 33 (2) (2022), DOI:10.1564/v33_apr_08.

77 Lerro, Catherine, et al., "Dicamba Use and Cancer Incidence in the Agricultural Health Study: An Updated Analysis," *International Journal of Epidemiology*, 2020, https://academic.oup.com/ije/article/49/4/1326/5827818.

78 Newman, Peter, "Mixing Herbicides Wins Again," Australian Herbicide Resistance Initiative, 2019, https://www.ahri.uwa.edu.au/mixing-herbicides-wins-again/?ref=hir.harvard.edu.

79 "Harvest Weed Seed Control as a Tactic for Weed Management," No-Till Farmer, 2019, https://www.no-tillfarmer.com/articles/8425-harvest-weed-seed-control-as-a-tactic-for-weed-management?ref=hir.harvard.edu.

80 "Harvest Weed Seed Control as a Tactic for Weed Management," No-Till Farmer, 2019, https://www.no-tillfarmer.com/articles/8425-harvest-weed-seed-control-as-a-tactic-for-weed-management?ref=hir.harvard.edu.

81 "Market Profile for Fresh Papaya from West Africa to Regional and European Markets," COLEAD, 2024, https://resources.colead.link/en/system/files/file_fields/2024/08/21/marketprofile-freshpapayaenweb.pdf.

82 Garrison, Cassandra, and Adriana Barrera, "EXCLUSIVE Mexico to Proceed with GMO Corn Ban, Seeks International Grain Deals –Official," Reuters, 2022, https://www.reuters.com/markets/commodities/exclusive-mexico-proceed-with-gmo-corn-ban-seeks-international-grain-deals-2022-10-27/.

83 "Mexican President Confident Congress Will Ban Planting GM Corn in 2025," Reuters, 2024, https://www.reuters.com/markets/commodities/mexican-president-confident-congress-will-ban-planting-gm-corn-2025-2024-12-21/.

84 Woźniak-Gientka, Ewa, Agata Tyczewska, and Tomasz Twardowski, "Public Opinion on Biotechnology and Genetic Engineering in the European Union: Polish Consumer Study," National Library of Medicine, 2022, https://pmc.ncbi.nlm.nih.gov/articles/PMC9642953/.

85 "Pesticide," Britannica, accessed: February 17, 2025, https://www.britannica.com/technology/pesticide.

86 Luna, Jessie, and Brian Dowd-Uribe, "Power and GM Cotton in Burkina Faso," The Ecologist, 2020, https://theecologist.org/2020/sep/02/power-and-gm-cotton-burkina-faso.

87 Qiao, Fangbin, "Fifteen Years of Bt Cotton in China: The Economic Impact and its Dynamics," World Development, 70 (2015), https://doi.org/10.1016/j.worlddev.2015.01.011.

88 "The Power of Orange: Catalyzing Delivery of Vitamin A Maize in Zambia," Harvest Plus, 2022, https://www.harvestplus.org/wp-content/uploads/2022/09/The-Power-of-Orange-Catalyzing-Delivery-of-Vitamin-A-Maize-in-Zambia.pdf.

89 "The Power of Orange: Catalyzing Delivery of Vitamin A Maize in Zambia," Harvest Plus, 2022, https://www.harvestplus.org/wp-content/uploads/2022/09/The-Power-of-Orange-Catalyzing-Delivery-of-Vitamin-A-Maize-in-Zambia.pdf.

90 "The Power of Orange: Catalyzing Delivery of Vitamin A Maize in Zambia," Harvest Plus, 2022, https://www.harvestplus.org/wp-content/uploads/2022/09/The-Power-of-Orange-Catalyzing-Delivery-of-Vitamin-A-Maize-in-Zambia.pdf.

91 Simpungwe, Eliab, Thanda Dhliwayo, Markus Palenberg, Victor Taleon, Ekin Birol, Adewale Oparinde, A Saltzman, and Michael Diressie, "Orange Maize in Zambia: Crop Development and Delivery Experience," Research Gate, 2017, https://www.researchgate.net/publication/318469363_Orange_maize_in_Zambia_Crop_development_and_delivery_experience.

92 Abbas, Qussai, Mathias Wilhelm, Bernhard Kuster, Brigitte Poppenberger, and Dmitrij Frishman, "Exploring Crop Genomes: Assembly Features, Gene Prediction Accuracy, and Implications for Proteomics," BMC Genomics, 2024, https://bmcgenomics.biomedcentral.com/articles/10.1186/s12864-024-10521-w.

93 Hsu, Charlotte, "Guacamole Lovers Rejoice! Avocado Genome Has Been Sequenced," University of Buffalo, 2019, https://arts-sciences.buffalo.edu/news-and-events/recent-news/2019/august/avocado-genome-sequenced.html#:~:text=We%2520needed%2520to%2520sequence%2520the,LANGEBIO%252C%2520where%2520he%2520is%2520emeritus.

9 Feeding a City

1 Steel, Carolyn, *Hungry City*, Chatto & Windus, June 2008, https://www.carolynsteel.com/hungrycitybook.

2 Steel, Carolyn, *Hungry City*, Chatto & Windus, June 2008, https://www.carolynsteel.com/hungrycitybook.

3 "Make Cities and Human Settlements Inclusive, Safe, Resilient and Sustainable," United Nations Sustainable Development Goals, accessed: August 20, 2025, https://unstats.un.org/sdgs/report/2023/goal-11/#:~:text=The%20world's%20population%20reached%208,exacerbating%20inequalities%20and%20urban%20poverty.

4 "2018 Revision of World Urbanization Prospects," United Nations, 2018, https://www.un.org/en/desa/2018-revision-world-urbanization-prospects.

5 "Ecological Threat Report 2022," Institute for Economics & Peace, 2022, https://www.visionofhumanity.org/wp-content/uploads/2022/10/ETR-2022-Web-V1.pdf.

6 "What You Need to Know about the Food Miles," ESG Report, January 2025, https://esgthereport.com/the-truth-about-food-miles-what-you-need-to-know-about-the-foods-you-eat/.

7 Shim, Jee-Seon, "Ultra-Processed Food Consumption and Obesity: A Narrative Review of Their Association and Potential Mechanisms," *Journal of Obesity & Metabolic Syndrome*, January 2025, https://www.jomes.org/journal/view.html?doi=10.7570/jomes24045.

8 Neuman, Melissa, Ichiro Kawachi, Steven Gortmaker, and S. V. Subramanian, "Urban-Rural Differences in BMI in Low- and Middle-Income Countries: The Role of Socioeconomic Status," Science Direct, February 2013, https://www.sciencedirect.com/science/article/pii/S0002916523054230?via%3Dihub.

9 Dutko, Paula, Michele Ver Ploeg, and Tracey Farrigan, "Characteristics and Influential Factors of Food Deserts," United States Department of Agriculture, August 2012, https://ers.usda.gov/sites/default/files/_laserfiche/publications/45014/30940_err140.pdf?v=86608

10 Dutko, Paula, Michele Ver Ploeg, and Tracey Farrigan, "Characteristics and Influential Factors of Food Deserts," United States Department of Agriculture, August 2012, https://ers.usda.gov/sites/default/files/_laserfiche/publications/45014/30940_err140.pdf?v=86608.

11 Corfe, Scott, "What Are the Barriers to Eating Healthily in the UK?," Social Market Foundation, October 2018, https://www.smf.co.uk/wp-content/uploads/2018/10/What-are-the-barriers-to-eating-healthy-in-the-UK.pdf.

12 Silviera, V. N. C., A. M. dos Santos, and A. K. T. C França, "Determinants of the Consumption of Ultra-Processed Foods in the Brazilian Population," Cambridge University Press, October 2024, https://www.cambridge.org/core/journals/brit

ish-journal-of-nutrition/article/determinants-of-the-consumption-of-ultraprocessed-foods-in-the-brazilian-population/C0A95F8C5C65B7B679837043DF6039E5.
13 "Snack Food," Statista, 2025, https://www.statista.com/outlook/cmo/food/confectionery-snacks/snack-food/worldwide.
14 "Pennsylvania Fresh Food Financing Initiative," Reinvestment Fund, accessed: August 20, 2025, https://www.reinvestment.com/insights/pennsylvania-fresh-food-financing-initiative/#:~:text=Share,the%20administrator%20of%20the%20program.
15 Dubowitz, Tamara, Madhumita Ghosh-Dastidar, Deborah Cohen, Robin Beckman, Elizabeth Steiner, Gerald Hunter, Karen Florez, Christina Huang, Christine Vaughan, Jennifer Sloan, Shannon N. Zenk, Steven Cummins, and Rebecca Collins, "Diet and Perceptions Change with Supermarket Introduction in a Food Desert, but Not Because of Supermarket Use," Health Affairs, November 2015, https://www.healthaffairs.org/doi/10.1377/hlthaff.2015.0667.
16 "SB 159," Kansas Legislative Sessions, 2025, https://www.kslegislature.gov/li_2024/b2023_24/measures/sb159/.
17 "App-Based Ride Service Launches for Columbus Community," SmartCitiesWorld, July 2019, https://www.smartcitiesworld.net/news/app-based-ride-service-launches-for-columbus-community-4390; Thompson, Kelsey, "CapMetro Launches New On-Demand Service in East Austin," Kxan, January 2025, https://www.kxan.com/traffic/traffic-projects/capmetro-launches-new-on-demand-service-in-east-austin/.
18 Westhagen, Amy, "Announcing Lyft's Grocery Access Program," Lyft, December 2018, https://www.lyft.com/blog/posts/announcing-lyfts-grocery-access-program.
19 "Mayor Cracks Down on Opening of New Hot-Food Takeaways around Schools," London Assembly, November 2017, https://www.london.gov.uk/press-releases/mayoral/mayor-cracks-down-on-new-takeaways-near-schools.
20 "Food Access Research Atlas—Go to the Atlas," USDA Economic Research Service, accessed: August 20, 2025, https://www.ers.usda.gov/data-products/food-access-research-atlas/go-to-the-atlas.
21 "Quito's Participatory Urban Agriculture Programme," Future Policy.org, https://www.futurepolicy.org/global/quito-agrupar/.
22 Halliday, J., and L. Platenkamp, "Nicolarea, Bioferia Food Markets—Quite, Educador," FoodActionCities, 2019, https://foodactioncities.org/case-studies/urban-markets/.
23 Halliday, J., and L. Platenkamp, "Nicolarea, Bioferia Food Markets—Quite, Educador," FoodActionCities, 2019, https://foodactioncities.org/case-studies/urban-markets/.
24 "COVID-19 Will Double Number of People Facing Food Crises Unless Swift Action is Taken," World Food Programme, April 2020, https://www.wfp.org/news/covid-19-will-double-number-people-facing-food-crises-unless-swift-action-taken.

25 Christian, Alex, "The Untold Story of the Big Boat That Broke the World," Wired, 2021, https://www.wired.com/story/ever-given-global-supply-chain/.

26 Hallstrom, Erin, "The Wake from the Suez Canal Blockage Is Hitting Parts of CPG Industry," Food Processing, April 2021, https://www.foodprocessing.com/supply-chain/logistics/news/11297669/the-wake-from-the-suez-canal-blockage-is-hitting-parts-of-cpg-industry; "Love Coffee? Your Favorite Beverage May Get Dearer if the Suez Canal Remains Blocked by Cargo Ship," News 18, March 2021, https://www.news18.com/news/buzz/love-coffee-your-favorite-beverage-may-get-dearer-if-the-suez-canal-remains-blocked-by-cargo-ship-3576809.html; Knowles, Mike, "Suez Shutdown Delays Fruit Supply," Eurofruit, March 2021, https://www.fruitnet.com/eurofruit/suez-shutdown-delays-fruit-supply/184793.article.

27 "H.R.3684—Infrastructure Investment and Jobs Act," Congress, 2022, https://www.congress.gov/bill/117th-congress/house-bill/3684.

28 "Investing in America: Biden-Harris Administration Announces Nearly $580 Million for Ports to Strengthen American Supply Chains and Lower Costs," US Department of Transportation Maritime Administration, November 2024, https://www.maritime.dot.gov/newsroom/investing-america-biden-harris-administration-announces-nearly-580-million-ports.

29 "Construction of Maasvlakte 2, Rotterdam," Port Economics, Management, and Policy, accessed: August 20, 2025, https://porteconomicsmanagement.org/pemp/contents/part4/port-terminal-construction/construction-maasvlakte-2-rotterdam/.

30 Rose, Nick, Elena Serrano, Kathy Hosig, Carola Haas, Dixie Reaves, and Sharon Nickols-Richardson, "The 100-Mile Diet: A Community Approach to Promote Sustainable Food Systems Impacts Dietary Quality," *Journal of Hunger & Environmental Nutrition*, October 2008, https://www.tandfonline.com/doi/full/10.1080/19320240802244082.

31 "Local Food Growth Plan," Local Food Plan, April 2025, https://www.sustainweb.org/assets/local-food-growth-plan-1748336069.pdf.

32 "Paul Baran and the Origins of the Internet," RAND, March 2018, https://www.rand.org/pubs/articles/2018/paul-baran-and-the-origins-of-the-internet.html#:~:text=%22On%20Distributed%20Communications%22%20Series,Communications%2C"%20published%20in%201964.

33 "Harvesting Opportunity," Federal Reserve Bank of St. Louis, 2017, https://www.thecommonmarket.org/assets/uploads/reports/Harvesting_Opportunity.compressed.pdf.

34 Thompson, Bella, "Why Local Food Can Restore Our Failing Food System," Sustainable Food Trust, September 2021, https://sustainablefoodtrust.org/news-views/local-food-restore-failing-food-system/#:~:text=Local%20food%20systems%20have%20the,each%20other%20and%20the%20land.

35 "Policy, Plates, and Planet: Actions to catalyze Urban Food System Transformation," World Wildlife Fund, accessed: August 20, 2025, https://wwfint.awsassets.panda.org/downloads/policy-plates-and-planet-actions-to-catalyze-urban-food-system-transformation_1.pdf.

36 "How Local Food Systems Support Sustainability and Food Access," Churchill Northern Studies Centre, accessed: August 20, 2025, https://churchillscience.ca/how-local-food-systems-support-sustainability-and-food-access/; "Field to Fork: Global Food Miles Generate Nearly 20% of All CO2 Emissions from Food," European Union Environment, January 2023, https://environment.ec.europa.eu/news/field-fork-global-food-miles-generate-nearly-20-all-co2-emissions-food-2023-01-25_en.

37 McCurdy, Madison, "Health Benefits of Eating Locally," University of New Hampshire, May 2022, https://extension.unh.edu/blog/2022/05/health-benefits-eating-locally.

38 "Supporting Local Farmers: The Importance of Buying Locally Grown Produce," Sustainable Agriculture Network, May 2024, https://www.sustainableagriculture.eco/post/supporting-local-farmers-the-importance-of-buying-locally-grown-produce; "Research Shows Significant Environmental Benefits of Local Food Production," Iowa State University, April 2022, https://www.cals.iastate.edu/news/2022/research-shows-significant-environmental-benefits-local-food-production; "How Local Food Systems Support Sustainability and Food Access," Churchill Northern Studies Centre, accessed: August 20, 2025, https://churchillscience.ca/how-local-food-systems-support-sustainability-and-food-access/.

39 "Research Shows Significant Environmental Benefits of Local Food Production," Iowa State University, April 2022, https://www.cals.iastate.edu/news/2022/research-shows-significant-environmental-benefits-local-food-production.

40 McCurdy, Madison, "Health Benefits of Eating Locally," University of New Hampshire, May 2022, https://extension.unh.edu/blog/2022/05/health-benefits-eating-locally.

41 "Our Mission," Growing Communities, accessed: August 20, 2025, https://growingcommunities.org/about-us.

42 Jha, Alok, "City Living Affects Your Brain, Researchers Find," *The Guardian*, June 2011, https://www.theguardian.com/science/2011/jun/22/city-living-afffects-brain.

43 "Cities for All Ages," OECD, April 2025, https://www.oecd.org/en/publications/cities-for-all-ages_f0c8fefa-en.html.

44 "Social Isolation and Loneliness in Older Adults: Opportunities for the Health Care System," National Academies Press, February 2020, https://www.ncbi.nlm.nih.gov/books/NBK557972/; "Mental Health of Older Adults," World Health Organization, October 2023, https://www.who.int/news-room/fact-sheets/detail/mental-health-of-older-adults.

45 Ainamani, Herbert, Wilson Bamwerinde, Godfrey Rukundo, Sam Tumwesigire, Rebecca Kalibwani, Evard Bikaitwaho, and Alexander Tsai, "Participation in Gardening Activity and Its Association with Improved Mental Health among Family Caregivers of People with Dementia in Rural Uganda," National Library of Medicine, May 2021, https://pmc.ncbi.nlm.nih.gov/articles/PMC8193614/#:~:text=Linear%20multivariable%20regression%20models%20estimated,depression%2C%20anxiety%2C%20and%20stress.

46 Chase, Spencer, "Study: Nearly 40% of US Adults Have Never Met a Farmer," AgriPulse, October 2022, https://www.agri-pulse.com/articles/18354-study-nearly-40-of-us-adults-have-never-met-a-farmer.

47 Whetstone, Holly, "Inflation, Pandemic, War Reflect Consumers' Wariness on Grocery Prices," Michigan State University, April 2022, https://www.canr.msu.edu/news/inflation-pandemic-war-reflect-consumers-wariness-on-grocery-prices.

48 Rutzerveld, Chloé, *Food Futures*, BIS Publishers, September 2018, https://www.bispublishers.com/food-futures.html.

49 Rutzerveld, Chloé, *Food Futures*, BIS Publishers, September 2018, https://www.bispublishers.com/food-futures.html.

50 Birkby, Jeff, "Vertical Farming," ATTRA, 2024, https://attra.ncat.org/publication/vertical-farming/.

51 Singh, Swapnil, Poonam Singh, Avdhesh Kumar, Ajay Kumar Baheliya, and Krishna Kumar Patel, "Promoting Environmental Sustainability through Vertical Farming: A Review," *Journal of Advances in Biology & Biotechnology*, 2024, https://journaljabb.com/index.php/JABB/article/view/780/1555.

52 Benke, Kurt, "Future Food-Production Systems: Vertical Farming and Controlled-Environment Agriculture," Research Gate, January 2017, https://www.researchgate.net/publication/321186650_Future_food-production_systems_Vertical_farming_and_controlled-environment_agriculture.

53 "Social Media," Sky Greens, accessed: August 20, 2025, https://www.skygreens.com; "Welcome to ComCrop," ComCrop, accessed: August 20, 2025, https://www.sfa.gov.sg/fromSGtoSG/farms/farm/Detail/comcrop; "Indoor Farming. Like Never Before," Go Farm, accessed: August 20, 2025, https://gofarm.sg.

54 Benke, Kurt, and Bruce Tomkins, "Future Food-Production Systems: Vertical Farming and Controlled-Environment Agriculture," Sustainability: Science, Practice and Policy, November 2017, https://www.tandfonline.com/doi/full/10.1080/15487733.2017.1394054#d1e185.

55 Despommier, Dickson, and Eric Ellingsen, "The Vertical Farm: The Sky-Scraper as Vehicle for Sustainable Urban Agriculture," CTBUH, 2008, https://global.ctbuh.org/resources/papers/download/1305-the-vertical-farm-the-sky-scraper-as-vehicle-for-a-sustainable-urban-agriculture.pdf.

56 "City Population 2050," Ontario Tech University, accessed: August 20, 2025, https://sites.ontariotechu.ca/sustainabilitytoday/urban-and-energy-systems/Worlds-largest-cities/population-projections/city-population-2050.php.

57 Santarelli, Marco, "Manhattan Real Estate Market Trends and Forecast 2025–2026," Norada, March 2025, https://www.noradarealestate.com/blog/manhattan-real-estate-market/.

58 "2023 Organic Industry Survey," Organic Trade, 2023, https://ota.com/resources/market-analysis/organic-industry-survey/2023-organic-industry-survey.

59 "Hydroponics," Organic Farmers Association, 2025 https://organicfarmersassociation.org/wp-content/uploads/2024/12/424-Hydroponics-Factsheet.pdf.

60 Steel, Carolyn, *Hungry City*, Chatto & Windus, June 2008, https://www.carolynsteel.com/hungrycitybook.

61 Gayle, Damien, "Waiting List for Allotments in England Almost Doubles in 12 Years," *The Guardian*, October 2023, https://www.theguardian.com/environment/2023/oct/11/waiting-list-for-allotments-in-england-almost-doubles-in-12-years?utm_source=chatgpt.com.

62 "Urban Agriculture: Laws, Zoning, and Regulations," University of California, accessed: August 20, 2025, https://ucanr.edu/site/urban-agriculture/urban-agriculture-incentive-zones-act-ab551#:~:text=In%202014%2C%20California%20implemented%20Assembly,create%20urban%20agriculture%20incentive%20zones.

63 "Nairobi City County Urban Agriculture Promotion and Regulation Act, 2015 (No. 4 of 2015)," Food and Agriculture Organization, 2015, https://www.fao.org/faolex/results/details/en/c/LEX-FAOC162509/#:~:text=The%20objects%20of%20this%20Act,safety%2C%20public%20health%20and%20the.

64 "Direct Payments," Fact Sheets on the European Union, accessed: August 20, 2025, https://www.europarl.europa.eu/factsheets/en/sheet/109/direct-payments#:~:text=All%20Member%20States%20set%20minimum,4).

65 "Shapiro Administration Invests $500,000 to Support Urban Farmers, Increase Availability of Fresh Food in Cities across Pennsylvania," Commonwealth of Pennsylvania | Department of Agriculture, March 2025, https://www.pa.gov/agencies/pda/newsroom/pa-invests--500k-to-support-urban-farmers--increase-fresh-food-a.

66 Sheldon, Marissa, "Urban Agriculture Ordinance Increases Local Food Access in Kampala, Uganda," Hunter College New York City Food Policy Center, May 2023, https://www.nycfoodpolicy.org/food-policy-snapshot-kampala-urban-agriculture-ordinance/#:~:text=Overview%3A%20Prior%20to%202006%2C%20urban,urban%20and%20peri%2Durban%20areas.&text=Program%20goals%3A%20To%20make%20public,to%20increase%20household%20food%20security.

67 Sheldon, Marissa, "Urban Agriculture Ordinance Increases Local Food Access in Kampala, Uganda," Hunter College New York City Food Policy Center, May

2023, https://www.nycfoodpolicy.org/food-policy-snapshot-kampala-urban-agri culture-ordinance/#:~:text=Overview%3A%20Prior%20to%202006%2C%20ur ban,urban%20and%20peri%2Durban%20areas.&text=Program%20goals%3A%20 To%20make%20public,to%20increase%20household%20food%20security.

68 Kwiringira, Japheth, Elizabeth Perry Mohling, Akugizibwe Mathias, Paulino Ariho, James Mugisha, Henry Zakumumpa, Joseph Rujumba, and Innocent Kamara Tumwebaze, "Urban Agriculture as an Evolved Sustainable Urban Livelihood: Evidence from Kampala City, Uganda," Agriculture & Food Security, December 2024, https://agricultureandfoodsecurity.biomedcentral.com/articles/10.1186/s40066-024-00507-y; Kirya, Michael, and Esau Galukande, "Improving Resilience of Cities through Urban Farming: The Case of Kampala City," Conference on International Research on Food Security, Natural Resource Management and Rural Development, September 2016, https://www.tropentag.de/2016/abstracts/full/782.pdf.

69 Sheldon, Marissa, "Urban Agriculture Ordinance Increases Local Food Access in Kampala, Uganda," Hunter College New York City Food Policy Center, May 2023, https://www.nycfoodpolicy.org/food-policy-snapshot-kampala-urban-agri culture-ordinance/#:~:text=Overview%3A%20Prior%20to%202006%2C%20ur ban,urban%20and%20peri%2Durban%20areas.&text=Program%20goals%3A%20 To%20make%20public,to%20increase%20household%20food%20security.

70 Kanosvamhira, T. P., and D. Tevera, "Urban Community Gardens in Cape Town, South Africa: Navigating Land Access and Land Tenure Security," *GeoJournal* 88 (3) (2023): 3105–20. doi: 10.1007/s10708-022-10793-3.

71 Kanosvamhira, T. P., and D. Tevera, "Urban Community Gardens in Cape Town, South Africa: Navigating Land Access and Land Tenure Security," *GeoJournal* 88 (3) (2023): 3105–20. doi: 10.1007/s10708-022-10793-3.

72 Haysom, Gareth, "Planning for Resilient, Nourishing and Generative African Urban Landscapes," *African Jounral of Landscape Architecture* 3 (2) (2022), https://www.ajlajournal.org/articles/planning-for-resilient-nourishing-and-generative-afri can-urban-landscapes.

73 Alexander Haywood, Cory, "Getting Dirty with Ron Finley: The 'Gangsta Gardener,'" *Our Weekly*, June 2017, https://www.ourweekly.com/2017/06/01/gett ing-dirty-ron-finley-gangsta-gardener/.

74 Eschner, Kat, "The Bizarre Story of Piggly Wiggly, the First Self-Service Grocery Store," *Smithsonian Magazine*, September 2017, https://www.smithso nianmag.com/smart-news/bizarre-story-piggly-wiggly-first-self-service-groc ery-store-180964708/.

75 "About," Faneuil Hall Marketplace, accessed: August 20, 2025, https://faneuilhall marketplace.com/about/history-of-faneuil-hall.

76 Bloom, Nicholas, "Merchant of Illusion: James Rouse, America's Salesman of Business Utopia," Ohio State University Press, 2004, https://www.csus.edu/indiv/s/shawg/courses/182/articles/rouse_festival_marketplaces.pdf.

77 "H.R.10339—Farmer-to-Consumer Direct Marketing Act of 1976," Congress, October 1976, https://www.congress.gov/bill/94th-congress/house-bill/10339.

78 "Farmers Market Promotion Program (FMPP)," Farmers Market Coalition, accessed: August 20, 2025, https://farmersmarketcoalition.org/advocacy/farmers-market-promotion-program/.

79 Byczynski, Lynn, "30 Years: A Brief History of Growing for Market," Growing for Market, January 2021, https://growingformarket.com/articles/30-years-brief-history-growing-market#:~:text=Although%20they%20have%20long%20been,read%20him%20yet%2C%20you%20should.

80 Witzling, Laura, Bret Shaw, and Darlene Wolnik, "U.S. Farmers Market Attendance and Experiences," *Journal of Agriculture, Food Systems, and Community Development*, 2025, https://www.foodsystemsjournal.org/index.php/fsj/article/view/1321.

81 "Fair Food Network's Double Up Food Bucks Model Expands Across U.S.," Fair Food Network, June 2016, https://fairfoodnetwork.org/press-release/fair-food-networks-double-up-food-bucks-model-expands-across-u-s/.

82 "Rose Vouchers for Early Years: How It Works," Alexandra Rose Charity, accessed: August 20, 2025, https://www.alexandrarose.org.uk/rose-vouchers/how-it-works/.

83 Wilton, Eleanor, "3 Million Rose Vouchers Spent by Families!," Alexandra Rose Charity, June 2024, https://www.alexandrarose.org.uk/3-million-rose-vouchers-spent-by-families/; "Our Impact," Alexandra Rose Charity, accessed: August 20, 2025, https://www.alexandrarose.org.uk/rose-vouchers/impact/.

84 "New Report Reveals Power of Fruit and Veg on Prescription," Sustain, accessed: August 20, 2025, https://www.sustainweb.org/news/may24-alexrose-impact.

10 Agri-Technology

1 "Drip Irrigation," University of Rhode Island, accessed: February 10, 2025, https://web.uri.edu/safewater/protecting-water-quality-at-home/sustainable-landscaping/drip-irrigation/#:~:text=Benefits%20of%20Drip%20Irrigation&text=In%20addition%2C%20drip%20irrigation%3A,the%20system%20is%20so%20efficient.

2 Al-Khoury, Laure, "Sprinklers and Drip Irrigation Help Iraqis Beat Drought," PhysOrg, 2024, https://phys.org/news/2024-03-sprinklers-irrigation-iraqis-drought.html.

3 "Spiio Soil Sensor," Spiio, accessed: February 10, 2025, https://spiio.greencastonline.com/?gclid=Cj0KCQjwxqayBhDFARIsAANWRnSY-I6hEb2NQUaiEMdQMioOrnf6HycHqnwqqO1Yk9DIbRvB_bAX6KUaAp3lEALw_wcB.

4 "Spiio Soil Sensor," Spiio, accessed: February 10, 2025, https://spiio.greencastonline.com/?gclid=Cj0KCQjwxqayBhDFARIsAANWRnSY-I6hEb2NQUaiEMdQMioOrnf6HycHqnwqqO1Yk9DIbRvB_bAX6KUaAp3lEALw_wcB.

5 "Case Study: 14% Yield Increase, 57% Water Use Reduction," CropX, accessed: February 10, 2025, https://cropx.com/2022/10/19/grapefruit/.

6 "Water Quality Instrumentation | Eureka Water Probes," Solinst Eureka, accessed: February 10, 2025.

7 "Water Usage in the Agricultural Industry," High Tide Technologies, accessed: February 10, 2025, https://www.htt.io/learning-center/water-usage-in-the-agricultural-industry.

8 Newton, Amanda Rose, and Rajesh Melaram, "Harmful Algal Blooms in Agricultural Irrigation: Risks, Benefits, and Management," Frontiers, 2023 https://www.frontiersin.org/journals/water/articles/10.3389/frwa.2023.1325300/full.

9 Trafton, Anne, "Plant Sensors Could Act as an Early Warning System for Farmers," MIT News, 2024, https://news.mit.edu/2024/plant-sensors-could-be-early-warning-system-farmers-0417#:~:text=In%202020%2C%20Strano%27s%20lab%20developed,carbon%20nanotubes%20wrapped%20in%20polymers.

10 "Livestock and Crop Health," AGXIO, accessed: February 10, 2025, https://agxio.com/traditional-farming/livestock-crop-health/#:~:text=Sensors%20placed%20directly%20in%20livestock,health%20to%20effectively%20increase%20yields.

11 "Edible Rates Are China's Latest Live-Streaming Stars," *The Economist*, 2025, https://www.economist.com/china/2025/05/01/edible-rats-are-chinas-latest-live-streaming-stars.

12 "Edible Rates Are China's Latest Live-Streaming Stars," *The Economist*, 2025, https://www.economist.com/china/2025/05/01/edible-rats-are-chinas-latest-live-streaming-stars.

13 "Edible Rates Are China's Latest Live-Streaming Stars," *The Economist*, 2025, https://www.economist.com/china/2025/05/01/edible-rats-are-chinas-latest-live-streaming-stars.

14 "Edible Rates Are China's Latest Live-Streaming Stars," *The Economist*, 2025, https://www.economist.com/china/2025/05/01/edible-rats-are-chinas-latest-live-streaming-stars.

15 Howe, Scott, "6 Barriers to Drone Adoption in Precision Agriculture and How to Overcome Them," Commercial UAV News, 2023, https://www.commercialuavnews.com/forestry/6-barriers-to-drone-adoption-in-precision-agriculture-and-how-to-overcome-them.

16 "Hyperspectral Imaging for Agriculture," IMEC, accessed: February 10, 2025, https://www.imechyperspectral.com/en/applications/hyperspectral-imaging-agri culture.

17 "Hyperspectral Imaging for Agriculture," IMEC, accessed: February 10, 2025, https://www.imechyperspectral.com/en/applications/hyperspectral-imaging-agri culture.

18 "Hyperspectral Imaging Technology," IMEC, accessed: February 10, 2025, https://www.imechyperspectral.com/en/applications/hyperspectral-imaging-agriculture.

19 "ULTRIS X20 Plus," Cubert, accessed: February 10, 2025, https://cubert-hyperspect ral.com/en/ultris-x20-plus/?gad_source=1&gclid=Cj0KCQjwxqayBhDFARIsAAN WRnQAaRVniJGovENRLFhWdEbe9iV8aZjreDDSSsZ3tJdcscJa9M1IfNkaAsVTE ALw_wcB.

20 Brooks, Kevin, "FAA Requirements for Farmers Operating a Drone," University of Illinois Urbana-Champaign, 2023, https://extension.illinois.edu/news-releases/faa-requirements-farmers-operating-drone.

21 Brooks, Kevin, "FAA Requirements for Farmers Operating a Drone," University of Illinois Urbana-Champaign, 2023, https://extension.illinois.edu/news-releases/faa-requirements-farmers-operating-drone.

22 "California Assembly Bill 1016 (Prior Session Legislation)," LegiScan, accessed: February 10, 2025, https://legiscan.com/CA/text/AB1016/2023.

23 Bohri, Andras, "Drone Spraying Regulation and Opportunities in Europe," Eurofins Agricultural Services, 2023, https://www.eurofins.com/agroscience-services/about-us/latest-news/drone-spraying-regulation-and-opportunities-in-europe/.

24 "Japan Drone Laws: Transforming to Save Agriculture," Viewpoints, 2019, https://viewpoints.dji.com/blog/japan-drone-laws-transforming-to-save-agriculture.

25 Obayashi, Yuka, "As Japan's Farmers Age, Drones Help With Heavy Lifting," The Christian Science Monitor, 2018, https://www.csmonitor.com/Technol ogy/2018/0823/As-Japan-s-farmers-age-drones-help-with-heavy-lifting.

26 Kesari, Ganes, "The Future of Farming: AI Innovations That Are Transforming Agriculture," Forbes, 2024, https://www.forbes.com/sites/ganeskesari/2024/03/31/the-future-of-farming-ai-innovations-that-are-transforming-agriculture/.

27 Kesari, Ganes, "The Future of Farming: AI Innovations That Are Transforming Agriculture," Forbes, 2024, https://www.forbes.com/sites/ganeskesari/2024/03/31/the-future-of-farming-ai-innovations-that-are-transforming-agriculture/.

28 "Carbon Robotics Introduces Faster, Lighter and Modular LaserWeeder G2 Product Line," Business Wire, 2025, https://www.businesswire.com/news/home/2025021 0556114/en/Carbon-Robotics-Introduces-Faster-Lighter-and-Modular-LaserWee der-G2-Product-Line.

29 "CGIAR Research Program on Climate Change, Agriculture and Food Security, Climate-Smart Agriculture In Kenya," International Center for Tropical Agriculture,

2015, https://climateknowledgeportal.worldbank.org/sites/default/files/2019-06/CSA%20KENYA%20NOV%2018%202015.pdf.

30 "Kenya Poverty and Equity Assessment 2023," From Poverty to Prosperity: Making Growth More Inclusive, World Bank, 2023, https://documents1.worldbank.org/curated/en/099121323073037589/pdf/P1773530a7eb3009308e3f08663aa95c826.pdf; "CGIAR Research Program on Climate Change, Agriculture and Food Security, Climate-Smart Agriculture in Kenya," International Center for Tropical Agriculture, 2015, https://climateknowledgeportal.worldbank.org/sites/default/files/2019-06/CSA%20KENYA%20NOV%2018%202015.pdf.

31 Kenya National Data Archive, Kenya Continuous Household Survey Programme 2022, Kenya National Bureau of Statistics, 2024, https://statistics.knbs.or.ke/nada/index.php/catalog/131.

32 Kenya National Data Archive, Kenya Continuous Household Survey Programme 2022, Kenya National Bureau of Statistics, 2024, https://statistics.knbs.or.ke/nada/index.php/catalog/131.

33 "Smart Farming Platform Improves Crop Yields, Minimizes Pollution," University of Texas at Austin, ScienceDaily, 2023, https://www.sciencedaily.com/releases/2023/06/230621164518.htm.

34 Chandler, David, "Reducing Pesticide Use While Increasing Effectiveness," MIT News, 2024, https://news.mit.edu/2024/reducing-pesticide-use-while-increasing-effectiveness-agzen-0312#:~:text=%E2%80%9CAcross%20the%20board%2C%20we%20were,pest%20control%2C%E2%80%9D%20Jayaprakash%20says.

35 Stokstad, Erik, "Ammonia Pollution from Farming May Exact Hefty Health Costs," Science, 2014, https://www.science.org/doi/full/10.1126/science.343.6168.238.

36 Ménesguen, Alain, Xavier Desmit, Valerie Duliere, Genevieve Lacroix,w Benedicte Thouveninin, Vincent Thieu, and Morgan Dussauze, "How to Avoid Eutrophication in Coastal Seas? A New Approach to Derive River-Specific Combined Nitrate and Phosphate Maximum Concentrations," Science Direct, 2018, https://www.sciencedirect.com/science/article/abs/pii/S0048969718304133.

37 Kutyrev, Alexey, Nikolay Kiktev, Marcin Jewiarz, Dmitriy Khort, Igor Smirnov, Valeria Zubina, Taras Hutsol, Marcin Tomasik, and Mykola Biliuk, "Robotic Platform for Horticulture: Assessment Methodology and Increasing the Level of Autonomy," National Library of Medicine, 2022, https://pmc.ncbi.nlm.nih.gov/articles/PMC9693234/.

38 Jurkovich, Viktor, Peter Hejel, and Leventa Kovas, "A Review of the Effects of Stress on Dairy Cattle Behaviour," *Animals, (Basel)* 11:14(14) (2024), doi: 10.3390/ani14142038.

39 Kutyrev, Alexey, Nikolay Kiktev, Marcin Jewiarz, Dmitriy Khort, Igor Smirnov, Valeria Zubina, Taras Hutsol, Marcin Tomasik, and Mykola Biliuk, "Robotic Platform for Horticulture: Assessment Methodology and Increasing the Level of

Autonomy," National Library of Medicine, 2022, https://pmc.ncbi.nlm.nih.gov/articles/PMC9693234/

40 Thomas, Noah, "The Robots Are Coming—to Pick Northwest Apples," Oregon Public Broadcasting, 2023, https://www.opb.org/article/2023/04/06/northwest-oregon-apple-washington-farm-harvest-robots-robotics-orchards-agriculture-technology/.

41 Akins, Anna, "Peachy Robot: A Glimpse into the Peach Orchard of the Future," Georgia Tech, https://fptd.gatech.edu/peachy-robot-glimpse-peach-orchard-future.

42 Bernier, Catherine, "Robot Vision: How to Maximize the Flexibility of Robot Technology," HowToRobot, 2022, https://howtorobot.com/expert-insight/robot-vision.

43 "Mechanical Weeding Robots," Odd.Bot, accessed: February 10, 2025, https://www.odd.bot/.

44 Chanarin, George, and Paul Silcock, "AgroEcoTech: How Can Technology Accelerate a Transition to Agroecology?," The Soil Association, 2021, https://www.soilassociation.org/media/22821/agroecotech-soil-association-report.pdf.

45 Chanarin, George, and Paul Silcock, "AgroEcoTech: How Can Technology Accelerate a Transition to Agroecology?," The Soil Association, 2021, https://www.soilassociation.org/media/22821/agroecotech-soil-association-report.pdf.

46 Abdulla Ahmed, Mohamed, Seydeh-Somayyeh Shafiei-Mosouleh, Riyadh Mannaa Mohsin, and Ziyad Khalf Salih, "Foliar Application of Iron Oxide Nanoparticles Promotes Growth, Mineral Contents, and Medicinal Qualities of Solidago virgaurea L," Springer Nature, 2023, https://link.springer.com/article/10.1007/s42729-023-01218-2; Baral Dola, Dipanjoli, Md Abdul Mannan, Umakanta Sarker, Md Abdullah Al Mamun, Tofazzal Islam, Sezai Ercisli, Muhammad Hamzah Saleem, Baber Ali, Oana Lelia Pop, and Romina Alina Marc., "Nano-Iron Oxide Accelerates Growth, Yield, and Quality of *Glycine Max* Seed in Water Deficits," National Library of Medicine, 2022, https://pmc.ncbi.nlm.nih.gov/articles/PMC9500458/.

47 Abdulla Ahmed, Mohamed, et al., "Foliar Application of Iron Oxide Nanoparticles Promotes Growth, Mineral Contents, and Medicinal Qualities of Solidago virgaurea L," Springer Nature, 2023, https://link.springer.com/article/10.1007/s42729-023-01218-2; Ullah, Asad, Sidra Munir, Syed Lal Badshah, Noreen Khan, Lubna Ghani, Benjamin Gabriel Poulson, Abdul-Hamid Emwas, and Mariusz Jaremko, "Important Flavonoids and Their Role as a Therapeutic Agent," National Library of Medicine, 2020, https://pmc.ncbi.nlm.nih.gov/articles/PMC7697716/#:~:text=Flavonoids%20are%20phytochemical%20compounds%20present,%2Dinflammatory%2C%20and%20antiviral%20properties.

48 Abdulla Ahmed, Mohamed, et al., "Foliar Application of Iron Oxide Nanoparticles Promotes Growth, Mineral Contents, and Medicinal Qualities of Solidago

virgaurea L," Springer Nature, 2023, https://link.springer.com/article/10.1007/s42729-023-01218-2.

49 Baral Dola, Dipanjoli, et al., "Nano-Iron Oxide Accelerates Growth, Yield, and Quality of *Glycine Max* Seed in Water Deficits," National Library of Medicine, 2022, https://pmc.ncbi.nlm.nih.gov/articles/PMC9500458/.

50 Chaud, Marco, Eliana Souto, Aleksandra Zielinska, Patricia Severino, Fernando Batain, Jose Oliveira-Junior, and Thais Alves, "Nanopesticides in Agriculture: Benefits and Challenge in Agricultural Productivity, Toxicological Risks to Human Health and Environment," National Library of Medicine, 2021, https://pmc.ncbi.nlm.nih.gov/articles/PMC8230079/.

51 Ahmad, Khurshid, Yanyan Li, Chaoxin Tu, Yueting Guo, Xia Yang, Chensi Xia, and Hu Hou, "Nanotechnology in Food Packaging with Implications for Sustainable Outlook and Safety Concerns," Science Direct, 2024, https://www.sciencedirect.com/science/article/abs/pii/S2212429224000555#bib139.

52 Chaud, Marco, et al., "Nanopesticides in Agriculture: Benefits and Challenge in Agricultural Productivity, Toxicological Risks to Human Health and Environment," National Library of Medicine, 2021, https://pmc.ncbi.nlm.nih.gov/articles/PMC8230079/.

53 "Homepage," Modern Farmer, accessed: February 10, 2025, https://modernfarmer.com/.

54 "8 Things to Know About Palm Oil," World Wildlife Fund, accessed: February 10, 2025, https://www.wwf.org.uk/updates/8-things-know-about-palm-oil.

55 Vijay, Varsha, Stuart Pimm, Clinton Jenkins, and Sharon Smith, "The Impacts of Oil Palm on Recent Deforestation and Biodiversity Loss," PLOS One, 2016, https://journals.plos.org/plosone/article?id=10.1371/journal.pone.0159668; "What's Wrong with Palm Oil?," Orangutan Foundation Interntional, accessed: February 10, 2025, https://orangutan.org/palmoil/.

56 Choiruzzad S. A. B., A. Tyson, and H. Varkkey, "The Ambiguities of Indonesian Sustainable Palm Oil Certification: Internal Incoherence, Governance Rescaling and State Transformation. *Asia Europe Journal* 19 (2) (2021) :189–208. doi: 10.1007/s10308-020-00593-0.

57 Ryan, Jennifer, "Getting Ready for Robotic Milking," Veterinary Advantage, 2020, https://vet-advantage.com/vet-advantage/getting-ready-for-robotic-milking/#:~:text=On%20average%2C%20it%20costs%20between,50%20to%2070%20cows%20each.

58 Rainey, Clint, "These Autonomous Tractors Are Electrifying Farming," Fast Company, 2024, https://www.fastcompany.com/91033164/monarch-tractor-most-innovative-companies-2024; Rhett, C., "How Much Do Tractors Cost? (New vs. Used vs. Rent with Examples)," Heavy Equipment Appraisal, 2025, https://heavyequ

ipmentappraisal.com/how-much-do-tractors-cost/#:~:text=Tractor%20costs%20typically%20range%20from,the%20price%20of%20the%20tractor.

59 Reynolds, Matthew, "The Future of Food: How to Feed the Planet without Destroying It," Penguin, 2021.

60 "John Deere Acquires Smart Apply," John Deere, 2023, https://www.deere.com/en/news/all-news/john-deere-acquires-smart-apply/.

61 "John Deere Acquires Bear Flag Robotics to Accelerate Autonomous Technology on the Farm," John Deere, 2021, https://www.deere.com/en/news/all-news/2021aug5-bear-flag-robotics/.

62 Saik, Robert, *Food 5.0: How We Feed the Future*, Lioncrest Publishing, 2019.

63 "Farm Producers," Census of Agriculture, 2022, https://www.nass.usda.gov/Publications/Highlights/2024/Census22_HL_FarmProducers_FINAL.pdf; "Agricultural Workforce in England at 1 June 2024," Gov.UK, 2024, https://www.gov.uk/government/statistics/agricultural-workforce-in-england-at-1-june/agricultural-workforce-in-england-at-1-june-2023#:~:text=In%202023%2C%20the%20highest%20proportion,years%20(see%20Figure%205.

64 "Agriculture in the UK Evidence Pack," Gov.UK, 2022, https://assets.publishing.service.gov.uk/media/6331b071e90e0711d5d595df/AUK_Evidence_Pack_2021_Sept22.pdf.

65 "Internet Access Still Denied to Many in the Developing World," The World Economic Forum, 2023, https://www.weforum.org/stories/2023/09/broadband-no-luxury-basic-necessity/#:~:text=Only%2035%25%20of%20people%20in%20developing%20nations%20have%20access%20to,create%20140%20million%20new%20jobs.

66 "A Case for Fortifying the BUILD Act: The U.S., China, and Internet Infrastructure in the Global South," Harvard Kennedy School Belfer Center, 2019, https://www.belfercenter.org/publication/case-fortifying-build-act-us-china-and-internet-infrastructure-global-south.

67 "Lessons for America's Broadband Buildout from the Global South," Broadband Breakfast, 2024, https://broadbandbreakfast.com/lessons-for-americas-broadband-buildout-from-the-global-south/; Goldman, David, "Broadband, Business Formation, and Economic Growth in the Global South: Assessing China's Impact," *American Affairs Journal*, 2023, https://americanaffairsjournal.org/2023/11/broadband-business-formation-and-economic-growth-in-the-global-south-assessing-chinas-impact/.

68 "Lessons for America's Broadband Buildout from the Global South," Broadband Breakfast, 2024, https://broadbandbreakfast.com/lessons-for-americas-broadband-buildout-from-the-global-south/.

69 "The Great Divide: Broadband Challenges in Rural America," Rajant, accessed: February 10, 2010, https://rajant.com/the-great-divide-broadband-challenges-in-rural-america/.

70 "Broadband Access Still a Challenge in Rural Affordable Housing," Pew Trusts, 2022, https://www.pewtrusts.org/en/research-and-analysis/articles/2022/12/08/broadband-access-still-a-challenge-in-rural-affordable-housing.
71 "Broadband Access Still a Challenge in Rural Affordable Housing," Pew Trusts, 2022, https://www.pewtrusts.org/en/research-and-analysis/articles/2022/12/08/broadband-access-still-a-challenge-in-rural-affordable-housing.
72 "The Great Divide: Broadband Challenges in Rural America," Rajant, accessed: February 10, 2010, https://rajant.com/the-great-divide-broadband-challenges-in-rural-america/.
73 Saik, Robert, *Food 5.0: How We Feed the Future*, Lioncrest Publishing, 2019.

11 Combating Food Waste with Technology

1 "How Much Food Do We Really Waste?," World Resource Institute, 2024, https://www.wri.org/insights/how-much-food-does-the-world-waste.
2 "Food Waste FAQs," USDA, accessed: February 23, 2025, https://www.usda.gov/about-food/food-safety/food-loss-and-waste/food-waste-faqs; "Food Loss and Waste," USDA NIFA, accessed: February 23, 2025.
3 "Food and Agriculture Organization, 'Global Food Losses and Food Waste,'" United Nations, 2011, https://www.fao.org/4/mb060e/mb060e00.htm.
4 "Wasted: How America Is Losing up to 40 Percent of Its Food from Farm to Fork to Landfill," NRDC, 2017, https://www.nrdc.org/sites/default/files/wasted-2017-report.pdf.
5 "Fight Climate Change by Preventing Food Waste," WWF, accessed: February 23, 2025, https://www.worldwildlife.org/stories/fight-climate-change-by-preventing-food-waste#:~:text=But%2520wasted%2520food%2520isn't,more%2520potent%2520than%2520carbon%2520dioxide.
6 "Fight Climate Change by Preventing Food Waste," World Wildlife Fund, https://www.worldwildlife.org/stories/fight-climate-change-by-preventing-food-waste.
7 "Sustainability Pathways," FAO, accessed: February 23, 2025, https://www.fao.org/nr/sustainability/food-loss-and-waste/en/.
8 "Food Waste & Food Loss," WFP USA, accessed: February 23, 2025, https://www.wfpusa.org/drivers-of-hunger/food-waste/.
9 "Food Waste & Food Loss," WFP USA, accessed: February 23, 2025, https://www.wfpusa.org/drivers-of-hunger/food-waste/.
10 Murdock, Larry, Dogo Seck, Georges Ntoukam, Laurie Kitch, and R. E. Shade, "Preservation of Cowpea Grain in Sub-Saharan Africa—Bean/Cowpea CRSP Contributions," Science Direct, 2003, https://www.sciencedirect.com/science/article/abs/pii/S0378429003000364.

11 Reynolds, Matthew, "The Future of Food: How to Feed the Planet without Destroying It," Wired, 2021, https://www.amazon.co.uk/Future-Food-WIRED-guides-Destroying/dp/1847943284#:~:text=In%2520this%2520brilliantly%2520insightful%252C%2520one,to%2520replace%2520inefficient%2520livestock%2520farming.

12 "Food Waste vs. Food Loss: Know the Difference and Help #StopTheWaste Today," WFP USA, accessed: February 23, 2025, https://www.wfpusa.org/articles/food-loss-vs-food-waste-primer/.

13 "Engaging Women's Producer Organizations in Senegal to Inspire Inclusion," AgResults, 2024, https://agresults.org/news-and-blog/10-blog/engaging-womens-producer-organizations-in-senegal-to-inspireinclusion/.

14 "Engaging Women's Producer Organizations in Senegal to Inspire Inclusion," AgResults, 2024, https://agresults.org/news-and-blog/10-blog/engaging-womens-producer-organizations-in-senegal-to-inspireinclusion/.

15 "Our Projects, AgResults Senegal Crop Storage Finance Challenge Project," Connexus, 2021, https://connexuscorporation.com/projects/agresults-senegal-crop-storage-finance-challenge-project/.

16 "Senegal Crop Storage Finance Challenge Project," AgResults, 2024, https://agresults.org/projects/senegal-storage-finance/.

17 "Engaging Women's Producer Organizations in Senegal to Inspire Inclusion," AgResults, 2024, https://agresults.org/news-and-blog/10-blog/engaging-womens-producer-organizations-in-senegal-to-inspireinclusion/.

18 "Engaging Women's Producer Organizations in Senegal to Inspire Inclusion," AgResults, 2024, https://agresults.org/news-and-blog/10-blog/engaging-womens-producer-organizations-in-senegal-to-inspireinclusion/.

19 Yaffe-Bellany, David, and Michael Corkery, "Dumped Milk, Smashed Eggs, Plowed Vegetables: Food Waste of the Pandemic," *New York Times*, 2020, https://www.nytimes.com/2020/04/11/business/coronavirus-destroying-food.html.

20 "Creating a Sustainable Food Future, Reducing Food Loss and Waste," World Resource Institute, 2013, https://www.wri.org/research/reducing-food-loss-and-waste.

21 White, Daniel, "American's Unrealistic Food Beauty Standards Cause Huge Amounts of Waste," *Time*, 2016, https://time.com/4404312/americans-farm-food-waste-beauty-standards/.

22 "Wasted: How America Is Losing up to 40 Percent of Its Food from Farm to Fork to Landfill," NRDC, 2017, https://www.nrdc.org/sites/default/files/wasted-2017-report.pdf.

23 "Wasted: How America Is Losing up to 40 Percent of Its Food from Farm to Fork to Landfill," NRDC, 2017, https://www.nrdc.org/sites/default/files/wasted-2017-report.pdf.

24 "Wasted: How America Is Losing up to 40 Percent of Its Food from Farm to Fork to Landfill," NRDC, 2017, https://www.nrdc.org/sites/default/files/wasted-2017-report.pdf.

25 Astill, Gregory, Steven Zahniser, and Belem Avendaño Ruiz, "Fruit and Vegetable Imports from Mexico Continue Upward Trend as Mexico's Growers Adopt U.S. Food Safety Rules," USDA Economic Research Service, 2023, https://www.ers.usda.gov/data-products/charts-of-note/chart-detail?chartId=108128.

26 The Wall Street Journal, "How a Steel Box Changed the World: A Brief History of Shipping," YouTube, 2018, https://www.youtube.com/watch?v=0MUkgDIQdcM.

27 "Agricultural Marketing Service, Implications of Rising Ocean Freight Rates for Agri-Food Product Markets (Summary)," USDA, 2023, https://www.ams.usda.gov/sites/default/files/media/ImplicationsofRisingOceanFreighRatesSummary.pdf.

28 Chambers, Sam, "80% of Ports Missing Out on the Benefits of Digitalisation, Creating Last Mile Risks," Splash, 2021, https://splash247.com/80-of-ports-missing-out-on-the-benefits-of-digitalisation-creating-last-mile-risks/.

29 "Wasted: How America Is Losing up to 40 percent of Its Food from Farm to Fork to Landfill," NRDC, 2017, https://www.nrdc.org/sites/default/files/wasted-2017-report.pdf.

30 Tatum, Megan, "Everything You Need to Know about Apeel," Wicked Leeks, 2023, https://wickedleeks.riverford.co.uk/features/everything-you-need-to-know-about-apeel/.

31 Tatum, Megan, "Everything You Need to Know about Apeel," Wicked Leeks, 2023, https://wickedleeks.riverford.co.uk/features/everything-you-need-to-know-about-apeel/.

32 Hayes, Wallace, and Saura C. Sahu, "Nanotechnology in the Food Industry: A Short Review," Food Safety, 2017, https://www.food-safety.com/articles/5193-nanotechnology-in-the-food-industry-a-short-review.

33 Julian McClements, David, "Future Foods: How Modern Science Is Transforming the Way We Eat," Copernicus, 2019, https://www.amazon.co.uk/Future-Foods-Modern-Science-Transforming/dp/3030129942?dplnkId=4fffdcfc-0dd1-463b-8202-9e77d5da1e21&nodl=1.

34 Julian McClements, David, "Future Foods: How Modern Science Is Transforming the Way We Eat," Copernicus, 2019, https://www.amazon.co.uk/Future-Foods-Modern-Science-Transforming/dp/3030129942?dplnkId=4fffdcfc-0dd1-463b-8202-9e77d5da1e21&nodl=1.

35 Jadhav, Rachana, Pallavi Pawar, Vishnu Choudhari, Niraj Topare, Sunita Raut-Jadhav, Shantini Bokil, and Anish Khan, "An Overview of Antimicrobial Nanoparticles for Food Preservation," Science Direct, 2023, https://www.sciencedirect.com/science/article/abs/pii/S2214785322046478.

36 Bajpai, Vivek, Madhu Kamle, Shruti Shukla, Dipendra Kumar Mahato, Pranjal Chandra, Seung Kyu Hwang, Pradeep Kumar, Yun Suk Huh, and Young-Kyu Han, "Prospects of Using Nanotechnology for Food Preservation, Safety, and Security," Science Direct, 2018, https://www.sciencedirect.com/science/article/pii/S1021949818301169#bib24.

37 Ahmad, Khurshid, Yanyan Li, Yueting Guo, Xia Yang, Chensi Xia, and Hu Hou, "Nanotechnology in Food Packaging with Implications for Sustainable Outlook and Safety Concerns," Science Direct, 2024, https://www.sciencedirect.com/science/article/abs/pii/S2212429224000555.

38 Ghosh, Tabli, G. V. S. Bhagya Raj, and Kshirod Kumar Dash, "A Comprehensive Review on Nanotechnology Based Sensors for Monitoring Quality and Shelf Life of Food Products," Science Direct, 2022, https://www.sciencedirect.com/science/article/pii/S2772275922000260#bib0006.

39 "Mori," King Philanthropies, 2025, https://kingphilanthropies.org/portfolio-organization/mori/.

40 "FAQs," Mori, 2025, https://mori.com.

41 "Wasted: How America Is Losing up to 40 Percent of Its Food from Farm to Fork to Landfill," NRDC, 2017, https://www.nrdc.org/sites/default/files/wasted-2017-report.pdf.

42 "The Global Food Donation Policy Atlas: Executive Summary: Argentina," Food Banking, accessed: February 23, 2025, https://www.foodbanking.org/wp-content/uploads/2023/01/Argentina-Executive-Summary-05.29.20.pdf.

43 "About Us: Fighting Food Waste Together since 2016," Too Good To Go, accessed: February 23, 2025, https://www.toogoodtogo.com/about-us?utm_medium=search&utm_source=google&utm_campaign=uk_b2b_paid_marketing_search_google_generic&utm_content=145160682893&utm_term=too+good+to+go&gad_source=1&gbraid=0AAAAADJNt-2ghZyKGP8z_LnOTqm5g8h6Z&gclid=EAIaIQobChMIrNqWvIyGiAMV95ODBx1pQhlZEAAYASADEgK-KfD_BwE.

44 Namkung, Victoria, "From Trash to Table: Will Upcycled Food Save the Planet?," *The Guardian*, 2024, https://www.theguardian.com/environment/2024/feb/22/us-food-waste-upcycling-snacks#:~:text=Pulp%2520Pantry%252C%2520along%2520with%2520other,edible%2520food%2520waste%2520and%2520byproducts.

45 "Growing the Upcycled Food Economy," Upcycled Food Association, accessed: February 23, 2025. https://www.upcycledfood.org/.

46 Namkung, Victoria, "From Trash to Table: Will Upcycled Food Save the Planet?," *The Guardian*, 2024, https://www.theguardian.com/environment/2024/feb/22/us-food-waste-upcycling-snacks#:~:text=Pulp%2520Pantry%252C%2520al

ong%2520with%2520other,edible%2520food%2520waste%2520and%2520byp roducts.
47 "Turning Overripe Bananas into Wine in Malawi," Fresh Plaza, 2024, https://www.freshplaza.com/north-america/article/9658151/turning-overripe-bananas-into-wine-in-malawi/.
48 "Food Waste—2025 Facts & Statistics," Waste Managed, 2025, https://www.wastemanaged.co.uk/our-news/food-waste/food-waste-facts-statistics/#:~:text=The%2520hospitality%2520and%2520food%2520service,tonnes%2520of%2520food%2520waste%2520annually; "Food Waste in Restaurants: What We Know," Fourth, 2023, https://uk.fourth.com/article/food-waste-in-restaurants-what-we-know.
49 "Most Wasted Food in UK Restaurants and Cafes," Willshees, accessed: February 23, 2025, https://www.willshees.co.uk/news/most-wasted-food-in-uk-restaurants-and-cafes/.
50 "Most Wasted Food in UK Restaurants and Cafes," Willshees, accessed: February 23, 2025, https://www.willshees.co.uk/news/most-wasted-food-in-uk-restaurants-and-cafes/.
51 "Most Wasted Food in UK Restaurants and Cafes," Willshees, accessed: February 23, 2025, https://www.willshees.co.uk/news/most-wasted-food-in-uk-restaurants-and-cafes/.
52 "Why Choose Winnow?," Winnow Solutions, accessed: February 23, 2025, https://www.winnowsolutions.com/why-winnow.
53 "Test Report 2023," Shelfy, 2023, https://vitesy.com/download/Shelfy_Test-Report_ENG.pdf.
54 "Wasted Food Scale," EPA, accessed: February 23, 2025, https://www.epa.gov/sustainable-management-food/wasted-food-scale.
55 "Bovine Spongiform Encephalopathy (BSE): Feed Controls," Scottish Agricultural and Rural Economy Directorate, 2018, https://www.gov.scot/publications/bovine-spongiform-encephalopathy-bse-feed- ontrols/.
56 Okayama, Tomoko, and Kohei Watanabe, "Performance of the Food Waste Recycling Law in Japan with Reference to SDG 12.3," Recycling, 2024, https://doi.org/10.3390/recycling9010018.
57 McBride, Monica, "Turning Food Waste into Feed: Benefits and Trade-offs for Nature," WWF, 2021, https://www.worldwildlife.org/blogs/sustainability-works/posts/turning-food-waste-into-feed-benefits-and-trade-offs-for-nature#:~:text=In%2520fact%252C%252030%2525%2520of%2520what's,from%2520manufacturing%2520or%2520grocery%2520stores.
58 "Report Submitted by the Special Rapporteur on the Right to Food, Olivier De Schutter," UN General Assembly, 2010, https://www2.ohchr.org/english/issues/food/docs/A-HRC-16-49.pdf.

59 Payton, Louise, "Saving Our Soils: Healthy Soils for Our Climate, Nature and Health," Soil Association, 2021, https://www.soilassociation.org/media/24941/saving-our-soils-report-dec21.pdf.

60 Payton, Louise, "Saving Our Soils: Healthy Soils for Our Climate, Nature and Health, Soil Association, 2021, https://www.soilassociation.org/media/24941/saving-our-soils-report-dec21.pdf.

61 Iacomino, Giuseppina, "Biochar and Compost Application either Alone or in Combination Affects Vegetable Yield in a Volcanic Mediterranean Soil," MDPI, 2022, https://www.mdpi.com/2073-4395/12/9/1996.

62 "Mitigation Pathways Compatible with 1.5°C in the Context of Sustainable Development," IPCC, 2018, https://www.ipcc.ch/sr15/chapter/chapter-2/.

Conclusion

1 Ritchie, Hannah, "Food Production Is Responsible for One-Quarter of the World's Greenhouse Gas Emissions," OurWorldinData.org, 2019, https://ourworldindata.org/food-ghg-emissions.

2 "Noncommunicable Disease Deaths in the Americas Have Risen 43% Since 2000, PAHO Report Shows," Pan American Health Organisation, 2025, https://www.paho.org/en/news/2-7-2025-noncommunicable-disease-deaths-americas-have-risen-43-2000-paho-report-shows.

3 Root, Skye, "Agriculture in the Red: Net Farm Income Drops Again in 2024 Forecast," Root Agricultural Advisory, November 2024, https://rootagadvisory.com/agriculture-in-the-red-net-farm-income-drops-again-in-2024-forecast/.

4 Chen, D. M. C., B. Bodirsky, Wang, X., J. Xuan, J. P. Dietrich, A. Popp, H. Lotze-Campen, "Future Food Prices Will Become Less Sensitive to Agricultural Market Prices and Mitigation Costs," *Nat Food* 6 (2025): 85–96, https://doi.org/10.1038/s43016-024-01099-3.

5 Food and Agriculture Organization of the United Nations, "Livestock Solutions for Climate Change," FAO 2017, https://openknowledge.fao.org/handle/20.500.14283/i8098en.

6 Liz, Godwin, and Brian Lipinski, "How Much Food Does the World Really Waste?," World Resource Institute, 2024, https://www.wri.org/insights/how-much-food-does-the-world-waste#:~:text=Around%20one%2Dthird%20of%20all,gas%20emissions%20warming%20the%20planet.

7 United Nations, "68% of the World Population Projected to Live in Urban Areas by 2050, Says UN," 2018, https://www.un.org/uk/desa/68-world-population-projected-live-urban-areas-2050-says-un.

Recommended Further Reading

Foundational Works on Food Systems

Holt-Giménez, E. (2017). *A Foodie's Guide to Capitalism: Understanding the Political Economy of What We Eat*. Monthly Review Press.

IPES-Food. (2016). *From Uniformity to Diversity: A Paradigm Shift from Industrial Agriculture to Diversified Agroecological Systems*. IPES-Food.

Lang, T., and Heasman, M. (2004). *Food Wars: The Global Battle for Mouths, Minds and Markets*. Earthscan.

Lappé, F. M. (1971, updated 2021). *Diet for a Small Planet*. Ballantine/Penguin Random House.

Nestle, M. (2002). *Food Politics: How the Food Industry Influences Nutrition and Health*. University of California Press.

Patel, R. (2007). *Stuffed and Starved: Markets, Power and the Hidden Battle for the World's Food System*. Melville House.

Pollan, M. (2006). *The Omnivore's Dilemma: A Natural History of Four Meals*. Penguin Press.

Sen, A. (1981). *Poverty and Famines: An Essay on Entitlement and Deprivation*. Oxford University Press.

Steel, C. (2008). *Hungry City: How Food Shapes Our Lives*. Chatto & Windus.

Weis, T. (2007). *The Global Food Economy: The Battle for the Future of Farming*. Zed Books.

Climate, Environment, and Agriculture

FAO. (2023). *The State of Food and Agriculture*. Food and Agriculture Organization of the UN.

FAO. (2022). *The State of World Fisheries and Aquaculture*. FAO.

IPCC. (2019). *Climate Change and Land: Special Report*. Intergovernmental Panel on Climate Change.

McMichael, P. (2011). *Development and Social Change: A Global Perspective*. SAGE Publications.

Pretty, J. (2002). *Agri-Culture: Reconnecting People, Land and Nature*. Earthscan.

Rockström, J., W. Steffen, K. Noone, A. Persson, S. Chapin, E. F. Lambin, T. M. Lenton, M. Scheffer, C. Folke, and H. J. Schellnhuber (2009). "Planetary

Boundaries: Exploring the Safe Operating Space for Humanity." *Ecology and Society*, 14 (2): 472–5.

Shiva, V. (1989). *Staying Alive: Women, Ecology and Development*. Zed Books.

Wise, T. (2019). *Eating Tomorrow: Agribusiness, Family Farmers, and the Battle for the Future of Food*. New Press.

Technology, Innovation, and Biotech

Bobo, J. (2021). *Why Smart People Make Bad Food Choices: The Invisible Influences That Guide Our Thinking*. Prometheus Books.

Clapp, J. (2012). *Food*. Polity Press.

Friends of the Earth. (2018). *Biotechnology and Agriculture: Promises and Pitfalls*. Friends of the Earth International.

IPBES. (2019). *Global Assessment Report on Biodiversity and Ecosystem Services*. Intergovernmental Science-Policy Platform on Biodiversity and Ecosystem Services.

ISAAA. (2022). *Global Status of Commercialized Biotech/GM Crops: ISAAA Brief*. International Service for the Acquisition of Agri-biotech Applications.

Saik, R. (2019). *Food 5.0: How We Feed the Future*. Lioncrest Publishing.

USDA. (2020). *Economic Research Service Reports on Biotechnology*. US Department of Agriculture.

Health, Nutrition, and Diet

Nestle, M. (2018). *Unsavory Truth: How Food Companies Skew the Science of What We Eat*. Basic Books.

Popkin, B. (2004). *The Nutrition Transition: Diet and Disease in the Developing World*. Academic Press.

Swinburn, B. A., V. I. Kraak, S. Allender, V. J. Atkins, P. I. Baker, J. R. Bogard, H. Brinsden, A. Calvillo, O. De Schutter, R. Devarajan, M. Ezzati, S. Friel, S. Goenka, R. A. Hammond, G. Hastings, C. Hawkes, M. Herrero, P. S. Hovmand, M. Howden, L. M. Jaacks, A. B. Kapetanaki, M. Kasman, H. V. Kuhnlein, S. K. Kumanyika, B. Larijani, T. Lobstein, M. W. Long, V. K. R. Matsudo, S. D. H. Mills, G. Morgan, A. Morshed, P. M. Nece, An Pan, D. W. Patterson, G. Sacks, M. Shekar, G. L. Simmons, W. Smit, A. Tootee, S. Vandevijvere, W. E. Waterlander, L. Wolfenden, and W. H. Dietz (2019). "The Global Syndemic of Obesity, Undernutrition, and Climate Change." *The Lancet Commission Report*.

Willett, W. (2001). *Eat, Drink, and Be Healthy: The Harvard Medical School Guide to Healthy Eating*. Free Press.

Willett, W., Rockström, J., B. Loken, M. Springmann, T. Lang, S. Vermeulen, T. Garnett, D. Tilman, F. DeClerck, A. Wood, M. Jonell, M. Clark, L. J. Gordon, J. Fanzo, C. Hawkes, R. Zurayk, J. A. Rivera, W. De Vries, L. M. Sibanda, A. Afshin, A. Chaudharyw, M. Herrero, R. Agustina, F. Branca, A. Lartey, S. Fan, B. Crona, E. Fox, V. Bignet, M. Troell, T. Lindahl, S. Singh, S. E. Cornell, K. S. Reddy, S. Narain, S. Nishtar, and Christopher J. L. Murray (2019). "Food in the Anthropocene: The EAT-Lancet Commission on Healthy Diets from Sustainable Food Systems." *The Lancet*, 393 (10170), 447–92.

Food Justice and Equity

Civil Eats. (2020). *Food Justice Reporting Anthology*. Civil Eats.

La Via Campesina. (2018). *La Via Campesina: International Peasant's Voice*. Zed Books.

Patel, R., and Moore, J. W. (2017). *A History of the World in Seven Cheap Things: A Guide to Capitalism, Nature, and the Future of the Planet*. University of California Press.

Penniman, L. (2018). *Farming While Black: Soul Fire Farm's Practical Guide to Liberation on the Land*. Chelsea Green Publishing.

White, M. (2019). *Freedom Farmers: Agricultural Resistance and the Black Freedom Movement*. University of North Carolina Press.

Global Trade and Policy

Clapp, J. (2016). *Food (Global Institutions Series)*. Polity Press.

Lang, T. (2020). *Feeding Britain: Our Food Problems and How to Fix Them*. Pelican.

OECD/FAO. (2023). *Agricultural Outlook 2023–2032*. OECD Publishing / FAO.

UNCTAD. (2023). *Trade and Development Report*. United Nations Conference on Trade and Development.

WTO. (2022). *World Trade Report*. World Trade Organization.

Index

About the Author

Honor May Eldridge is a policy and advisory expert in food and environment impact, with over a decade of experience advocating for resilient, equitable food systems in both the UK and the United States. She has worked to advise NGOs and businesses in the agriculture and food retail sectors. Her expertise spans international trade, climate-resilient agriculture, and food system transformation—always with an eye on equity and long-term viability. Honor was previously the senior legislative advisor to the UK's Food Standards Agency, working on how innovation can be delivered through regulation. She previously worked at the Soil Association, Sustainable Food Trust, and the Center for Food Safety in Washington, DC. Honor is the author of *The Avocado Debate* (2023), an incisive exploration of the environmental, ethical, and social issues tied to one of the world's most popular superfoods. Through her writing and policy work, she challenges readers and policymakers alike to reconsider the impact of their food choices and to push for a more sustainable future.